SEMI-CLASSICAL APPROXIMATION IN QUANTUM MECHANICS

MATHEMATICAL PHYSICS AND
APPLIED MATHEMATICS

Editors:

M. FLATO, *Université de Dijon, Dijon, France*

R. RĄCZKA, *Institute of Nuclear Research, Warsaw, Poland*

with the collaboration of:

M. GUENIN, *Institut de Physique Théorique, Geneva, Switzerland*

D. STERNHEIMER, *Collège de France, Paris, France*

VOLUME 7

V. P. MASLOV and M. V. FEDORIUK

SEMI-CLASSICAL APPROXIMATION IN QUANTUM MECHANICS

Translated from the Russian by J. Niederle and J. Tolar

D. REIDEL PUBLISHING COMPANY

DORDRECHT : HOLLAND / BOSTON : U.S.A.
LONDON : ENGLAND

Library of Congress Cataloging in Publication Data

CIP

Maslov, V. P.
  Semi-classical approximation in quantum mechanics.

  (Mathematical physics and applied mathematics; v. 7)
  Translation of: Kvaziklassicheskoe priblizhenie dlîa uravnenii
kvantovoï mekhaniki.
  Bibliography: p.
  Includes index.
  1. Approximation theory. 2. Quantum theory. I. Fedoriuk,
Mikhail Vasil'evich. II. Title. III. Series.
QC 174.17.A66M3713   520.1′2   81–2529
                            AACR2

Published by D. Reidel Publishing Company
P.O. Box 17, 3300 AA Dordrecht, Holland

Sold and distributed in the U.S.A. and Canada
by Kluwer Boston Inc.,
190 Old Derby Street, Hingham, MA 02043, U.S.A.

In all other countries, sold and distributed
by Kluwer Academic Publishers Group
P.O. Box 322, 3300 AH Dordrecht, Holland

D. Reidel Publishing Company is a member of the Kluwer Group

ISBN 1-4020-0306-4
Transferred to Digital Print 2001

TABLE OF CONTENTS

PART I

QUANTIZATION OF VELOCITY FIELD
(THE CANONICAL OPERATOR)

PART II

SEMI-CLASSICAL APPROXIMATION FOR
NON-RELATIVISTIC AND RELATIVISTIC
QUANTUM MECHANICAL EQUATIONS

PROFILE

In this volume the multi-dimensional semi-classical approximation to equations of quantum mechanics is discussed. The first part is devoted to quantization of the velocity field for general Hamiltonians. The second part is concerned with the semi-classical approximation of the Cauchy problem for the initial data satisfying the correspondence principle, with the scattering problem, and with the asymptotics of spectral series for relativistic, as well as non-relativistic, equations of quantum mechanics.

The book is intended for undergraduate and graduate students and for physicists and mathematicians engaged in research in theoretical and mathematical physics.

PREFACE

This volume is concerned with a detailed description of the canonical operator method – one of the asymptotic methods of linear mathematical physics. The book is, in fact, an extension and continuation of the authors' works [59], [60], [65].

The basic ideas are summarized in the Introduction. The book consists of two parts. In the first, the theory of the canonical operator is developed, whereas, in the second, many applications of the canonical operator method to concrete problems of mathematical physics are presented.

The authors are pleased to express their deep gratitude to S. M. Tsidilin for his valuable comments.

THE AUTHORS

INTRODUCTION

1. Various problems of mathematical and theoretical physics involve partial differential equations with a small parameter at the highest derivative terms. For constructing approximate solutions of these equations, asymptotic methods have long been used.

In recent decades there has been a renaissance period of the asymptotic methods of linear mathematical physics. The range of their applicability has expanded: the asymptotic methods have been not only continuously used in traditional branches of mathematical physics but also have had an essential impact on the development of the general theory of partial differential equations. It appeared recently that there is a unified approach to a number of problems which, at first sight, looked rather unrelated. Thus identical methods are applied to the study of singularities of elementary solutions of differential equations; to the problem of local solvability of differential equations; to the investigation of asymptotics of a solution of the Cauchy problem with rapidly oscillating initial data; to the problem of propagation of discontinuities of fundamental solutions of a hyperbolic equation; to the scattering problem for equations of quantum mechanics in semi-classical approximation; to the construction of high-frequency asymptotics in the problems of wave diffraction theory; to the problems of asymptotic behaviour of eigenvalues and eigenfunctions of differential operators; to the investigation of singularities of solutions of partial differential equations with analytic coefficients, and to other problems.

In the present monograph one of the basic asymptotic methods of linear mathematical physics – the method of canonical operator – is discussed and a number of its concrete applications presented. This method yields the global asymptotics for a solution of the Cauchy problem with rapidly oscillating initial data, for the short-wave approximation of a solution of the field equation with a point source in an inhomogeneous medium, and for solutions of many other problems. We hope that the capability of the canonical operator method is by no means exhausted; without any doubt this method will surely be applied to the boundary value problems of mathematical physics.

1

As a matter of fact, the canonical operator method reflects the profound wave-particle duality: in semi-classical approximation, the quantum mechanical quantities are described in the framework of classical mechanics, and the laws of wave optics are replaced by those of geometrical optics. This fact is mathematically expressed in the duality between the Fourier and the contact transformations.

The canonical operator method is based on ideas introduced by P. Debye, G. D. Birkhoff [6], V. A. Fock [22], S. L. Sobolev [78], J. Leray [53], and by many other mathematicians and physicists.

In the Introduction we would like to summarize all the basic ideas, methods and results of the monograph without going into necessary but cumbersome technical details.

2.   First consider a typical example of the problems studied in the present book – the Cauchy problem with rapidly oscillating initial data for the Schrödinger equation

$$ih\frac{\partial\psi}{\partial t} = -\frac{h^2}{2m}\Delta\psi + V(x)\psi, \tag{0.1}$$

$$\psi|_{t=0} = \psi_0(x)\exp\left[\frac{i}{h}S_0(x)\right]. \tag{0.2}$$

Here $x \in \mathbf{R}^n$, functions $V(x)$, $S_0(x)$ are real-valued and infinitely differentiable, and function $\psi_0(x) \in C_0^\infty(\mathbf{R}^n)$, i.e. $\psi_0(x)$ is infinitely differentiable and with compact support. We are looking for an asymptotic solution of the problem (0.1), (0.2) as $h \to +0$ and $x \in \mathbf{R}^n$, $0 \le t \le T$, i.e. within an arbitrary finite time T.

The corresponding asymptotic formulae are said to be a 'semi-classical approximation' or 'semi-classical asymptotics'.

Consider now the equation

$$L(x, \lambda^{-1}D_x)u(x) = 0, \qquad x \in \mathbf{R}^n, \tag{0.3}$$

over the whole space $\mathbf{R}^n$, i.e. a generalization of the previous case. Here $\lambda > 0$ is a large parameter and L is a differential operator of the form

$$L(x, \lambda^{-1}D_x) = \sum_{|\alpha|=0}^{m} a_\alpha(x)(\lambda^{-1}D_x)^\alpha$$

with the following standard notation: $\alpha = (\alpha_1, \alpha_2, \ldots, \alpha_n)$ for a multi-index, $\alpha_j \geqq 0$ – integers, $|\alpha| = \alpha_1 + \alpha_2 + \ldots + \alpha_n$ and

$$D_x = \left(\frac{1}{i}\frac{\partial}{\partial x_1}, \ldots, \frac{1}{i}\frac{\partial}{\partial x_n}\right),$$

$$D_x^\alpha = \left(\frac{1}{i}\frac{\partial}{\partial x_1}\right)^{\alpha_1} \ldots \left(\frac{1}{i}\frac{\partial}{\partial x_n}\right)^{\alpha_n}.$$

The function

$$L(x, p) = \sum_{|\alpha| = 0}^{m} a_\alpha(x)p^\alpha$$

is said to be a *symbol* of operator L(here $p = (p_1, p_2, \ldots, p_n)$).

Schrödinger equation (0.1) is apparently of the form (0.3) with $\lambda = h^{-1}$ and

$$L(t, x, E, p) = E + \frac{h^2}{2m}\langle p, p \rangle + V(x)$$

(E is the variable conjugate to t). Here and later on, the bracket $\langle ., . \rangle$ denotes a (non-hermitian) scalar product.

Many equations of mathematical physics are of the form (0.3), too, for example the Helmholtz equation, i.e. the equation of wave optics or wave acoustics:

$$(\Delta + k^2 n^2(x)) u(x) = 0.$$

In this case $\lambda = k$ and the symbol L (after dividing the equation by k^2) is of the form

$$L(x, p) = -\langle p, p \rangle + n^2(x).$$

The asymptotic solutions of the Helmholtz equation (as $k \to +\infty$) are called *short-wave* or (*high-frequency*) *asymptotics*.

Notice that the Maxwell equations, the Dirac equation, the system of equations of linear elasticity theory as well as many others are also of the form (0.3).

We are interested in an asymptotic solution of Equation (0.3) in the whole space $\mathbf{R}_x^n$ as $\lambda \to +\infty$. At this stage we shall not specify our problem in more detail; the scattering problem for the Helmholtz equation can, however, serve as an example.

We begin with a simpler problem, i.e. with the construction of an asymptotic solution of Equation (0.3) as $\lambda \to +\infty$ in some domain of space $\mathbf{R}_x^n$. Then, after solving this local problem, we shall try to obtain a global solution by patching up the local ones.

As is well known, the construction of asymptotic solutions consists of two steps: (1) construction of a formal asymptotic solution; (2) derivation of the obtained asymptotic solution on a rigorous basis.

These two steps are, as a rule, solved by completely different methods; in the Introduction we shall consider only the first one. Let us also note that in the Introduction we shall only discuss the main features of facts, i.e. we shall write down statements, mathematical rigour of which will be achieved by adding natural assumptions. The exact explanations provide ·the content of the rest of this volume.

Finally let us state the assumption to be valid throughout the book.

ASSUMPTION. The coefficients of operator L are infinitely differentiable for all $x \in \mathbf{R}^n$, and, under various circumstances, with compact support or rapidly tending to constants as $|x| \to \infty$. Symbol $L(x, p)$ is real-valued.

3. We shall look for a solution of Equation (0.3) in the form of a formal series

$$u(x, \lambda) = \exp\left[i\lambda S(x)\right] \sum_{j=0}^{\infty} (i\lambda)^{-j} \varphi_j(x), \tag{0.4}$$

where $S(x)$ and $\varphi_j(x)$ are unknown functions. If the coefficients of operator L are constant, Equation (0.3) has the exact plane-wave solutions

$$u(x, \lambda) = \exp\left[i\lambda \langle x, p \rangle\right],$$

where p is a constant vector. The function (0.4) can be interpreted as a distorted plane wave.

The Ansatz (0.4) for solving Equation (0.3) has been known in the theory of differential equations since the time of Green and Liouville (1837). For partial differential equations this Ansatz was introduced by famous English physicist and mathematician P. Debye in 1911, and for quantum mechanical problems by Wentzel, Kramers, and Brillouin (the WKB method).

Now insert series (0.4) in Equation (0.3), divide both sides of the resulting equation by the exponential $\exp(i\lambda S)$ and, finally, set the coefficients at the powers $(i\lambda)^0$ and $(i\lambda)^{-1}$ equal to zero. The coefficient at $(i\lambda)^0$ is equal to

$L(x, \partial S(x)/\partial x)\varphi_0(x)$; hence

$$L\left(x, \frac{\partial S(x)}{\partial x}\right) = 0. \tag{0.5}$$

For Schrödinger equation (0.1), Equation (0.5) represents the Hamilton–Jacobi equation of classical mechanics:

$$\frac{\partial S}{\partial t} + \frac{1}{2m}(\nabla_x S)^2 + V(x) = 0. \tag{0.6}$$

Its solution is a classical *action*.

The Equation (0.5) will be called the *Hamilton–Jacobi equation* or the *characteristic equation* corresponding to Equation (0.3). Notice that the characteristic equation for the Helmholtz equation:

$$(\nabla S(x))^2 = n^2(x)$$

is the *eikonal* equation known from geometrical optics. Function $S(x)$, the solution of Hamilton–Jacobi equation (0.5), will be called a *characteristic* of operator L or an *action* [in analogy with Equation (0.6)]. Only real characteristics will be considered. In other words, the equation

$$L(x, p) = 0$$

with $p = (p_1, \ldots, p_n)$ has to determine a C^∞-manifold of maximal possible dimension $2n - 1$ in space $\mathbf{R}^n_x \times \mathbf{R}^n_p$.

The Hamilton–Jacobi equation (0.5) is a non-linear first order partial differential equation. It might happen that such an equation either admits no solutions without singularities in the whole space $\mathbf{R}^n_x$, or the class of these solutions is rather poor. For example, any solution of the eikonal equation $(\nabla S(x))^2 = 1$ belonging to $C^\infty(\mathbf{R}^n_x)$ is just a linear function of x. This is exactly the main difficulty we meet with when trying to construct a global asymptotic solution, i.e., a solution in the whole space. Global solutions cannot be constructed in the form (0.4) – for this purpose new ideas have to be introduced.

However, let us begin with the study of Hamilton–Jacobi equation (0.5). In order to obtain a unique solution, the Cauchy problem on an $(n-1)$-dimensional manifold $x = f(y)$, $y \in U$, with U being a domain in $\mathbf{R}^{n-1}_y$, is specified by the condition

$$S(x) = S_0(y). \tag{0.7}$$

One of the most remarkable achievements of mathematical analysis consists in reducing the integration of Cauchy problem (0.6), (0.7) for a non-linear first order partial differential equation to the integration of the Cauchy problem for the system of ordinary differential equations

$$\frac{dx}{dt} = \frac{\partial L(x, p)}{\partial p}, \qquad \frac{dp}{dt} = -\frac{\partial L(x, p)}{\partial x} \tag{0.8}$$

with the Cauchy data

$$x\big|_{t=0} = x^0(y), \qquad p\big|_{t=0} = p^0(y), \qquad y \in U. \tag{0.9}$$

Here $x^0(y) = f(y)$, and $p^0(y)$ will be specified later. The system (0.8) will be called a *Hamiltonian system* associated with operator L, or a *bicharacteristic system*. The phase trajectories of the Hamiltonian system $\{x = x(t, y),$ $p = p(t, y)\}$ in *phase space* $\mathbf{R}^{2n}_{x, p}$ will be called *bicharacteristics* of Equation (0.3) or simply *trajectories*, and their projections $x = x(t, y)$ on $\mathbf{R}^n_x$ – *rays*. The variable x and p will be called *coordinates* and *momenta*, respectively. We shall also use a number of known facts from the theory of partial differential equations and classical mechanics (see [4], [82], [31], [11], [77], [78]). For instance, along any bicharacteristic we have

$$dS = \langle p, dx \rangle. \tag{0.10}$$

As already mentioned, the solution $S(x)$ of Cauchy problem (0.6), (0.7) might be a non-smooth function. This is a consequence of the following fact. The phase trajectories of the Hamiltonian system do not intersect in accordance with the existence and uniqueness theorem for a normal system of ordinary differential equations (since for simplicity, we assume that a solution of Cauchy problem (0.8), (0.9) does exist and is infinitely differentiable for all $t \in \mathbf{R}$). The projections of bicharacteristics on x-space, i.e. the rays $x = x(t, y)$, might, however, intersect, touch or meet in one point, etc. Let us clarify the consequences in detail. The function S, by virtue of (0.7), (0.10), has the form

$$S(x(t, y)) = S_0(y) + \int_0^t \left\langle p, \frac{dx}{d\tau} \right\rangle d\tau,$$

where the integral is taken along the trajectories $x = x(t, y)$, $p = p(t, y)$. Consequently, S is a smooth function of 'curvilinear coordinates' (t, y). We may introduce coordinates $(t, y_1, \ldots, y_{n-1})$ instead of coordinates

Fig. 1.

$(x_1, \ldots, x_n)$ whenever the correspondence $(t, y) \leftrightarrow x(t, y)$ is one-to-one. Then necessarily the Jacobian

$$J(t, y) = \det \frac{\partial x(t, y)}{\partial(t, y)}$$

must not vanish.

As will be shown later, the Jacobian differs from zero for $t = 0$, but not necessarily for all (t, y). For the points in which $J = 0$, the variables $(t, y_1, \ldots, y_{n-1})$ appear to be non-smooth or multi-valued functions of $(x_1, \ldots, x_n)$ and, consequently, the function $S(x)$, too. In optics, the manifolds on which $J(t, y) = 0$ are known as *caustics* of the family of rays $x = x(t, y)$ $(y \in U, t \in \mathbf{R})$ and the points on the ray $x = x(t, y^0)$ in which $J(t, y^0) = 0$ as *focal* points (Figure 1).

Thus we see that the existence of caustics for a family of rays $x = x(t, y)$ presents an obstacle in constructing global asymptotics via (0.4).

4. Notice that even if the family of rays $x = x(t, y)$ has caustics, there are no caustics for the family of trajectories $x = x(t, y)$, $p = p(t, y)$. Let g^t be a *displacement* along trajectories *during the time t*; in particular, let $g^t(x^0(y), p^0(y)) = (x(t, y), p(t, y))$. The Cauchy data (0.9) determine an $(n-1)$-dimensional manifold Λ_0^{n-1} in the phase space. This manifold is by no means arbitrary: indeed, there is a necessary condition of compatibility of the two Cauchy problems: that for the Hamilton–Jacobi equation and that for the Hamilton system. By virtue of (0.10) this so called 'strip condition' is given by

$$\langle p^0(y), dx^0(y) \rangle = dS_0(y), \qquad y \in U.$$

This condition means that the differential 1-form

$$\omega^1 = \langle p, dx \rangle = \sum_{j=1}^{n} p_j \, dx_j \tag{0.11}$$

is closed (i.e. $d\omega^1 \equiv 0$) on the initial manifold Λ_0^{n-1}.

Manifolds in phase space $\mathbf{R}^{2n}_{x,p}$ on which form ω^1 is closed, i.e.

$$d\omega^1 = \sum_{j=1}^{n} dp_j \wedge dx_j = 0, \qquad (0.12)$$

are said to be *Lagrangian manifolds*. In other words, the Lagrangian manifolds are manifolds for which the integral $\int \langle p, dx \rangle$ is locally *path independent*. They represent an important object in analytical mechanics. As is known from there, the displacement g^t preserves the form $\omega^2 = d\omega^1$. Consequently, under a displacement along phase trajectories Lagrangian manifolds are transformed among themselves. In particular, the manifold $\Lambda_t^n = g^t \Lambda_0^{n-1}$ is Lagrangian again. Moreover, a 'tube of trajectories', i.e. the set

$$\Lambda^n = \bigcup_{-\infty < t < \infty} g^t \Lambda_0^{n-1},$$

is an n-dimensional Lagrangian manifold. For simplicity it will be assumed simply connected (Figure 2).

    Lagrangian manifolds are embedded in $2n$-dimensional space $\mathbf{R}^{2n}_{x,p}$ – the *phase space*. Its points will be denoted by $r = (x, p)$.

    Lagrangian manifold Λ^n is just that fundamental classical object which is associated with the quantum object – the operator L:

$$L(x, \lambda^{-1} D_x) \leftrightarrow \Lambda^n$$

The emphasis will be not on the Hamilton–Jacobi equation but on its solution $S(x)$ and the Lagrangian manifold Λ^n.

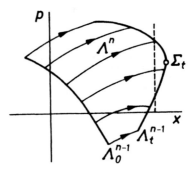

Fig. 2.

The function

$$S(r) = S(r^0) + \int_{r^0}^{r} \langle p, dx \rangle \tag{0.12'}$$

is uniquely defined on a simply connected Lagrangian manifold Λ^n. (Here $r^0, r \in \Lambda^n$, point r^0 is fixed, and the integral is taken along an arbitrary path on Λ^n connecting points r^0 and r.)

If point $r = (x, p) \in \Lambda^n$ lies near the initial manifold Λ_0^{n-1}, then p is uniquely determined by $x : p = p(x)$. Therefore $S(r) = S(x)$ $(r = (x, p(x)))$; the letter S is used in shorthand notation for function $S(r)$ defined on manifold Λ^n as well as for action $S(x)$. However, function $S(r)$, in contrast to $S(x)$, is determined globally. Function $S(r)$ turns out to be a kind of an 'analytic continuation' of function $S(x)$, and manifold Λ^n – an analogue of the Riemann surface for function $S(x)$.

Let us denote by Λ_t^n the piece $\Lambda^n \cap \{0 \leq t \leq T\}$ of the tube of trajectories. Provided the projection of Λ^n on $\mathbf{R}_x^n$ is a diffeomorphism, function $S(x)$ is smooth and the function $S(r)$ can be regarded as a function of x.

There is a remarkable property possessed by a Lagrangian manifold Λ^n: its local coordinates can always be chosen to be either x, or p, or the set

$$(p_{(\alpha)}, x_{(\beta)}), (\alpha) = (\alpha_{i_1}, \ldots, \alpha_{i_k}), \quad (\beta) = (\beta_{j_1}, \ldots, \beta_{j_{n-k}}),$$

not containing conjugate coordinates and momenta (i.e. pairs of the form (x_j, p_j)). Hence locally $S(r)$ is a function either of x or of p, or of $(p_{(\alpha)}, x_{(\beta)})$. Thus in a neighbourhood of point $r \in \Lambda^n$ for which the projection on $\mathbf{R}_x^n$ is not a diffeomorphism one may introduce either a *p-representation* (i.e. to take momenta p as coordinates), or a mixed (x, p)-*representation* $(p_{(\alpha)}, x_{(\beta)})$. The passage to p-representation is realized by means of the *Legendre transformation*.

Now our task is to express the initial Equation (0.3) in p-representation or in mixed coordinate-momentum representation.

5. In order to determine an approximate solution of Equation (0.3) with accuracy, say, to $O(\lambda^{-1})$, one has to find the function $\varphi_0(x)$ (see (0.4)). This function $\varphi_0(x)$ has to fulfil the equation

$$\left\langle \frac{\partial \varphi_0(x)}{\partial x}, \frac{\partial L(x, p)}{\partial p} \right\rangle + \frac{1}{2} \sum_{j,k=1}^{n} \frac{\partial^2 L(x, p)}{\partial p_j \partial p_k} \frac{\partial^2 S(x)}{\partial x_j \partial x_k} \varphi_0(x) = 0,$$

$$\tag{0.13}$$

where $p = \partial S(x)/\partial x$. It is not difficult to see that this equation considered along a ray $x = x(t, y)$ leads to

$$\left\langle \frac{\partial \varphi_0}{\partial x}, \frac{\partial L}{\partial p} \right\rangle = \frac{d\varphi_0}{dt}$$

with d/dt being the time derivative with respect to Hamilton system (0.8). Consequently (0.13) is an ordinary differential equation along the ray $x = x(t, y)$ (for fixed y). It is called a *transport equation*. By using the known Liouville formula, the transport equation takes the form

$$\frac{1}{\sqrt{J}} \frac{d}{dt} (\sqrt{J} \, \varphi_0) - \frac{1}{2} \sum_{j=1}^{n} \frac{\partial^2 L(x, p)}{\partial x_j \partial p_j} \varphi_0 = 0, \tag{0.14}$$

where $x = x(t, y)$, $p = p(t, y)$, and $J = J(t, y)$.

There are two important facts which immediately follow from (0.14):

(i) The *commutation formula*

$$L(x, \lambda^{-1} D_x) \left[\frac{\varphi}{\sqrt{J}} \exp (i\lambda S) \right] =$$

$$= \exp (i\lambda S) \frac{1}{\sqrt{J}} \left[L(x, p)\varphi + \frac{1}{i\lambda} R_1 \varphi + O(\lambda^{-2}) \right], \quad (\lambda \to + \infty) \tag{0.15}$$

holds. Here R_1 is the first order operator

$$R_1 = \frac{d}{dt} - \frac{1}{2} \sum_{j=1}^{n} \frac{\partial^2 L(x, p)}{\partial x_j \partial p_j}, \tag{0.16}$$

and p in (0.15) and (0.16) is equal to $\partial S(x)/\partial x$.

(ii) The leading term of asymptotic solution $u(x, \lambda)$ is of the form

$$v(x, \lambda) = \varphi_0(y) \sqrt{\frac{J(0, y)}{J(t, y)}} \exp \left[i\lambda \left(S_0(y) + \int_0^t \langle p, dx \rangle \right) - \right.$$

$$\left. - \frac{1}{2} \int_0^t \sum_{j=1}^{n} \frac{\partial^2 L(x, p)}{\partial x_j \partial p_j} dt \right]. \tag{0.17}$$

Here $x = x(t, y)$, $p = p(t, y)$.

More precisely, formula (0.17) determines a *formal asymptotic solution* of Equation (0.3) (i.e. $Lv = O(\lambda^{-2})$ where v is the right-hand side of formula (0.17) and of order $O(1)$).

As seen from formula (0.17), the leading term of an asymptotic solution of partial differential Equation (0.3) is expressible in terms of classical mechanics; namely, we can determine it provided the solution of Cauchy problem (0.3) for Hamilton system (0.8) is known. For this reason asymptotic formulae of type (0.17) carry a special name – 'semi-classical approximation'.

Previously we have specified the connection between function S and Lagrangian manifold Λ^n; now we shall specify an analogous connection between Jacobian $J(t, y)$ and manifold Λ^n. The points $r = (x, p) \in \Lambda^n$ are parametrized by variables (t, y); we set $r(t, y) = x(t, y)$, $p(t, y)$. Let volume element $d\sigma^n(r)$ on Λ^n be defined by

$$d\sigma^n(r(t, y)) = dt \, dy.$$

It is invariant with respect to dynamical system (0.8). We have

$$|J(t, y)| = \left| \frac{dx}{d\sigma^n(r)} \right|.$$

Assume that Λ^n admits a diffeomorphic projection on R_x^n; then the operator

$$(K_{\Lambda^n} \varphi(r))(x) = \sqrt{\left| \frac{d\sigma^n(r)}{dx} \right|} \exp \left[i\lambda \int_{r^0}^{r} \langle p, dx \rangle \right] \varphi(x) \qquad (0.18)$$

on Λ^n can be introduced; here $r = (x, p)$, and $r^0 \in \Lambda^n$ is a fixed point. Consequently, according to (0.15), (0.16), the commutation formula

$$[L(x, \lambda^{-1} D_x) K_{\Lambda^n} \varphi(r)](x) =$$
$$= K_{\Lambda^n} \left[L(x, p) \varphi(r) + \frac{1}{i\lambda} R_1 \varphi(r) \right](x) + O(\lambda^{-2}) \qquad (0.19)$$

holds, or, in a shorthand notation,

$$\mathscr{L} K = K \left(L + \frac{1}{i\lambda} R_1 \right) \qquad (0.19')$$

with accuracy to $O(\lambda^{-2})$. Thus, by commuting $\mathscr{L}$ with K, we obtain the operator $L + (1/i\lambda) R_1$. It is substantially more simple than the initial operator $\mathscr{L}$, since L is a multiplication operator by function $L(x, p)$ and R_1 is a first-order differential operator.

Thus our construction of the asymptotic solution of Equation (0.3) consists of the following steps. First we look for a solution of the form

$$u = K\left(\varphi_0 + \frac{1}{i\lambda} \varphi_1 + \ldots \right). \tag{0.20}$$

Then by applying commutation formula (0.19) we get the equations

$$L(x, p) = 0, \qquad R_1 \varphi_0 = 0, \text{ etc.}$$

The first equation is algebraic, the second one is the transport equation, i.e. a first-order ordinary differential equation along a phase trajectory of the Hamilton system.

Summarizing, whenever Λ^n admits a diffeomorphic projection on R_x^n, the asymptotic solution is constructed via the chain

$$\mathscr{L} \to \Lambda^n \to K_{\Lambda^n},$$

where $\mathscr{L}, \Lambda^n$, and K_{Λ^n} are quantum, classical, and semi-classical objects, respectively.

These results will be discussed in detail in §§ 1–4.

6. Consider now Equation (0.3) in p-representation. First of all let us remark that operator L can be represented as

$$L(x, \lambda^{-1} D_x) u(x) = F_{\lambda, p \to x}^{-1} L(x, p) F_{\lambda, x \to p} u(x). \tag{0.20'}$$

Here, $F_{\lambda, x \to p}$ is the λ-Fourier transformation

$$F_{\lambda, x \to p} u(x) = \left(\frac{\lambda}{2\pi i} \right)^{n/2} \int\limits_{R^n} \exp\left[- i\lambda \langle x, p \rangle \right] u(x) \, dx,$$

$$\sqrt{i} = e^{i\pi/4}.$$

Following Feynman, we shall write operator (0.20') in the form $L(\overset{2}{x}, \lambda^{-1} \overset{1}{D}_x)$ to indicate that we perform first differentiation and then multiplication by functions of x (see (0.3)).

We take formula (0.20') as the definition of a λ-pseudo-differential operator (λ-p.d. operator). The operator $\mathscr{L} = L(\overset{2}{x}, \lambda^{-1} \overset{1}{D}_x)$ is a differential operator, whenever symbol $L(x, p)$ is a polynomial in variables p. In other cases $\mathscr{L}$ is not a differential operator, but, nevertheless, it exhibits a number of important properties of differential operators, provided the class of symbols L is properly chosen (see § 2). In particular, the asymptotic expansion

$$\mathscr{L}\left[\varphi(x)\exp\left(i\lambda S(x)\right)\right] \sim \exp\left[i\lambda S(x)\right]\sum_{j=0}^{\infty}(i\lambda)^{-j}\varphi_j(x)$$

holds for $\lambda \to +\infty$, where $\varphi \in C_0^{\infty}(\mathbf{R}^n)$, $S \in C^{\infty}(\mathbf{R}^n)$, and S is a real-valued function. If $\mathscr{L}$ is a differential operator, the series is finite. Thus we may look for an asymptotic solution of Equation (0.3) in the form (0.4) even in the case when $\mathscr{L}$ is a λ-p.d. operator.

The present book treats the λ-p.d. operators, the symbols of which belong to class T_+^m. Let us describe them in the simpler case when symbol $L(x, p)$ is independent of λ. Then a function $L(x, p)$ belongs to T_+^m, if it is infinitely differentiable for all $(x, p) \in \mathbf{R}_x^n \times \mathbf{R}_p^n$ and if the estimates

$$\left|D_x^{\alpha}D_p^{\beta}L(x, p)\right| \leq C_{\alpha\beta}(1 + |x|)^m(1 + |p|)^m, (x, p) \in \mathbf{R}_x^n \times \mathbf{R}_p^n,$$

hold for arbitrary multi-indices α, β.

The union $\bigcup_{m \in R} T_+^m$ of all classes T_+^m will be denoted by T_+. In particular, if $L(x, p)$ is a polynomial of degree not greater than m, then L belongs to T_+^m. Hence all differential operators with polynomial coefficients belong to T_+. Moreover, class T_+ contains also differential operators with coefficients tending to constants sufficiently rapidly as $|x| \to \infty$, as well as shift operators, i.e. operators of the form

$$\mathscr{L}f(x) = f(x + h),$$

where $h \in \mathbf{R}^n$ is a constant vector. Let us remark that the symbol L of a shift operator is equal to $\exp\left[i\langle x, h\rangle\right]$, and therefore belongs to class T_+^0.

It will be proved in § 2, Theorem 2.12, that composition of λ-p.d. operators (with the symbols of class T_+) yields a λ-p.d. operator with the symbol of class T_+ again.

We shall look for a solution u of Equation (0.3) in the form of the λ-Fourier transform of a function $v(p, \lambda)$:

$$u(x, \lambda) = F_{\lambda, p \to x}^{-1}v(p, \lambda).$$

Inserting u into Equation (0.3) and then applying operator $F_{\lambda, p \to x}$ from the left, we obtain Equation (0.3) in p-representation:

$$\tilde{\mathscr{L}}v(p, \lambda) = 0,$$

where

$$\tilde{\mathscr{L}} = F_{\lambda, x \to p} L(\overset{2}{x}, \lambda^{-1} \overset{1}{D}_x) F^{-1}_{\lambda, p \to x}. \tag{0.21}$$

Operator $\tilde{\mathscr{L}}$ is said to be the *Fourier transform of operator* $\mathscr{L}$. It is not difficult to see that

$$\tilde{\mathscr{L}} = L(-\lambda^{-1}\overset{1}{D}_p, \overset{2}{p}).$$

If $\mathscr{L}$ has form (0.3), then

$$\tilde{\mathscr{L}}v = \sum_{|\alpha|=0}^{m} (-\lambda^{-1} D_p)^\alpha (a_\alpha(p)v).$$

Operator $\tilde{\mathscr{L}}$ is again a λ-p.d. operator, i.e. the class of λ-p.d. operators is invariant with respect to the λ-Fourier transformations. Thus we can try to construct a particular asymptotic solution of Equation (0.3) in the form of the λ-Fourier transform of a rapidly oscillating exponential function u of type (0.4):

$$u(x, \lambda) = F^{-1}_{\lambda, p \to x}\left[\exp(i\lambda \tilde{S}(p)) \sum_{j=0}^{\infty} (i\lambda)^{-j} \tilde{\varphi}_j(p) \right]. \tag{0.22}$$

Obviously, one might construct solutions of (0.3) with the aid of λ-Fourier transformations over some of the variables.

The new class of asymptotic solutions (0.22) enables us to find out global asymptotic solutions of Equation (0.3). This construction procedure can be briefly described as follows. If over domain $\Omega \subset \mathbf{R}^n$ there lie only such domains ('leaves') of Lagrangian manifold Λ^n which admit diffeomorphic projections on Ω, solution u is equal to the sum (over the number of leaves) of exponentials (0.4). If domain Ω contains a caustic, solution u (in a neighbourhood of the caustic) has the form of an exponential in p-representation (i.e. it has the form (0.22) or a similar one in which the λ-Fourier transformation is made over a subset of variables). The global asymptotic solution will be patched up just from these local solutions.

Let us investigate now the structure of the solution in p-representation in more detail. Since $\tilde{\mathscr{L}}$ is again a λ-p.d. operator, it is clear that the formula for solution v of the equation $\tilde{\mathscr{L}}v = 0$ is of the form (0.17). It is obvious that a commutation formula of type (0.19) is also valid in p-representation.

Using p-representation, and others, we immediately encounter the problem of their selection. More specifically: let Λ^n admit diffeomorphic projections on both $\mathbf{R}^n_x$ and $\mathbf{R}^n_p$. Then one can look for an asymptotic solution in form (0.4) and in form (0.22). How are these two representations of the solution related?

To answer this question, let us evaluate the asymptotics of the λ-Fourier transform of a rapidly oscillating exponential. Let function $\varphi(x) \in C_0^\infty(\mathbf{R}^n)$, function $S(x) \in C^\infty(\mathbf{R}^n)$, and function $S(x)$ be real-valued. By applying the stationary phase method (see § 1) we obtain

$$F_{\lambda, x \to p}\left[\varphi(x)\left|\det\frac{\partial p(x)}{\partial x}\right|^{1/2}\exp\left(i\lambda S(x)\right)\right](p)=$$

$$= \exp\left(-\frac{i\pi m}{2}\right)\varphi(x(p))\left|\det\frac{\partial x(p)}{\partial p}\right|^{1/2}\exp\left(i\lambda\tilde{S}(p)\right)+$$

$$+ O(\lambda^{-1}). \tag{0.23}$$

In this formula $p = \partial S(x)/\partial x$, $\tilde{S}(p)$ is a function dual to function $S(x)$ in the Young sense (so that $x = \partial\tilde{S}(p)/\partial p$), and $m = $ inerdex $\partial x(p)/\partial p$. The matrix $\partial x(p)/\partial p = \partial^2\tilde{S}(p)/\partial p^2$ is symmetric, and inerdex A denotes the negative inertial index (the number of negative eigenvalues) of real symmetric matrix A.

In deriving asymptotic formula (0.23) we have assumed that matrix $S''_{xx}(x)$ is non-degenerate on supp φ.

Formula (0.23) expresses still another interesting fact: the λ-Fourier transform of a localized rapidly oscillating exponential is a localized rapidly oscillating exponential again. Indeed, the λ-Fourier transform of the function $\varphi \exp(i\lambda S)$, where φ is a smooth function with compact support, and $\det S''_{xx} \neq 0$ on supp φ, is given by $\tilde{\varphi} \exp(i\lambda\tilde{S}) + O(\lambda^{-1})$, where $\tilde{\varphi}$ is a function with compact support.

In view of the result we introduce the operator on Λ^n:

$$(\tilde{K}_{\Lambda^n}\varphi(r))(x)=$$

$$= F_{\lambda, p \to x}^{-1}\left(\sqrt{\left|\frac{d\sigma^n(p)}{dp}\right|}\exp\left[i\lambda\left(\int_{r^0}^{r}\langle p, dx\rangle-\right.\right.\right.$$

$$\left.\left.\left. - \langle x(p), p\rangle\right)\right]\varphi(r)\right)(x). \tag{0.24}$$

Here $r = (x, p)$, and $x = x(p)$ is the equation of manifold Λ^n. It is not hard to check that the commutation formula

$$\tilde{\mathscr{L}}\tilde{K} = \tilde{K}\left(L + \frac{1}{i\lambda}R_1\right) \tag{0.19''}$$

holds (with accuracy to $O(\lambda^{-2})$). Evaluation of the asymptotics of integral (0.24) by the stationary phase method yields the connection between K and $\tilde{K}$:

$$\tilde{K}\varphi = e^{-i\pi m/2} K\varphi + O(\lambda^{-1}). \qquad (0.25)$$

Here m is an integer,

$$m = \text{inerdex} \, \frac{\partial x(p)}{\partial p}.$$

Thus, we get $K = \tilde{K}$ up to the multiplier $e^{i\pi m/2}$ and apart from a summand of order $O(\lambda^{-1})$. A similar situation arises also in the case when Λ^n is diffeomorphically projected on both $\mathbf{R}^n_x$ and $\mathbf{R}^n_{p_{(\alpha)}, x_{(\beta)}}$. This completes our survey of the main results of §§ 5 and 6 of this volume.

Finally let us observe that it is sufficient to know number m in (0.25) modulo 4, since $e^{2\pi i} = 1$. Then the comparison of various representations of the solution yields a certain integer-valued invariant mod 4. Namely, if Λ^n admits diffeomorphic projections on Lagrangian coordinate planes $(p_{(\alpha)}, x_{(\beta)})$ and $(p_{(\tilde{\alpha})}, x_{(\tilde{\beta})})$, we obtain

$$\text{inerdex} \, \frac{\partial x_{(\alpha)}(r)}{\partial p_{(\alpha)}} \equiv \text{inerdex} \, \frac{\partial x_{(\tilde{\alpha})}(r)}{\partial p_{(\tilde{\alpha})}} \, (\text{mod } 4)$$

in all non-singular points $r \in \Lambda^n$ (i.e. all points, the neighbourhoods of which are diffeomorphically projected on $\mathbf{R}^n_x$).

7. Now we construct a global asymptotic solution of Equation (0.3). We set up a covering of manifold Λ^n by charts $\{\Omega_j\}$ and take as their local coordinates either x, or p, or a special set $(p_{(\alpha)}, x_{(\beta)})$ not containing conjugate coordinates and momenta. Take a C^∞-partition of unity: $1 \equiv \sum_j e_j(r)$, $\text{supp} e_j \subset \Omega_j$, and consider the function

$$(K^{r_0}_{\Lambda^n} \varphi(r))(x) = \sum_j c_j K(\Omega_j)(e_j(r) \, \varphi(r))(x). \qquad (0.26)$$

Here $\varphi \in C^\infty_0 (\Lambda^n)$ and c_j are constants. Operator $K(\Omega_j)$ is given by (0.18), (0.24), and (6.3), whenever the coordinates in Ω_j are x, p, and $(p_{(\alpha)}, x_{(\beta)})$, respectively. Manifold Λ^n is assumed to be simply connected (since otherwise $\int \langle p, dx \rangle$ can be path dependent).

It remains to clarify constants c_j. They are taken in such a way that for operator $K = K^{r_0}_{\Lambda^n}$ the commutation formula

$$\mathscr{L}K\varphi = K\left(L\varphi + \frac{1}{i\lambda} R_1 \varphi\right) + O(\lambda^{-2}) \qquad (0.27)$$

holds. For simplicity, let atlas $\{\Omega_j\}$ consist of two charts Ω_1, Ω_2, and let their intersection $\Omega_{12} = \Omega_1 \cap \Omega_2$ be a domain, and function $\varphi \in C_0^\infty(\Omega_{12})$. Further let $L(x, p) = 0$, $L''_{x_j p_j}(x, p) = 0$ on Λ^n, $1 \leq j \leq n$, and denote $K_j = K(\Omega_j)$. From (0.19), (0.19') we obtain

$$\mathscr{L} K \varphi = \mathscr{L}(c_1 K_1(e_1 \varphi) + c_2 K_2(e_2 \varphi)) =$$

$$= \frac{1}{i\lambda} \left(c_1 K_1 \left(\frac{d}{dt}(e_1 \varphi) \right) + c_2 K_2 \left(\frac{d}{dt}(e_2 \varphi) \right) \right) + O(\lambda^{-2}) =$$

$$= \frac{1}{i\lambda} K \left(\frac{d\varphi}{dt} \right) + \frac{1}{i\lambda} \left[c_1 K_1 \left(\frac{de_1}{dt} \varphi \right) + \right.$$

$$\left. + c_2 K_2 \left(\frac{de_2}{dt} \varphi \right) \right] + O(\lambda^{-2}).$$

The commutation formula (0.27) is valid if and only if the square bracket term is of order $O(\lambda^{-1})$. If $\psi \in C_0^\infty(\Omega_{12})$, (0.25) implies

$$K_2 \psi = e^{i\pi m/2} K_1 \psi + O(\lambda^{-1}),$$

so that the square bracket term is equal to

$$K_1 \left[\left(c_1 \frac{de_1}{dt} + c_2 e^{i\pi m/2} \frac{de_2}{dt} \right) \varphi \right] + O(\lambda^{-1}).$$

Consequently, the identity

$$c_1 \frac{de_1}{dt} + c_2 e^{i\pi m/2} \frac{de_2}{dt} \equiv 0$$

has to be satisfied in Ω_{12}. Since $e_1 + e_2 \equiv 1$ in Ω_{12}, we have $(de_1/dt) + (de_2/dt) \equiv 0$ and, therefore,

$$c_1 - c_2 e^{i\pi m/2} = 0,$$

which is nothing else than the necessary condition for patching.

Examination of the patching procedure shows that coefficient c_j is expressed in terms of the quantity

$$\text{ind } l[r^0, r].$$

Here $l[r^0, r]$ is a path joining non-singular points $r^0 \in \Lambda^n$ and $r \in \Omega_j$, and its index $\text{ind } l$ is an integer. The exact definition and further properties of index ind that was introduced by one of the authors of this work in [59], [60], will be given in § 7.

If $\operatorname{ind} l = 0$ for an arbitrary closed path l in Λ^n, then formula (0.26) determines the operator

$$K_{\Lambda^n}^{r^0} : C_0^\infty(\Lambda^n) \to C^\infty(\mathbf{R}_x^n) \, (\operatorname{mod} O(\lambda^{-1})),$$

which is called a *canonical operator*. Its most important property is expressed by commutation formula (0.27).

Up to now, canonical operator K has been defined with accuracy to order $O(\lambda^{-1})$ (for $\lambda \to +\infty$). Thus, whenever the atlas, the partition of unity and the local coordinates in charts are changed, the function $(K\varphi(r))(x)$ is replaced by the function $(\tilde{K}\varphi(r))(x) = (K\varphi(r))(x) + O(\lambda^{-1})$. However, it is possible to improve our construction of the canonical operator in such a way that it is determined with accuracy to $O(\lambda^{-N})$, where $N \geq 1$ is an arbitrary fixed number (see § 9).

Now we shall discuss the structure of the canonical operator a bit in detail. Let us fix manifold Λ^n, distinguished point $r^0 \in \Lambda^n$, and also charts Ω_j, their coordinates and the partition of unity. Furthermore, denote the *set of singular points* (with respect to the projection on $\mathbf{R}_x^n$) of manifold Λ^n by $\Sigma(\Lambda^n)$. In the neighbourhoods of the singular points we cannot take the coordinates $x_1, \ldots, x_n$ as local coordinates. Further let $\pi_x \Sigma(\Lambda^n)$ (or shortly $\pi_x \Sigma$) be the projection of set $\Sigma(\Lambda^n)$ on $\mathbf{R}_x^n$; it is called a *caustic*. Consider the function

$$u(x, \lambda) = (K\varphi(r))(x, \lambda),$$

where $\varphi \in C_0^\infty(\Lambda^n)$, $K = K_{\Lambda^n}^{r^0}$, for three particular cases.

(1) Let $\pi_x \operatorname{supp} \varphi$ be the projection of the support of function φ on $\mathbf{R}_x^n$ and let $U \subset \mathbf{R}_x^n$ be a domain that does not intersect $\pi_x \operatorname{supp} \varphi$. Then

$$u(x, \lambda) = O(\lambda^{-1}) \quad (\lambda \to +\infty)$$

uniformly in $x \in U$. Consequently, function $K\varphi$ is localized in a neighbourhood of $\pi_x \operatorname{supp} \varphi$ for $\lambda \gg 1$.

(2) Let x^0 be a point belonging to $\pi_x \operatorname{supp} \varphi$ but not to the caustic. For simplicity let point x^0 be the image (under projection π_x) of only a finite number of points $r^1, r^2, \ldots, r^N$ of manifold Λ^n. Then

$$u(x^0, \lambda) = \sum_{j=1}^N c_j \exp\left(i\lambda \int_{r^0}^{r^j} \langle p, dx \rangle \varphi(r^j) + O(\lambda^{-1}) \right),$$

where $c_j \neq 0$. Consequently, $(K\varphi)(x^0)$, for our x^0, is a sum of rapidly oscillating exponentials.

(3) Finally, let point $x^0 \in \pi_x$ supp φ and belong to the caustic. For simplicity, suppose there is only one point $r^0 \in \Lambda^n$ which is projected into x^0. If, in a neighbourhood of point r^0, we can introduce local coordinates $(p_{(\alpha)}, x_{(\beta)})$ on Λ^n, then, up to a factor of modulus one, we obtain

$$u(x, \lambda) = F^{-1}_{\lambda, p_{(\alpha)} \to x_{(\alpha)}} \left[\sqrt{\left| \frac{d\sigma^n}{dp_{(\alpha)} \, dx_{(\beta)}} \right|} \, \varphi(r) \times \right.$$

$$\left. \times \exp\left(i\lambda \int_{r^0}^{r} \langle p, dx \rangle - \langle x_{(\alpha)}(p_{(\alpha)}, x_{(\beta)}), p_{(\alpha)} \rangle \right) \right], \quad (0.28)$$

for x close to x^0. Here the manifold Λ^n is defined by the equations

$$x_{(\alpha)} = x_{(\alpha)}(p_{(\alpha)}, x_{(\beta)}), \qquad p_{(\beta)} = p_{(\beta)}(p_{(\alpha)}, x_{(\beta)}).$$

Consequently, function $K\varphi$ is the Fourier transform of a rapidly oscillating exponential (i.e. the continual sum of rapidly oscillating exponentials) in a neighbourhood of the caustic point x^0.

The integral (0.28) can be written as

$$u(x, \lambda) = \lambda^{k/2} \int_{R^k} \psi(p_{(\alpha)}, x) \exp (i\lambda S) \, dp_{(\alpha)}, \qquad (0.28')$$

where variables x play the role of parameters, and $S \equiv S(p_{(\alpha)}, x)$ is a real function. The stationary points of phase S (considered as a function of variables $p_{(\alpha)}$ with fixed x) are found from the equation

$$x_{(\alpha)} = x_{(\alpha)}(p_{(\alpha)}, x_{(\beta)}).$$

It is easily verified that, whenever point x does not lie on the caustic, the corresponding stationary points $p_{(\alpha)} = p_{(\alpha)}(x)$ of phase S are non-degenerate, and the asymptotics of integral (0.28) is evaluated by the stationary phase method. If point x lies on the caustic, the corresponding stationary point of phase S is degenerate. In this case it is difficult to determine the asymptotics of the integral explicitly; one can only state that

$$u(x, \lambda) \sim \text{const.} \; \lambda^r (\ln \lambda)^m \qquad (\lambda \to +\infty),$$

where $r > 0$ is a rational number and m an integer. In particular, if point x belongs to the caustic, the function $(K\varphi)(x, \lambda)$ is increasing for $\lambda \to +\infty$. This fact (i.e. the steep increase of light intensity in the focal points of light rays) has been known in optics for long. Recall Archimedes who burned the Roman fleet in Syracuse by using a parabolic mirror made of polished copper sheets.

However, we are interested not so much in the behaviour of function $u(x, \lambda)$ in the fixed caustic point x^0, as in the asymptotic formulae which are uniform with respect to x (for $\lambda \to + \infty$) in a neighbourhood of the caustic point x^0. At present such formulae can be derived only in the simplest cases. For instance, assume that, in a neighbourhood of the point $p = 0$, $x = 0$, the manifold Λ^n is given by the equations

$$x_1 = p_1^2, \qquad p_j = \partial S / \partial x_j \qquad (j \geq 2),$$

where $S = S(p_1, x_2, \ldots, x_n) \in C^\infty$ for small $|p_1|, |x_j|$. Then integral (0.28′) is one-dimensional, $p_{(\alpha)} = p_1$, and the phase function S has the form

$$S = -\tfrac{2}{3} p_1^3 + x_1 p_1.$$

In this case the leading term of the asymptotics of integral $u(x, \lambda)$ (for small $|x|$) is expressed in terms of the Airy–Fock function

$$v(\alpha) = \frac{1}{2\sqrt{\pi}} \int_{-\infty}^{\infty} \exp\left[i\left(\alpha t + \frac{t^3}{3} \right) \right] dt.$$

However, the asymptotics of the integral $u(x, \lambda)$ in a neighbourhood of the caustic cannot be expressed by means of known special functions even for $n = 1$. To see this, it is sufficient to consider the canonical operator corresponding to the curve $x = p^k$ ($k \geq 3$) in (x, p)-plane (see § 8).

The study of asymptotics of the canonical operator near the caustic points leads to the following question: is it possible to replace the evaluation of asymptotics of integrals (0.28′) by that of some standard integrals? However, before answering this question one has to investigate singularities of Lagrangian manifolds (i.e. the structure of set $\Sigma(\Lambda^n)$). For dimensions $n \leq 5$ the singularities of Lagrangian manifolds were completely classified in [3]. In this case, the asymptotics is expressed in terms of standard integrals – i.e. by means of new special functions (see §§ 7 and 8). These special functions are practically unknown at present.

8. Let us summarize the content of the second part of this volume. There the canonical operator is used to construct asymptotic solutions of the following problems:

(I) The Cauchy problem with rapidly oscillating initial data for λ-pseudo-differential equations.

(II) The scattering problem in semi-classical approximation for $h \to +0$ for the stationary Schrödinger equation.

(III) The determination of the asymptotics for certain series of eigenvalues and eigenfunctions of self-adjoint λ-p.d. operators.

Thus, in § 10, the Cauchy problem

$$\hat{L}\psi \equiv \lambda^{-1}D_t\psi + H(t, x, \lambda^{-1}D_x)\psi = 0,$$
$$\psi|_{t=0} = \psi_0(x)\exp[i\lambda S_0(x)], \tag{0.29}$$

is considered, where $x \in \mathbf{R}^n$. For simplicity, it is assumed that the symbol $H(t, x, p)$ does not depend on λ. Function $\psi_0(x) \in C_0^\infty(\mathbf{R}^n)$, function $S_0(x) \in C^\infty(\mathbf{R}^n)$ and is real-valued, and symbol H is a real C^∞-function. For convenience, the reader may regard $\hat{L}$ to be a differential operator the coefficients of which (at the derivatives with respect to x_j) are polynomials in x or rapidly tend to constants at infinity. An example of the Cauchy problem of type (0.29) is the Cauchy problem (0.1), (0.2) for the Schrödinger equation with potential $V(x) \in \mathcal{S}(\mathbf{R}^n)$.

Next we have to find out the global asymptotic solution of Cauchy problem (0.29) for an arbitrary finite time T (i.e. for $0 \le t \le T, x \in \mathbf{R}^n$).

There is an n-dimensional Lagrangian manifold Λ_0^n in phase space $\mathbf{R}_{x,p}^{2n}$, naturally associated with the problem (0.29). This manifold is given by the equations

$$x = y, \qquad p = \partial S_0(y)/\partial y, \qquad y \in \mathbf{R}^n,$$

and is diffeomorphically projected on the coordinate space $\mathbf{R}_x^n$. The Hamilton system corresponding to problem (0.29) is given by

$$\frac{dx}{dt} = \frac{\partial H}{\partial p}, \qquad \frac{dp}{dt} = -\frac{\partial H}{\partial x}. \tag{0.30}$$

Let g^t be the displacement along trajectories of this system within time t. The solution of the Hamilton system with Cauchy data $(x, p)|_{t=0} \in \Lambda_0^n$ will be denoted by $\{x(t, y), p(t, y)\}$.

Displacement g^t transforms Lagrangian manifold Λ_0^n into the Lagrangian manifold $\Lambda_t^n = g^t\Lambda_0^n$. If $T > 0$ is sufficiently small, manifold Λ_t^n is diffeomorphically projected on $\mathbf{R}_x^n$ for $0 \le t \le T$. In this case (i.e. when constructing a local asymptotic solution of the Cauchy problem) we can just suppose an approximate solution in the form of series (0.4), i.e.

$$\psi(t, x, \lambda) = \exp\left[i\lambda S(t, x)\right] \sum_{j=0}^{N} (i\lambda)^{-j}\psi_j(t, x).$$

Function $S(t, x)$ satisfies the Hamilton–Jacobi equation

$$\frac{\partial S}{\partial t} + H\left(t, x, \frac{\partial S}{\partial x}\right) = 0,$$

and the Cauchy data are determined by (0.29):

$$S|_{t=0} = S_0(x).$$

Function $S(t, x)$ is of the form

$$S(t, x) = S_0(y) + \int_0^t \left[\left\langle p, \frac{dx}{dt} \right\rangle - H \right] dt. \tag{0.31}$$

In this formula $x = x(t, y)$ and the integral is taken along the trajectory of Hamilton system (0.30).

For function ψ_0 we obtain the transport equation

$$\frac{1}{\sqrt{J}} \frac{d}{dt}(\sqrt{J}\,\psi_0) = \frac{1}{2} \frac{\partial^2 H}{\partial x\, \partial p} \psi_0$$

(cf. (0.14)). Here, J is the Jacobian

$$J(t, y) = \det \frac{\partial x(t, y)}{\partial y}$$

and

$$\frac{\partial^2 H}{\partial x\, \partial p} = \sum_{j=1}^n \frac{\partial^2 H}{\partial x_j\, \partial p_j}.$$

We choose the Cauchy data in the following way:

$$\psi_0|_{t=0} = \psi_0(x).$$

Then

$$\psi_0(t, x) = \sqrt{\frac{J(0, y)}{J(t, y)}} \exp\left[\frac{1}{2} \int_0^t \frac{\partial^2 H}{\partial x\, \partial p} dt \right] \psi_0(y), \tag{0.32}$$

where $x = x(t, y)$, and the integral is taken along the phase trajectory.

Thus the function

$$v(t, x, \lambda) = \exp\left[i\lambda S(t, x) \right] \psi_0(t, x),$$

where S and ψ_0 are determined by relations (0.31), (0.32) exactly fulfils Cauchy data (0.29) and satisfies Equation (0.29) with accuracy to $O(\lambda^{-2})$ (i.e. $\hat{L}v = O(\lambda^{-2})$). Taking higher approximations into account, we can obtain function v^N which exactly fulfils the Cauchy data and satisfies Equation (0.29) with accuracy to $O(\lambda^{-N})$, for an arbitrary $N \geq 2$.

In the case when symbol H is independent of t and generates a self-adjoint operator in $L_2(\mathbf{R}^n)$, we can rigorously prove that approximate solution v^N is close to the exact solution in the norm of $L_2(\mathbf{R}^n)$ (see Proposition 10.3). Thus we see that the derivation of the asymptotic solution of Cauchy problem (0.29) for small t reduces to the integration of a system of ordinary differential equations (i.e. the Hamilton system). At the same time, asymptotic formulae for the solutions are expressed in the framework of classical mechanics.

Let us mention one important point. Denote by M a subset of phase space $R^{2n}_{x,p}$ and by $\Pi_T M$ the set consisting of the trajectories within time $0 \leq t \leq T$, which initiate in M at $t = 0$. In other words, $\Pi_T M = \bigcup_{0 \leq t \leq T} g^t M$ (a tube of trajectories). Now, let M be the set

$$M = \{(x,p) \in \mathbf{R}^{2n} : x \in \mathrm{supp}\,\psi_0, p = \partial S_0 / \partial x\}.$$

Then $\mathrm{supp}\,v(t,x,\lambda)$ is contained in the *tube of rays* $\Pi_{T,x}(M)$, i.e., in the projection of the tube of trajectories $\Pi_T(M)$ on x-space. This fact is true also for all higher approximations v^N.

The initial data

$$\psi|_{t=0} = \psi_0(x) \exp\left[i\lambda S(x)\right]$$

can be interpreted as a wave packet localized on $\mathrm{supp}\,\psi_0$. For $\lambda \gg 1$, this wave packet evolves according to the laws of classical mechanics; in particular, the support $\mathrm{supp}\,\psi(t,x,\lambda)$ displaces gradually with t along the tube of rays (provided all quantities of order $O(\lambda^{-1})$ are neglected). This is the meaning of the frequently used phrase "quantum mechanics passes into classical mechanics when $h \to 0$".

The formula (0.32) is true whenever the Jacobian $J(t,y)$ does not vanish, or, in other words, as long as the Lagrangian manifolds $\Lambda^n_t, 0 \leq t \leq T$ admit diffeomorphic projections on $\mathbf{R}^n_x$. The validity of approximation (0.32) is violated in the points for which Jacobian J vanishes.

Now we shall use the canonical operator technique to construct the global asymptotic solution, i.e. the solution within an arbitrary finite time T. First of all we realize that, taking the volume element $d\sigma^n(x) = dx$ on the initial manifold Λ^n_0, we can express Cauchy data (0.29) by means of the canonical operator:

$$\psi|_{t=0} = e^{i\gamma}(K_{\Lambda_0^n}\psi_0(r))(x).$$

Here $\psi_0(r) = \psi_0(x)$, $r = (x, p(x))$, and $\gamma = \lambda S(x^0)$, where $r^0 = (x^0, p(x^0))$ is a distinguished point on manifold Λ_0^n. Actually,

$$\int_{r^0}^{r} \langle p, dx \rangle = \int_{r^0}^{r} \langle \partial S_0(x)/\partial x, dx \rangle = S_0(x) - S_0(x^0).$$

(Here again $r = (x, p(x))$ and the representation of the initial data in terms of canonical operator K_{Λ_0} is obtained by virtue of (0.18).) On the Lagrangian manifold Λ_t^n we take volume element $d\sigma_t^n$ which is induced by phase flow g^t from the volume element $d\sigma^n$ on Λ_0^n. By using commutation formula (0.27) it is not hard to show that the function

$$v(t, x, \lambda) = e^{i\lambda\delta} K_{\Lambda_t^n}\tilde{\psi} \tag{0.33}$$

exactly fulfils Cauchy data (0.29) and satisfies Equation (0.29) with accuracy to $O(\lambda^{-2})$. Functions δ and $\tilde{\psi}$ are easily expressed in terms of ψ_0 and L (see § 10); the formula (0.33) corresponding to the Schrödinger equation will be written in full detail later on.

This result can be briefly formulated in the following way: if the initial data have the form

$$\psi|_{t=0} = K_{\Lambda_0}\psi_0$$

the asymptotic solution at time t is given (up to a multiplier) by the formula

$$\psi|_t = K_{\Lambda_t^n}\psi.$$

For non-focal points, there exist simpler asymptotic formulae. We shall write them down for the Schrödinger equation, i.e. for problem (0.1), (0.2). We fix a non-focal point (t^0, x^0). Let $y^1, \ldots, y^N$ be all points such that $x(t^0, y^j) = x^0$ (i.e. $x(t, y^j)$ is a ray which begins in point y^j at time $t = 0$ and reaches point x^0 at the time $t = t^0$). Then, for $h \to +0$,

$$\psi(t^0, x^0) = \sum_{j=1}^{N} \psi_0(y^j)\sqrt{\left|\frac{J(0, y^j)}{J(t^0, y^j)}\right|} \times$$

$$\times \exp\left(\frac{i}{h}S_j - \frac{i\pi}{2}\mu_j\right) + O(h). \tag{0.34}$$

Here,

$$S_j = S_0(y^j) + \int_0^{t^0} \left(\frac{\dot{x}^2}{2} - V(x) \right) dt,$$

and μ_j is the *Morse index* of the ray $x = x(t, y^j)$, $0 \leq t \leq t^0$. In our particular case, μ_j is equal to the number of zeros of the Jacobian $J(t, y^j)$ on the ray (including their multiplicities). Thus μ_j is equal to the number of the contact points of ray $x = x(t, y^j)$ with the caustic of ray family $\{x = x(t, y)\}$ within time $0 \leq t \leq t^0$. In the simplest case, when $N = 2$ and when, on one of the rays, Jacobian J has just one simple zero, while on the other J is non-vanishing, the usual interpretation of formula (0.34) states: "when passing through a caustic the phase changes by $-\pi/2$".

As before (and again for small time intervals) the asymptotic solution $v(t, x, \lambda)$ (see (0.33)) of Cauchy problem (0.29) is localized in a neighbourhood of the tube of rays $\Pi_{T,x}(M)$. Namely,

$$v(t, x, \lambda) = O(\lambda^{-1}) \qquad (\lambda \to +\infty)$$

outside of an arbitrary fixed neighbourhood of the tube. We stress once more that the asymptotic solution of the Cauchy problem constructed in § 10, is global, i.e. for all $x \in \mathbf{R}^n$ and for an arbitrary finite time T.

The Cauchy problem for the Schrödinger equation is discussed in § 12.

In § 11, Cauchy problem (0.29) is considered for the system of equations (ψ is an N-vector, H is a Hermitian ($N \times N$)-matrix) under the assumption that the eigenvalues h_α of matrix H have constant multiplicities. The general construction procedure of the asymptotics is the same as for scalar equations; we mention only some new features. There are not one but $m \geq 1$ Hamilton–Jacobi equations (in accordance with the number of distinct eigenvalues h_α of matrix H)

$$\frac{\partial S_\alpha}{\partial t} + h_\alpha \left(t, x, \frac{\partial S_\alpha}{\partial x} \right) = 0,$$

associated with system (0.29). Consequently, there are m Hamilton systems and m Lagrangian manifolds $\Lambda_{t,\alpha}^n$ corresponding to Cauchy problem (0.29). The transport equations are substantially more complicated, especially in the case when matrix H has multiple eigenvalues. These equations are written down in § 11.

Notice that in the case of the system of Equations (0.29), *polarization* of the initial wave packet takes place. Namely, let all eigenvalues of matrix H be simple for all (t, x, p). Then space $\mathbf{R}^N$ decomposes into a direct sum of N invariant subspaces of matrix H for any (t, x, p); correspondingly, the initial vector ψ_0 is represented by the sum $\sum_{j=1}^{n} \psi^{0j}$ of the eigenvectors of matrix $H(0, x, p)$, where $p = \partial S_0(x)/\partial x$. In the first approximation, the asymptotics of the solution is equal to a sum of rapidly oscillating exponentials for short time intervals, whereas, for long time intervals, it is equal to a sum of canonical operators acting on vectors ψ^{0j} of the form $\exp(i\lambda S_j)\psi^j$, where ψ^j are the eigenvectors of matrix H. Thus, the original wave packet $\psi|_{t=0}$ splits into the N wave packets, each of which is moving along the trajectories of its own Hamilton system.

§ 14 is devoted to the Dirac equation (i.e. to the system of four first-order partial differential equations in t and x, involving matrix H with two double eigenvalues). Thus, with the aid of the results contained in § 11, the asymptotic solution of Cauchy problem (0.29) for the Dirac system is derived. It is shown that spin polarization has the classical limit for $h \to 0$. Moreover, in § 14, we derive the transport equations by a method different from that in § 11.

The scattering problem for the Schrödinger equation when $h \to 0$ is considered in § 12; we present there only a survey of the results obtained in [42].

The problem (III) (see p. 21) is discussed in § 13. Namely, the equation

$$\mathscr{A}(h)\psi = E\psi \tag{0.35}$$

is investigated where $\mathscr{A}(h)$ is a self-adjoint operator in $L_2(\mathbf{R}^n)$ (with a purely discrete spectrum) generated by the formally symmetric h^{-1}-pseudo-differential operator

$$\mathscr{L} = \tfrac{1}{2}[L(\overset{2}{x}, h\overset{1}{D}_x) + L(\overset{1}{x}, h\overset{2}{D}_x)]$$

with symbol $L(x, p)$ smooth and real-valued.

It is well known that there is a tremendous difference between one- and multi-dimensional eigenvalue problems. For illustration, let us consider two cases:

$$-\psi''(x) + V(x)\,\psi(x) = E\,\psi(x), \quad x \in \mathbf{R},$$
$$-\Delta\psi(x) + V(x)\,\psi(x) = E\,\psi(x), \quad x \in \mathbf{R}^n (n \geq 2).$$

If $V(x) \in C^\infty(\mathbf{R}^n)$, $V(x) \to +\infty$ for $|x| \to \infty$, then the spectra in both cases are purely discrete and the eigenfunctions belong to $L_2(\mathbf{R}^n)$. If $n = 1$, the asymptotics of both eigenvalues E_m and eigenfunctions ψ_m for $m \to +\infty$ were evaluated for fairly large classes of potentials. The following asymptotic formula holds for the eigenvalues:

$$\int\limits_{V(x) < E_m} \sqrt{E_m - V(x)}\, dx = m\pi + (\pi/2) + o(1) \quad (m \to \infty).$$

For $n \geq 2$, an analogous asymptotic formula for the eigenvalues can be derived, too, but more general results on asymptotics of eigenfunctions are missing. Moreover, even in the cases in which separation of variables is possible, the semi-classical asymptotic formulae can be derived only for some special series of eigenfunctions. Recall that the *semi-classical asymptotics* is understood to be an asymptotic formula written in terms of classical mechanics (i.e. all quantities appearing in the formula are expressed with the aid of the parameters of a certain family of phase trajectories).

It turns out that not all eigenfunctions (with large quantum numbers) possess the semi-classical asymptotics, even for the Schrödinger equation.

Let us discuss again problem (0.35). In order to get information on the corresponding spectrum it is sufficient to construct an *approximate eigenfunction* (see Lemma 13.1), i.e., the approximate solution of Equation (0.35). It is found via the canonical operator method. Let Λ^n be the compact Lagrangian manifold which belongs to the set of the level $L(x, p) = E = $ const. of symbol L, and function $\varphi \in C^\infty(\Lambda^n)$. Let this manifold be invariant with respect to displacements along trajectories of the dynamical system associated with symbol L. Let us consider the expression $(K\varphi(r))(x)$ for fixed $h = \lambda^{-1}$ where $K = K_{\Lambda^n}$. The expression $K\varphi$ is, in general, a multi-valued function of x, since the integral $\int_{r^0}^r \langle p, dx \rangle$ may depend on integration path. In addition, the index ind $l[r^0, r]$ may also be different from zero.

Further, it is shown that the necessary and sufficient condition for expression $u = (K\varphi)(x)$ to be a single-valued function of x (for h fixed) is given by

$$h^{-1} \oint\limits_\gamma \langle p, dx \rangle - \frac{\pi}{2} \text{ind } \gamma = 2\pi m, \quad m \in \mathbf{Z}, \tag{0.36}$$

for an arbitrary closed path γ lying on manifold Λ^n. Let $H_1(\Lambda^n, \mathbf{Z})$ be the first homology group of manifold Λ^n with integral coefficients, and p_1 its dimension. Then it suffices if relations (0.36) are fulfilled only for all basis cycles $\gamma_1, \ldots, \gamma_{p_1}$ of group $H_1(\Lambda^n, \mathbf{Z})$. If $p_1 = 1$, relations (0.36) can be satisfied only for some discrete set of values of h. If $p_1 \geqq 2$, it may happen that no value of h exists for which relations (0.36) are satisfied. Obviously, all relations (0.36) (for $\gamma = \gamma_1, \ldots, \gamma_{p_1}$) can be fulfilled, if there exists a family of compact Lagrangian manifolds $\{\Lambda^n(\alpha)\}$ depending smoothly on p_1 parameters $\alpha = (\alpha_1, \ldots, \alpha_{p_1})$. In this case a series of approximate eigenfunctions of operator $\mathscr{A}(h)$ can be constructed. The series allows to prove the existence of asymptotics for the corresponding series of eigenvalues of operator $\mathscr{A}(h)$ (Theorem 13.3). However, it is not possible rigorously to prove the proximity of these approximate and exact eigenfunctions. In Section 13, the eigenvalue problem for the Laplace–Beltrami operator on the sphere in $\mathbf{R}^n$ is presented as an example.

The boundary-value problems for partial differential equations are not treated in this volume. The asymptotic series of eigenvalues and eigenfunctions of operator $-c^2(x)\Delta$ in a bounded domain in $\mathbf{R}^n$, with vanishing Dirichlet or Neumann conditions on the boundary, are investigated in the works [5], [38].

In the present monograph there is no discussion of various diffraction problems for electromagnetic, acoustic and other waves; these topics are treated in the monographs [1], [5], [7], [8], [85].

Recently the canonical operator method was further developed in a series of works (see [9], [42]–[45], [64]–[67]). Except the above mentioned cases, this method finds application in studying the propagation of discontinuities of a solution of the Cauchy problem for hyperbolic equations; it is briefly mentioned in § 10 together with the references.

This volume cannot be regarded as a complete exposition or even a review of the whole variety of asymptotic methods of linear mathematical physics; its main purpose is just to explain one of these methods – the method of canonical operator. In particular, it does not treat the algebraic ideas of J. Leray which essentially extended the scope of the canonical operator method.

PART I

QUANTIZATION OF VELOCITY FIELD
(THE CANONICAL OPERATOR)

§ 1. The Method of Stationary Phase. The Legendre Transformation

The semi-classical approximation yields integrals of rapidly oscillating functions. The asymptotics of such integrals is evaluated by means of the stationary phase method. Its basic formulae are summarized in this paragraph together with the definition and the main properties of the Legendre transformation.

1. *The Method of Stationary Phase*

We shall review the main results on the asymptotics of oscillatory integrals of the form

$$\int \varphi(x) \exp{(i\lambda\, S(x))}\, dx,$$

where λ is a large real parameter. Let $\mathbf{R}_\lambda^\pm$ denote the half-axes $\lambda \geq 1$, $\lambda \leq -1$. Here function $\varphi(x)$ has compact support, function $S(x)$ is real-valued, and both belong to $C^\infty(\mathbf{R}^n)$. Point x^0 is said to be a *critical* or *stationary* point of function $S(x)$, if $\partial S(x^0)/\partial x = 0$; the critical point is non-degenerate, whenever $\det{(\partial^2 S(x^0)/\partial x^2)} \neq 0$.

THEOREM 1.1. *Let function $S(x)$ have a single and moreover non-degenerate critical point x^0 on supp φ. Then for λ real and $|\lambda| \geq 1$, and for any integer $N \geq 1$*

$$I(\lambda) \equiv \int \varphi(x) \exp{(i\lambda\, S(x))}\, dx =$$

$$= \lambda^{-n/2} \exp{(i\lambda\, S(x^0))} \sum_{j=0}^{N-1} a_j(\varphi, S)\lambda^{-j} + R_N(\lambda). \tag{1.1}$$

Here $a_j(\varphi, S) = (P_j\varphi)(x^0)$, where P_j is a linear differential operator of order $2j$ with C^∞-coefficients. The estimate of the remainder is given by

$$\left| R_N(\lambda) \right| \leq C_n \lambda^{-(n/2)-N} \|\varphi\|_{C^\beta(\mathbf{R}^n)}, \tag{1.2}$$

29

where $\beta = \beta(N) < \infty$. Here

$$\sqrt{\lambda} > 0, \quad \lambda > 0; \qquad \sqrt{\lambda} = i|\sqrt{-\lambda}|, \quad \lambda < 0. \tag{1.3}$$

For the proof see Theorem 2.2 in [20], where a review of the literature on the stationary phase method is also given.

As well known, it is just the stationary points which give the leading contribution to the asymptotics of an oscillatory integral.

Now suppose A be a real symmetric non-degenerate $(n \times n)$-matrix. We introduce the following notation: sgn A for the signature of the quadratic form associated with matrix A, inerdex A for the inertial index of matrix A. If v_+ and v_- are the numbers of positive and negative eigenvalues of matrix A, respectively, then

$$\text{sgn}\, A = v_+ - v_-, \qquad \text{inerdex}\, A = v_-,$$

$$\text{sgn}\, A + 2\, \text{inerdex}\, A = n.$$

Formula (1.1) expresses the asymptotic expansion of integral $I(\lambda)$ for $\lambda \to \pm \infty$. The leading term of the asymptotics has the form

$$I(\lambda) \sim \left(\frac{2\pi}{\lambda}\right)^{n/2} \left|\det \frac{\partial^2 S(x^0)}{\partial x^2}\right|^{-1/2} \times$$

$$\times \varphi(x^0) \exp\left[i\lambda S(x^0) + \frac{i\pi}{4} \text{sgn}\, \frac{\partial^2 S(x^0)}{\partial x^2}\right], \qquad (\lambda \to +\infty). \tag{1.4}$$

The expansion coefficients in (1.1) are determined by the formula ([20], Theorem 2.3)

$$I(\lambda) = b_n \lambda^{-n/2} \exp\left[i\lambda S(x^0)\right] \sum_{j=0}^{N} \left(\frac{\lambda^{-j}}{j!}\right) L^j(\varphi(x) \times$$

$$\times \exp\left[i\lambda S(x, x^0)\right])\big|_{x=x^0} + \lambda^{-\alpha_N} R_N^*(\lambda), \qquad \lambda \in \mathbf{R}_\lambda^+. \tag{1.5}$$

Here

$$L = \frac{i}{2}\left\langle \left(\frac{\partial^2 S(x^0)}{\partial x^2}\right)^{-1} \frac{\partial}{\partial x}, \frac{\partial}{\partial x}\right\rangle,$$

$$S(x, x^0) = S(x) - S(x^0) - \frac{1}{2}\left\langle \frac{\partial^2 S(x^0)}{\partial x^2}(x - x^0), (x - x^0)\right\rangle,$$

$$b_n = (2\pi)^{n/2} \left|\det \frac{\partial^2 S(x^0)}{\partial x^2}\right|^{-1/2} \exp\left(\frac{i\pi}{4} \text{sgn}\, \frac{\partial^2 S(x^0)}{\partial x^2}\right),$$

$$\alpha_N = \frac{n}{2} + N + 1 - \left[\frac{2N+2}{3}\right] \tag{1.6}$$

and the estimate of the remainder for $\lambda \geq 1$ is given by

$$|\mathbf{R}_N^*(\lambda)| \leq C_N \|\varphi\|_{C^\gamma(\mathbf{R}^n)}, \tag{1.7}$$

where $\gamma = \gamma(N) < \infty$.

The asymptotic expansions (1.1) and (1.5) can be differentiated any number of times with respect to λ, with an analogous estimate for the remainder.

In the one-dimensional case, there exist more precise formulae and estimates ([20], Theorem 2.1).

Since, in applications, $I(\lambda)$ usually depends on supplementary slowly varying parameters ω, $\omega \in \mathbf{R}^k$, Theorem 1.1 can rarely be applied in its original form. Let $\Omega \subset \mathbf{R}_\omega^k$ be a finite domain. Analogously to E. Landau's symbol 'O' we introduce

DEFINITION 1.2. *Function* $u(\omega, \lambda)$ *belongs to the class* $O_m^+(\Omega)$, *if the following conditions are satisfied*:

(1) $u(\omega, \lambda) \in C^\infty(\Omega \times \mathbf{R}_\lambda^+)$.
(2) For any compact subset K of Ω, any multi-index α and any integer $\beta \geq 0$

$$|D_\omega^\alpha D_\lambda^\beta u(\omega, \lambda)| \leq C_{\alpha, \beta, K} \lambda^{m+|\alpha|}, \quad (\omega \in K, \lambda \in \mathbf{R}_\lambda^+). \tag{1.8}$$

The class $O_m^-(\Omega)$ $(\lambda \in \mathbf{R}_\lambda^-)$ is defined analogously. As a typical example of the function of class $O_m^\pm(\Omega)$ one may consider the function

$$u(\omega, \lambda) = \lambda^m \varphi(\omega) \exp(i\lambda S(\omega)),$$

where $\varphi, S \in C^\infty(\Omega)$ and function S is real-valued.

We set $O_{-\infty}^\pm(\Omega) = \bigcap_{m=-\infty}^{+\infty} O_m^\pm(\Omega)$.

PROPOSITION 1.3. *The following inclusions are true*:

(1) $O_{m_1}^+(\Omega) + O_{m_2}^+(\Omega) \subset O_m^+(\Omega)$, $m = \text{Max}(m_1, m_2)$.
(2) $O_{m_1}^+(\Omega) O_{m_2}^+(\Omega) \subset O_{m_1+m_2}^+(\Omega)$.
(3) $\lambda O_m^+(\Omega) \subset O_{m+1}^+(\Omega)$.
(4) $D_\omega^\alpha O_m^+(\Omega) \subset O_{m+|\alpha|}^+(\Omega)$.
(5) $D_\lambda^\beta O_m^+(\Omega) \subset O_m^+(\Omega)$.

Analogous inclusions hold for the classes $O_m^-(\Omega)$.
The proof is obvious.
Now we shall consider the integral

$$I(\lambda, \omega) = \int \varphi(x, \omega, \lambda) \exp\left[i\lambda S(x, \omega)\right] dx, \tag{1.9}$$

and formulate differential conditions for functions φ and S:

(1) $\varphi \in C^\infty(G \times \Omega \times \mathbf{R}_\lambda^+)$, $S \in C^\infty(G \times \Omega)$, and function S is real-valued.

(2) There exists a compact subset $K \subset G$ such that the projection $\pi_x(\operatorname{supp}\varphi)$ on $\mathbf{R}_x^n$ is contained in K.

(3) For arbitrary multi-indices α, β. and any integer $\gamma \geq 0$

$$\left|D_x^\alpha D_\omega^\beta D_\lambda^\gamma \varphi(x, \omega, \lambda)\right| \leq C_{\alpha, \beta, \gamma} \lambda^{m-\gamma}$$

for $(x, \omega, \lambda) \in G \times \Omega \times \mathbf{R}_\lambda^+$, where m is a fixed number.

The conditions on a stationary point, i.e. on a solution of the equation

$$\frac{\partial S(x, \omega)}{\partial x} = 0, \tag{1.10}$$

state:

(4) For each $\omega \in \Omega$, function $S(x, \omega)$ has a single critical point $x = x^0(\omega) \in \Omega$.

(5) $\inf_\Omega |\mu_j(\omega)| \geq \delta > 0$, $1 \leq j \leq n$, where $\mu_j(\omega)$ are the eigenvalues of the matrix $\partial^2 S(x^0(\omega), \omega)/\partial x^2$.

THEOREM 1.4. *Suppose conditions (1)–(5) are satisfied. Then the expansion*

$$I(\lambda, \omega) = \lambda^{-n/2} \exp\left[i\lambda S(x^0(\omega), \omega)\right] \times$$
$$\times \sum_{j=0}^{N-1} a_j(\varphi, S; \omega)\lambda^{-j/2} + R_N(\lambda, \omega) \tag{1.11}$$

holds for an arbitrary integer $N \geq 1$ and for $(\lambda, \omega) \in \mathbf{R}_\lambda^+ \times \Omega$. Here $a_j = [P_j(\omega, x, D_x)\varphi]$ for $x = x^0(\omega)$, where P_j are linear differential operators of order $2j$ with C^∞-coefficients. The remainder $R_N(\omega, \lambda) \in O_{m-(n/2)-N}^+(\Omega)$.

The leading term of asymptotics (1.11) is of the form (provided $\varphi(x^0(\omega), \omega, \lambda) \neq 0$ for $\lambda \gg 1$)

$$I(\lambda, \omega) \sim \left(\frac{2\pi}{\lambda}\right)^{n/2} \varphi(x, \omega, \lambda) \exp\left[i\lambda S(x, \omega)\right] \times$$

$$\times \left| \det \frac{\partial^2 S(x, \omega)}{\partial x^2} \right|^{-1/2} \exp\left[\frac{i\pi}{4} \operatorname{sgn}\left(\frac{\partial^2 S(x, \omega)}{\partial x^2} \right) \right] \Bigg|_{x = x^0(\omega)} \quad (1.12)$$

Moreover a formula analogous to (1.5) with an estimate of the remainder à la (1.7) holds, too.

The proof follows from [20], Theorem 2.4. Notice that formula (1.11) is also true if the assumptions of Theorem 1.4 are satisfied for $\lambda \in \mathbf{R}_\lambda^-$ and the branch of $\sqrt{\lambda}$ is chosen in accordance with (1.3).

2. *The λ-Fourier Transformation*

We shall use the following Fourier transformation depending on the real parameter $\lambda \neq 0$:

$$(F_{\lambda, x \to p} u(x))(p) = \left(\frac{\lambda}{2\pi i} \right)^{n/2} \int \exp\left[-i\lambda \langle x, p \rangle \right] u(x) \, dx. \quad (1.13)$$

The inverse transformation is given by

$$(F_{\lambda, p \to x}^{-1} v(p))(x) = \left(\frac{\lambda}{-2\pi i} \right)^{n/2} \int \exp\left[i\lambda \langle x, p \rangle \right] v(p) \, dp. \quad (1.14)$$

Here $\sqrt{i} = e^{i\pi/4}$, $\sqrt{-i} = e^{-i\pi/4}$, and $\sqrt{\lambda}$ is defined in (1.3). The λ-Fourier transformation with respect to some of the variables is defined similarly.

The following lemma will be useful in § 2.

LEMMA 1.5. *Suppose $\varphi(x) \in C_0^\infty(\mathbf{R}^n)$, $S(x) \in C^\infty(\mathbf{R}^n)$, and function $S(x)$ to be real-valued. Moreover, let $M(\varphi, S) = \{p: p = \partial S(x)/\partial x, x \in \operatorname{supp} \varphi\}$, and $G(p) \subset \mathbf{R}_p^n$ be a domain such that $G(p) \cap M(\varphi, S) = \varnothing$. Then we have*

$$\left| D_p^\alpha \left[(F_{\lambda, x \to p}(\varphi(x) \exp(i\lambda S(x))))(p) \right] \right| \leq$$
$$\leq C_{N, \alpha} |\lambda|^{-N} (1 + |p|)^{-N}, \quad \lambda \in \mathbf{R}_\lambda^\pm, p \in G(p), \quad (1.15)$$

for arbitrary integer $N \geq 0$ and for any multi-index α.

Proof. Take the formula

$$e^{i\lambda S(x)} = \frac{1}{i\lambda} L(e^{i\lambda S(x)}), \quad (1.16)$$

with

$$L = \sum_{j=1}^{n} a_j(x) \frac{\partial}{\partial x_j}, \qquad a_j(x) = \frac{\partial S(x)}{\partial x_j} \left| \frac{\partial S(x)}{\partial x} \right|^{-2},$$

where $S(x)$ is a sufficiently smooth function and $\nabla S(x) \neq 0$. Then

$$F_{\lambda, x \to p}(\varphi(x) \exp(i\lambda S(x))) =$$
$$= \left(\frac{\lambda}{2\pi i} \right)^{n/2} \frac{i}{\lambda} \int \varphi_1(x, p) \exp[i\lambda(S(x) - \langle x, p \rangle)] \, dx, \quad (1.17)$$

where

$$\varphi_1 = {}^t L \varphi, \qquad {}^t L = - \sum_{j=1}^{n} \frac{\partial}{\partial x_j} a_j(x).$$

For $x \in \operatorname{supp} \varphi, p \in G, |\alpha| > 0$, we have

$$0 < C_1(1 + |p|) \leq \left| \frac{\partial S(x)}{\partial x} - p \right| \leq C_2(1 + |p|),$$

$$\left| D_x^\alpha \left(\frac{\partial S(x)}{\partial x} - p \right) \right| \leq C.$$

Consequently, the modulus of the left-hand side of (1.17) is not larger than $C |\lambda|^{n/2 - 1} (1 + |p|)^{-1}$ for $\lambda \in \mathbf{R}_\lambda^\pm$, $p \in G$. Applying (1.16) once again we obtain (1.15) for $|\alpha| = 0$. The differentiation with respect to p yields an integral of the same form.

The usual Fourier transformation maps the Schwartz space S into itself; the λ-Fourier transformation preserves the Schwartz space S_λ^+ (or S_λ^-) depending on parameter λ, instead.

Let us recall that the *space* $S_\lambda^+ (\mathbf{R}_x^n)$, by definition, consists of functions $u(x, \lambda) \in C^\infty(\mathbf{R}_x^n \times \mathbf{R}_x^n)$ for which the estimate

$$|x^\beta D_x^\alpha u(x, \lambda)| \leq C_{\alpha\beta} \lambda^{|\alpha|}$$

holds for arbitrary multi-indices α, β, and $x \in \mathbf{R}^n$, $\lambda \in \mathbf{R}_\lambda^+$. The *space* $S_\lambda^- (\mathbf{R}_x^n)$ is introduced similarly.

PROPOSITION 1.6. *If* $u(x, \lambda) \in S_\lambda^+ (\mathbf{R}_x^n)$, *then*

$$\lambda^{-n/2} F_{\lambda, x \to p} u(x, \lambda) \in S_\lambda^+ (\mathbf{R}_p^n).$$

An analogous statement is true for $u \in S_\lambda^- (\mathbf{R}_x^n)$ and for $F_{\lambda, p \to x}^{-1}$. The proposition is proved by using the standard integration by parts.

3. *The Legendre Transformation*

First we shall consider the one-dimensional case. Let $\Gamma: y = S(x)$ be a convex sufficiently smooth curve in the plane. Thus to each point of Γ there corresponds a pair of numbers $(x, S(x))$. At each point of the curve consider the tangent. The obtained family of straight lines in the plane uniquely determines curve Γ, which turns out to be the envelope of this family. The pair of numbers which uniquely determine the tangent of Γ at each point, can, therefore, be taken as new coordinates of this point. The equation of the tangent of Γ in point $(x_0, S(x_0))$ is given by

$$y = S'(x_0)x + (S(x_0) - x_0 S'(x_0)),$$

so that the pair $(S'(x_0), S(x_0) - x_0 S'(x_0))$ determines the point $(x_0, S(x_0)) \in \Gamma$ uniquely. Let us introduce now a new independent variable,

$$p = S'(x), \tag{1.18}$$

and a new function

$$\tilde{S}(p) = xp - S(x). \tag{1.19}$$

Here, $x = x(p)$ is the function of p determined by (1.18). The transformation

$$L: (x, S(x)) \rightarrow (p, \tilde{S}(p)), \tag{1.20}$$

given by (1.18) and (1.19) is called a *Legendre transformation* or a *contact transformation*. Function $\tilde{S}(p)$ is said to be *Young-dual* to function $S(x)$.

Due to the convexity of function $S(x)$, the correspondence between x and p is one-to-one (for simplicity let $S(x) \in C^\infty(\mathbf{R})$). By virtue of (1.18) and (1.19) we have

$$d\tilde{S}(p) = \tilde{S}'(p)\,dp = x\,dp + p\,dx - S'(x)\,dx = x\,dp,$$

so that $\tilde{S}'(p) = x$ and the obtained formulae are symmetric:

$$p = S'(x), \qquad x = \tilde{S}'(p), \qquad S(x) + \tilde{S}(p) = xp. \tag{1.21}$$

If curve Γ is not strictly convex, the Legendre transformation has singularities.

EXAMPLE 1. $S(x) = x^3/3$. Function $\tilde{S}(p) = \pm (\frac{2}{3})p^{3/2}$ is double-valued (with a beak-like graph, Figure 3).

EXAMPLE 2. $S(x) = x^4/4$, $\tilde{S}(p) = \frac{3}{4}p^{4/3}$. Function S is convex every-

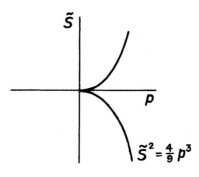

Fig. 3.

where, but not strictly convex at the point $x = 0$; function $\tilde{S}$ has a singularity at the point $p(0) = 0$, namely, $\tilde{S}''(0) = \infty$.

The Legendre transformation is *involutive*, i.e. $L^2 = I$ (I being the identity transformation), cf. (1.21). Further, if the curve $y = S(x)$ is convex, then the curve $q = \tilde{S}(p)$ is convex, too, since $d\tilde{S}'(p)/dp = 1/S''(x) > 0$. Moreover, $S''(x)\,\tilde{S}''(p) = 1$.

In the many-variable case the Legendre transformation is defined analogously to (1.18) and (1.19) where $x, p \in \mathbf{R}^n$, i.e. by

$$p = \frac{\partial S(x)}{\partial x}, \qquad \tilde{S}(p) = \langle x, p \rangle - S(x), \tag{1.22}$$

or in the symmetric form

$$p = \frac{\partial S(x)}{\partial x}, \qquad x = \frac{\partial \tilde{S}(p)}{\partial p},$$
$$S(x) + \tilde{S}(p) = \langle x, p \rangle. \tag{1.23}$$

Now we shall specify the main properties of the Legendre transformation (see [11], [23], [30]). Let G be a domain in $\mathbf{R}_x^n$, the function $S(x) \in C^\infty(G)$, and the manifold $x_{n+1} = S(x), x \in G$, have non-zero Gauss curvature, i.e.

$$\det \left\| \frac{\partial^2 S(x)}{\partial x^2} \right\| \neq 0, \qquad x \in G. \tag{1.24}$$

Then, according to the inverse function theorem, the equation $p = \partial S(x)/\partial x$ yields a diffeomorphism of some sufficiently small neighbourhoods U_0, V_0

of the points $x^0, p^0 = \partial S(x^0)/\partial x$, respectively; hence $\tilde{S}(p) \in C^{\infty}(V_0)$. With $x \in U_0$ and $p \in V_0$, we have

(1) The Legendre transformation is involutive (see (1.23)).

(2) The following identity holds:

$$\frac{\partial^2 S(x)}{\partial x^2} \frac{\partial^2 \tilde{S}(p)}{\partial p^2} = I. \tag{1.25}$$

Indeed,

$$dx = d(\partial \tilde{S}(p)/\partial p) = \partial^2 \tilde{S}(p)/\partial p^2 \, dp,$$

$$dp = \partial^2 S(x)/\partial x^2 \, dx.$$

(3) The Gauss curvature of the manifold $p_{n+1} = \tilde{S}(p)$, $p \in V_0$, differs from zero:

$$\det \left\| \frac{\partial^2 \tilde{S}(p)}{\partial p^2} \right\| \neq 0, \qquad p \in V_0, \tag{1.26}$$

and this manifold is convex, whenever such is the manifold $x_{n+1} = S(x)$, $x \in U_0$. This property follows from (2).

EXAMPLE 3. $S(x) = \frac{1}{2} \langle Ax, x \rangle$, where A is a real non-degenerate symmetric matrix. Then

$$p = Ax, \qquad \tilde{S}(p) = \frac{1}{2} \langle A^{-1} p, p \rangle.$$

The Legendre transformation with respect to some of the variables can be defined in a similar way.

Let us establish now the connection between the λ-Fourier transformation and the Legendre transformation. Suppose the following conditions are true:

(1) $S(x) \in C^{\infty}(U)$, $\varphi(x) \in C_0^{\infty}(U)$, and function $S(x)$ is real-valued (U being a domain in $\mathbf{R}_x^n$).

(2) The mapping $x \to p = \partial S(x)/\partial x$ is a diffeomorphism of U on $V \subset \mathbf{R}_p^n$.

Then the following theorem holds:

THEOREM 1.7. *For an arbitrary integer* $N \geq 1, p \in \mathbf{R}^n, \lambda \in \mathbf{R}_\lambda^+$, *we have the formula*

$$[F_{\lambda, x \to p}(\varphi(x) \exp(i\lambda S(x)))](p) =$$

$$= \exp\left[i\lambda\tilde{S}(p) - \frac{i\pi}{2} \text{ inerdex } \frac{\partial^2 S(x(p))}{\partial x^2} \right] \left| \det \frac{\partial^2 S(x(p))}{\partial x^2} \right|^{-1/2} \times$$

$$\times \left(\varphi(x(p)) + \sum_{k=1}^{N} \lambda^{-k}(R_k \varphi)(p) \right) + R_{-N}(p, \lambda). \tag{1.27}$$

The remainder R_{-N} belongs to $O^{+}_{-N}(\mathbf{R}^n_p)$.

REMARK 1.8. Operators R_k were described in Theorem 1.4. Expansion (1.27) can be differentiated with respect to p and λ any number of times, preserving the uniform estimate of the remainder in p and λ. Moreover, if G is the complement of a domain G' which contains $M(\varphi, S)$ (defined in Lemma 1.5), then more precise estimate (1.15) of the remainder is true.

Proof of Theorem 1.7. The left-hand side of (1.27) is equal to

$$\left(\frac{\lambda}{2\pi i} \right)^{n/2} \int \varphi(x) \exp\left[i\lambda(S(x) - \langle x, p \rangle) \right] dx.$$

The stationary points of the function in the exponent can be found from the equation $\partial S(x)/\partial x = p$; by virtue of condition (2), the stationary point $x = x(p)$ is unique for $x \in U$, and is non-degenerate. In this point, $S(x(p)) - \langle p, x(p) \rangle = \tilde{S}(p)$.

Now we shall apply the stationary phase method (Theorem 1.4); p plays the role of ω, V that of Ω. Theorem 1.4 yields the existence of an expansion in the powers of λ^{-1} and the estimate of the remainder. We evaluate the factor in front of the exponential in the leading term of the asymptotics. For $\lambda > 0$ it is given by

$$\left(\frac{\lambda}{2\pi i} \right)^{n/2} \left(\frac{2\pi}{\lambda} \right)^{n/2} \varphi(x) \left| \det \frac{\partial^2 S(x)}{\partial x^2} \right|^{-1/2} \times$$

$$\times \exp\left[\frac{i\pi}{4} \text{ sgn } \frac{\partial^2 S(x)}{\partial x^2} \right] \Bigg|_{x=x(p)} =$$

$$= \varphi(x) \left| \det \frac{\partial^2 S(x)}{\partial x^2} \right|^{-1/2} \exp\left[\frac{i\pi}{4} \left(\text{sgn } \frac{\partial^2 S(x)}{\partial x^2} - n \right) \right] \Bigg|_{x=x(p)},$$

which leads to (1.27).

COROLLARY 1.9. *Under the assumptions of Theorem 1.7 we have*

$$\left[F_{\lambda, x \to p}(\varphi(x)) \left| \det \frac{\partial^2 S(x)}{\partial x^2} \right|^{1/4} \exp(i\lambda S(x)) \right](p) =$$

$$= \varphi(x(p)) \exp\left[i\lambda \tilde{S}(p)\right] \left|\det \frac{\partial^2 \tilde{S}(p)}{\partial p^2}\right|^{1/4} \times$$

$$\times \exp\left[-\frac{i\pi}{2} \text{ inerdex } \frac{\partial^2 S(x(p))}{\partial x^2}\right] + R_{-1}(p, \lambda). \qquad (1.28)$$

The estimate of the remainder is the same as that in Theorem 1.7.

The proof follows from (1.27) and (1.25).

The formula (1.28) admits an interesting interpretation. Namely, let us consider the manifold

$$\Lambda^n = \left\{(x, p): x \in U, p = \frac{\partial S(x)}{\partial x}\right\}$$

in the space $\mathbf{R}^{2n} = \mathbf{R}^n_x \times \mathbf{R}^n_p$. Under the assumptions of Theorem 1.7 the equation of Λ^n can be written in the form

$$\Lambda^n = \left\{(x, p): x = \frac{\partial \tilde{S}(p)}{\partial p}, p \in V\right\},$$

so that either x, or p can be taken as the coordinates on Λ^n. Let x be the coordinates on Λ^n; then

$$\det \frac{\partial p(x)}{\partial x} = \det \frac{\partial^2 S(x)}{\partial x^2}$$

and, similarly, if p are the coordinates on Λ^n, then

$$\det \frac{\partial x(p)}{\partial p} = \det \frac{\partial^2 \tilde{S}(p)}{\partial p^2}.$$

Then formula (1.28) takes the symmetric form

$$\left[F_{\lambda, x \to p}(\varphi(x)) \left|\det \frac{\partial p(x)}{\partial x}\right|^{1/2} \exp(i\lambda S(x))\right](p) \sim$$

$$\sim \varphi(x(p)) \left|\det \frac{\partial x(p)}{\partial p}\right|^{1/2} \times$$

$$\times \exp\left[i\lambda \tilde{S}(p)\right] \exp\left(-\frac{i\pi}{2} \text{ inerdex } \frac{\partial x(p)}{\partial p}\right) \qquad (1.29)$$

with accuracy to $O(\lambda^{-1})$.

Symmetry of formula (1.29) is caused by unitarity of the Fourier transformation. Actually,

$$F_{\lambda, x \to p}(\varphi(x) \exp{(i\lambda S(x))}) = \tilde{\varphi}(p) \exp{(i\lambda \tilde{S}(p))}$$

where $\lambda \gg 1$ and terms of order $O(\lambda^{-1})$ were omitted. Consequently,

$$\int |\varphi(x)|^2 \, dx = \int |\tilde{\varphi}(p)|^2 \, dp.$$

If function $\varphi(x)$ is localized in a neighbourhood of the point x, then function $\tilde{\varphi}(p)$ is localized in a neighbourhood of the point $p = p(x)$, i.e.

$$|\varphi(x)|^2 \, dx = |\tilde{\varphi}(p)|^2 \, dp.$$

This implies that

$$|\varphi(x)| \sqrt{\left|\frac{dx}{dp}\right|} = |\tilde{\varphi}(p)| \sqrt{\left|\frac{dp}{dx}\right|}$$

(cf. (1.29)).

4. The Stationary Phase Method for Abstract Functions

The results of Section 1 concerning the asymptotics of oscillatory integrals can be generalized for the integrals of the form

$$\Phi(A) = \int_{\mathbf{R}^n} \exp{[iAS(x)]} \, \varphi(x) \, dx. \qquad (1.30)$$

Here A is an operator, $A : B \to B$, where B is a Banach space; the precise assumptions about operator A will be given later on. Furthermore, we shall assume that the following conditions are always satisfied:

$\Phi 1.1$. Function $S(x) \in C^\infty(\mathbf{R}^n)$, and is real-valued.
$\Phi 1.2$. Function $\varphi(x) \in C_0^\infty(\dot{B})$.

Here $C_0^\infty(B)$ is the class of functions $\varphi : \mathbf{R}_x^n \to B$ with values in Banach space B which are infinitely differentiable for $x \in \mathbf{R}^n$ and have compact support.

For the integrals of the form (1.30) asymptotic expansions will be derived, analogous to those obtained for the integrals of form (1.1). All the proofs are carried out via the scheme for the integrals of form (1.1) (cf. [15], [20]).

We shall denote by: $D(A)$ – the domain of definition of operator A, $\sigma(A)$ – its spectrum, and $\rho(A)$ – its resolvent set. We suppose the property

A1.1. Operator A is a generator of the strongly continuous group $\{\exp(itA)\}$ of bounded operators on the real axis $-\infty < t < +\infty$.

This assumption implies the following facts ([14], [87]). There exists $\omega > 0$ such that

$$\|\exp(itA)\| \le M e^{\omega|t|}, \quad t \in \mathbf{R}, \tag{1.31}$$

where constant M is independent of t. Spectrum $\sigma(A)$ is contained in the strip $|\operatorname{Im} \mu| < \omega$ of the complex plane μ. The resolvent $R_\mu(A) = (\mu I - A)^{-1}$ of operator A (I being the unit operator) is a holomorphic operator-valued function of μ outside the strip $|\operatorname{Im} \mu| \le \omega$, and satisfies the estimates

$$\|R_\mu(A)\| \le C_\varepsilon (1 + |\mu|)^{-1} \tag{1.32}$$

in the domain $|\operatorname{Im} \mu| \ge \omega + \varepsilon$ where $\varepsilon > 0$. Moreover, the following formulae hold:

$$\int_0^\infty \exp(-\mu t + iAt)\, dt = -iR_{-i\mu}(A) \quad (\operatorname{Re} \mu > \omega), \tag{1.33}$$

$$\int_{-\infty}^0 \exp(-\mu t + iAt)\, dt = iR_{-i\mu}(A) \quad (\operatorname{Re} \mu < -\omega). \tag{1.33'}$$

In particular, the operator $(A + i\varepsilon I)^{-1}$ exists and is bounded for $|\operatorname{Re} \varepsilon| > \omega$.

EXAMPLE 4. Consider the operator $A = (-i)d/d\tau$ in the Hilbert space $B = L_2(-\infty, \infty)$. Group $\{\exp(itA)\}$ appears to be the unitary group of the shifts

$$\exp\left(t \frac{d}{d\tau}\right) f(\tau) = f(\tau + t).$$

Spectrum $\sigma(A)$ coincides with the real axis $-\infty < \mu < +\infty$. Domain $D(A)$ of the operator $A = (-i)d/d\tau$ consists of functions $f(\tau) \in L_2(-\infty, \infty)$, which are absolutely continuous and such that $f'(\tau) \in L_2(-\infty, \infty)$. The operator $(A + i\varepsilon I)^{-1}$ exists and is bounded for arbitrary complex ε except purely imaginary.

Let $N \ge 1$ be an integer; then functions $f(\tau) \in D(A^N)$ are smooth, i.e. all their derivatives $f^{(k)}(\tau)$ to the order $k = N - 1$ included, are absolutely continuous. Moreover,

$$D((A + i\varepsilon I)^N) = D(A^N).$$

LEMMA 1.10. *Suppose the assumptions* $\Phi 1.1$, $\Phi 1.2$ *and* A1.1 *are satisfied, and*

$$\text{grad } S(x) \neq 0, \qquad x \in \text{supp } \varphi. \tag{1.34}$$

Then

$$\Phi(A) \in D(A^N) \tag{1.35}$$

for arbitrary integer $N \geq 1$.
    Proof. We use the identity

$$\exp\left[i(A + i\varepsilon I)S(x)\right] = (A + i\varepsilon I)^{-1} L \exp\left[i(A + i\varepsilon I)S(x)\right],$$

$$L = -i\,|\nabla S(x)|^{-2} \sum_{j=1}^{n} \frac{\partial S(x)}{\partial x_j}\frac{\partial}{\partial x_j},$$

where $|\text{Re } \varepsilon| > \omega$. Then integrating (1.30) by parts and taking into account that function $\varphi(x)$ has compact support, we get

$$\Phi(A) = (A + i\varepsilon I)^{-N} \int \exp\left[i(A + i\varepsilon I)S(x)\right] {}^t L^N(e^{\varepsilon S(x)}\varphi(x)),$$

where ${}^t L$ is the operator formally adjoint to L. The last integral, by virtue of A1.1, belongs to the space B so that $\Phi(A) \in D(A^N)$, q.e.d.
    In particular, if $A = (-i)d/d\tau$, $B = L_2(-\infty, \infty)$, then $\Phi(A)$ is an infinitely differentiable function of τ under the assumptions of Lemma 1.10.
    Now consider the case when phase $S(x)$ has non-degenerate stationary points on supp φ. In accordance with Lemma 1.10 it is sufficient to consider the case in which only one stationary point is present. The validity of formula (1.1) is based on two facts: on the Morse lemma and on the formula

$$\int_{0}^{\infty} \exp\left(ix^2 A\right) dx = \tfrac{1}{2}\, e^{i\pi/4}\sqrt{\frac{\pi}{A}} \qquad (\text{Im } A \geq 0). \tag{1.36}$$

Here A is a complex number different from zero, and the branch of $\sqrt{A}$ is taken in the upper half-plane $\text{Im } A \geq 0$ and in such a way that $\sqrt{A} > 0$ for real positive A.

MORSE LEMMA [66]. *Let function S(x) be real-valued and infinitely differentiable in a neighbourhood of the point x^0 which is a non-degenerate critical point of function S(x). Then there exist neighbourhoods U, V of the points $x = x^0$, $y = 0$, respectively, and a diffeomorphism $\varphi : V \to U$ of class C^∞ such that*

$$(S \circ \varphi)(y) = S(x^0) + \tfrac{1}{2} \sum_{j=1}^{n} \mu_j y_j^2. \tag{1.37}$$

Here, μ_j are the eigenvalues of the matrix $S''_{xx}(x^0)$, and det $\varphi'(0) = 1$.

In order to derive the analogue of formula (1.1) for integrals of the form (1.30), it is necessary to prove the operator analogue of formula (1.36).

LEMMA 1.11. *Let operator A satisfy assumption A1.1. Then the operator*

$$(A + i\varepsilon I)^{-1/2} = \frac{2e^{-i\pi/4}}{\sqrt{\pi}} \int_0^\infty \exp\left[it^2(A + i\varepsilon I)\right] dt \tag{1.38}$$

is bounded for Re $\varepsilon > \omega$, maps B into B, and fulfils the relation

$$\left[(A + i\varepsilon I)^{-1/2}\right]^2 g = (A + i\varepsilon I)^{-1} g \tag{1.39}$$

for any $g \in B$.

Proof. Since, according to the assumption,

$$\left\| \exp\left[it^2(A + i\varepsilon I)\right] \right\| \leq M \exp\left[t^2(\omega - \mathrm{Re}\,\varepsilon)\right]$$

and Re $\varepsilon > \omega$, integral (1.38) converges strongly and is a bounded operator (from B into B). By multiplying two integrals of the form (1.38) we obtain the strongly convergent double integral

$$\left[(A + i\varepsilon I)^{-1/2}\right]^2 = -\frac{4i}{\pi} \int_0^\infty \int_0^\infty \exp\left[i(t^2 + \tilde{t}^2)(A + i\varepsilon I)\right] dt \, d\tilde{t}.$$

Expressing it in polar coordinates (r, φ) and taking formula (1.33) into account, we obtain

$$\left[(A + i\varepsilon I)^{-1/2}\right]^2 = -\frac{4i\varepsilon}{\pi} \int_0^{\pi/2} d\varphi \int_0^\infty \exp\left[ir^2(A + i\varepsilon I)\right] r \, dr =$$

$$= -i \int_0^\infty \exp\left[i(A + i\varepsilon I)\rho\right] d\rho = (A + i\varepsilon I)^{-1},$$

q.e.d.

COROLLARY 1.12. *Under the assumptions of Lemma* 1.11 *and for* Re $\varepsilon < -\omega$, *the following formula holds*:

$$\int_0^\infty \exp\left[-i(A + i\varepsilon I)t^2\right] dt = \frac{\sqrt{\pi}}{2} e^{i\pi/4}(A + i\varepsilon I)^{-1/2}. \qquad (1.40)$$

REMARK 1.13.   Let assumption A1.1 be satisfied,

$$\|\exp(itA)\| \leqq M, \qquad t \in \mathbf{R},$$

and $\sigma(A)$ do not contain the half-axis $(-\infty, \delta]$, $\delta > 0$. Then the formula

$$A^{-1/2} = \frac{2}{\sqrt{\pi}} e^{-i\pi/4} \int_0^\infty \exp(it^2 A) dt \qquad (1.41)$$

is true. Operator $A^{-1/2}: B \to B$ is bounded, and $(A^{-1/2})^2 g = A^{-1}g$ for any $g \in B$.

In the same way one proves

LEMMA 1.14. *Let assumption* A1.1 *be satisfied,* Re $\varepsilon > \omega$, *and* $j \geqq 0$ *be an integer. Then*

$$\int_0^\infty \exp\left[i(A + i\varepsilon I)t^2\right] t^j dt =$$

$$= \tfrac{1}{2}\Gamma\left(\frac{j+1}{2}\right) \exp\left[\frac{i\pi}{4}(j+1)\right](A + i\varepsilon I)^{-(j+1)/2}. \qquad (1.42)$$

LEMMA 1.15. *Suppose assumption* A1.1 *be satisfied,* Re $\varepsilon > \omega, \delta > 0$, $j \geqq 0$ *be an integer, and* $g \in B$. *Then the expansion*

$$\int_0^\delta \exp\left[i(A + i\varepsilon I)t^2\right] t^j dt\, g =$$

$$= \tfrac{1}{2} \Gamma \left(\frac{j+1}{2} \right) \exp \left[\frac{i\pi}{4}(j+1) \right] (A + i\varepsilon I)^{-(j+1)/2} g +$$

$$+ \exp \left[i(A + i\varepsilon I)\delta^2 \right] \sum_{k=1}^{N} c_k (A + i\varepsilon I)^{-k} g + g_N \qquad (1.43)$$

holds for any integer $N \geq 0$. *Here* $g_N \in D(A^{N+1})$ *and* c_k *are constants.*

Proof. Represent the integral over the interval $[0, \delta]$ as a difference of two integrals over the half-axes $[0, \infty)$ and $[\delta, \infty)$. Then for the former formula (1.42) holds; the latter will be integrated by parts:

$$\int_\delta^\infty \exp \left[i(A + i\varepsilon I)t^2 \right] t^j \, dt \, g =$$

$$= -\frac{i}{2}(A + i\varepsilon I)^{-1} \exp \left[i(A + i\varepsilon I)t^2 \right] t^{j-1} g \Big|_\delta^\infty =$$

$$+ i(A + i\varepsilon I)^{-1} \int_\delta^\infty \exp \left[i(A + i\varepsilon I)t^2 \right] \frac{d}{dt}(t^{j-1}) \, dt \, g. \qquad (1.44)$$

Since

$$\| \exp \left[i(A + i\varepsilon I)t^2 \right] \| \leq M \exp \left[(-\operatorname{Re} \varepsilon + \omega)t^2 \right]$$

and $\operatorname{Re} \varepsilon > \omega$, the first term vanishes for $t = +\infty$. The integral on the right-hand side of (1.44) is strongly convergent; consequently, the last term in (1.44) belongs to $D(A)$. By repeating the integration by parts we obtain (1.43).

REMARK 1.16. Let $A = d/d\tau$, $B = L_2(-\infty, \infty)$. Then the expansion (1.43) is an asymptotic one with respect to the smoothness of function $\Phi(A)$ ($\Phi(A)$ being the integral in (1.43)), since remainder g_N belongs to the domain of definition of the operator $(d/d\tau)^{N+1}$.

Now we shall derive the fundamental formula of the stationary phase method in one-dimensional case.

LEMMA 1.17. *Let assumptions* $\Phi 1.1$, $\Phi 1.2$, $A 1.1$ *be satisfied, and* $\operatorname{Re} \varepsilon > \omega$. *Then the expansion*

$$\int_{-\infty}^\infty \exp(iAx^2)\varphi(x) \, dx =$$

$$= (A + i\varepsilon I)^{-1/2} \sum_{k=0}^{N} (A + i\varepsilon I)^{-k} a_k + g_N \qquad (1.45)$$

holds for any integer $N \geq 1$, where $a_k \in B$, $g_N \in D(A^{N+1})$.
  Proof. Consider the integral

$$\Phi(A) = \int_0^\infty \exp\left[ix^2(A + i\varepsilon I)\right] \varphi_\varepsilon(x)\, dx,$$

where the function $\varphi_\varepsilon(x) = \exp(\varepsilon x^2)\varphi(x)$ has compact support.
  Suppose function $\varphi(x)$ has a zero of order m at the origin of coordinates. Integrating by parts, we get

$$\Phi(A) = \frac{1}{2i}(A + i\varepsilon I)^{-1} \left.\frac{\varphi_\varepsilon(x)}{x}\right|_{x=0} -$$
$$- \frac{1}{2i}(A + i\varepsilon I)^{-1} \int_0^\infty \exp\left[ix^2(A + i\varepsilon I)\right]\frac{d}{dx}\left(\frac{\varphi_\varepsilon(x)}{x}\right)dx.$$

If $m \geq 2$, then $(\varphi_\varepsilon(x)/x)|_{x=0} = 0$ and the last integral belongs to space B, so that $\Phi(A) \in D(A)$. Repeating the integration by parts we obtain that $\Phi(A) \in D(A^k)$, where $k = [m/2] - 1$.
  Representing function $\varphi_\varepsilon(x)$ in the form

$$\varphi_\varepsilon(x) = \sum_{j=0}^{N} \varphi_j x^j + \psi_N(x), \qquad \varphi_j \in B,$$

we find that function $\psi_N \in C^\infty(B)$ has a zero of order $N + 1$ at the origin. We introduce a scalar cut-off function $\chi(x) \in C_0^\infty(\mathbf{R})$, identically equal to 1 in the neighbourhood $|x| < \delta$ of the origin, and consider the integral

$$\tilde{\Phi}(A) = \int_0^\infty \exp(iAx^2)\, \varphi(x)\, \chi(x)\, dx.$$

By virtue of Lemma 1.10 we have

$$\Phi(A) - \tilde{\Phi}(A) \in D(A^N)$$

for arbitrary integer $N \geq 0$. Actually,

$$\Phi(A) - \tilde{\Phi}(A) = \int_0^\infty \exp(iAx^2)\, \varphi(x)(1 - \chi(x))\, dx,$$

function $\varphi(x)(1 - \chi(x)) \in C_0^\infty(B)$ and is identically equal to zero for $|x| < \delta$ so that the phase $S(x) = x^2$ has no stationary points on the support of this function. We have

$$\tilde{\Phi}(A) = \sum_{j=0}^{N} \int_0^\infty \exp\left[i(A + i\varepsilon I)x^2\right] x^j \chi_\varepsilon(x) \varphi_j \, dx +$$

$$+ \int_0^\infty \exp\left[i(A + i\varepsilon I)x^2\right] x^j \psi_N(x) \chi_\varepsilon(x) \, dx, \qquad (1.46)$$

where $\chi_\varepsilon(x) = \chi(x) \exp(\varepsilon x^2)$. Suppose $M \geq 1$ is fixed; then, according to the previous results, the last integral in (1.46) belongs to $D(A^M)$ for sufficiently large N (of order $2M$).

Further,

$$\int_0^\infty \exp\left[i(A + i\varepsilon I)x^2\right] x^j \chi_\varepsilon(x) \, dx = \int_0^\delta \exp\left[i(A + i\varepsilon I)x^2\right] x^j \, dx +$$

$$+ \int_\delta^\infty \exp\left[i(A + i\varepsilon I)x^2\right] x^j \chi_\varepsilon(x) \, dx.$$

The integral over the interval $[0, \delta]$ is evaluated according to formula (1.43). The second integral over the half-axis $[\delta, \infty)$ is integrated by parts, yielding the expression

$$-\frac{i}{2}(A + i\varepsilon I)^{-1} x^j \exp\left[i(A + i\varepsilon I)x^2\right]\big|_{x=\delta} +$$

$$+\frac{i}{2}(A + i\varepsilon I)^{-1} \int_\delta^\infty \exp\left[i(A + i\varepsilon I)x^2\right] \frac{d}{dx}(x^{j-1} \chi_\varepsilon(x)) \, dx.$$

Repeating the integration by parts, we notice that the first term is cancelled by the terms of the form

$$c_k \exp\left[i(A + i\varepsilon I)\delta^2\right]((A + i\varepsilon I)\delta^2)^{-k}$$

entering in the expansion (1.43) of the integral over the interval $[0, \delta]$. Recall that function $\chi(x)$ has compact support and that $\chi(x) \equiv 1$ for $|x| \leq \delta$.

Thus we have proved that

$$\int_0^\infty \exp\left[i(A + i\varepsilon I)x^2\right]x^j\varphi_j\chi_\varepsilon(x)\,dx = c(A + i\varepsilon I)^{-(j+1)/2}\varphi_j + \tilde{g}_M,$$

where $\tilde{g}_M \in D(A^M)$ and $M \geq 1$ is arbitrary, $c = \text{const}$. Finally we obtain the expansion

$$\Phi(A) = \sum_{k=0}^M (A + i\varepsilon I)^{-(k+1)/2}\psi_k + h_M,$$

where $h_M \in D(A^{M+1})$ and $M > 0$ is arbitrary.

Explicit formulae for coefficients ψ_k will not be given; they are exactly the same as in the case of the integrals

$$\int_0^\infty \varphi(x)\exp(i\lambda x^2)\,dx$$

where $\lambda \to +\infty$.

An analogous expansion holds for an integral of the form $\Phi(A)$ over the half-axis $(-\infty, 0]$. Moreover, it is not hard to verify that, in the corresponding expansions, the summands containing integral powers of the operator $(A + i\varepsilon I)^{-1}$ cancel each other, and the expansion (1.45) is obtained.

Similarly one can prove

LEMMA 1.18.   *Under the assumptions of Lemma 1.17, and for* $\text{Re }\varepsilon < -\omega$, *expansion* (1.45) *holds for the integral*

$$\int_{-\infty}^\infty \exp(-iAx^2)\varphi(x)\,dx = \sum_{k=0}^N (A - i\varepsilon I)^{-k-1/2}b_k + g_N, \qquad (1.47)$$

where $b_k \in B$, $g_N \in D(A^{N+1})$.

Now we shall consider the n-dimensional integral (1.30). We obtain an analogue of Theorem 1.1.

THEOREM 1.19.   *Suppose assumptions* $\Phi1.1$, $\Phi1.2$ *and* $A1.1$ *are satisfied,* *and function S(x) has just one non-degenerate stationary point x^0 on supp φ. Then, for $\mathrm{Re}\,\varepsilon > \omega$ and for arbitrary integer $N \geq 0$ the expansion*

$$\Phi(A) = \exp\left[iAS(x^0)\right](A + i\varepsilon I)^{-v_+/2}(E - i\varepsilon I)^{-v_-/2} \times$$

$$\times \left[\sum_{0 \leq k+l \leq N} (A + i\varepsilon I)^{-k}(A - i\varepsilon I)^{-l} a_{kl}\right] + g_N \qquad (1.48)$$

holds where $a_{kl} \in B$, and the remainder $g_N \in D(A^{N+[n/2]+1})$.

Here v_+ and v_- are the numbers of positive and negative eigenvalues of matrix $S_{xx}''(x^0)$, respectively.

Proof. Due to Lemma 1.10 we may assume that supp φ is contained in an arbitrary small (but fixed) neighbourhood of point x^0. According to the Morse lemma, a smooth change of variables $x = f(y)$ can be performed such that $x^0 = f(0)$, and the function $(S \circ f)(y)$ will become the quadratic form (1.37). In this way

$$\Phi(A) = \exp\left[iAS(x^0)\right] \int_V \exp\left(\frac{iA}{2}\sum_{j=1}^{n}\mu_j y_j^2\right)\tilde{\varphi}(y)\,dy, \qquad (1.49)$$

where V is a neighbourhood of the point $y = 0$, and

$$\tilde{\varphi}(y) = (\varphi \circ f)(y)\det f'(y).$$

Now we repeatedly apply the stationary phase method (i.e. Lemmas 1.17 and 1.18) with respect to the variables $y_1, y_2, \ldots, y_n$, to the integral (1.49). First we consider the integral

$$\Phi_1(A) = \int_{-\infty}^{\infty} \exp\left(\frac{iA}{2}\mu_1 y_1^2\right)\tilde{\varphi}(y)\,dy_1.$$

If $\mu_1 > 0$, then according to Lemma 1.17 this integral admits expansion (1.45), the coefficients of which are compactly supported smooth functions of the variables $y_2, \ldots, y_n$. In the case $\mu_1 < 0$, we can apply Lemma 1.18. Then we multiply the obtained expansion by $\exp\left((i/2)\mu_2 Ay_2^2\right)$, integrate it over y_2, and, according to the sign of μ_2, apply Lemma 1.17 or Lemma 1.18. By repeating this procedure, we obtain expansion (1.48).

The leading term of expansion (1.48) is of the form

$$\Phi(A) = c\exp\left[iAS(x^0)\right](A + i\varepsilon I)^{-v_+/2}(A - i\varepsilon I)^{-v_-/2}\varphi(x^0) + g_1,$$

where $g_1 \in D(A^{[n/2]+1})$, and

$$c = (2\pi)^{n/2} \left| \det S''_{xx}(x^0) \right|^{-1/2} \exp\left[\frac{i\pi}{4} \operatorname{sgn} S''_{xx}(x^0) \right].$$

The stationary phase method for abstract functions developed above, was applied to investigate the propagation of discontinuities of solutions of the Cauchy problem [59], [61].

§ 2. Pseudodifferential Operators

Let L be a differential operator depending on a large parameter λ, namely, an operator involving derivatives $\partial/\partial x_j$, each of which is multiplied by a small parameter λ^{-1}. Applying operator L to the function $\varphi \exp(i\lambda S)$ we get

$$L[\varphi(x) \exp(i\lambda S(x))] = \exp(i\lambda S(x)) Q(x, \lambda^{-1}),$$

where Q is a polynomial in λ^{-1}.

In this paragraph we introduce the notion of a λ-pseudodifferential operator which generalizes the notion of a differential operator. The necessity of this generalization will be clarified in § 5. The above formula which expresses the action of a differential operator on function $\varphi \exp(i\lambda S)$ for large λ will be valid for λ-pseudodifferential operators, too (if S is real-valued and φ compactly supported), except that Q will not be a polynomial in λ^{-1} but an asymptotic series in powers of λ^{-1}.

1. *The Definition of a λ-Pseudodifferential Operator*

Let $L(x, p)$ and $u(x)$ be scalar functions. In Feynman's notation

$$\begin{aligned} L(\overset{2}{x}, \overset{1}{D}_x)u(x) &= F^{-1}_{p \to x} L(x, p) F_{x \to p} u(x), \\ L(\overset{1}{x}, \overset{2}{D}_x)u(x) &= F^{-1}_{p \to x} (F_{x \to p} L(x, p) u(x)). \end{aligned} \tag{2.1}$$

The indices 1 and 2 indicate which of the operators x, D_x acts first in the expression $L(x, D_x)$, and which acts second. For example, if $L(x, p) = a(x)p_j$, then

$$L(\overset{2}{x}, \overset{1}{D}_x)u(x) = -ia(x)\frac{\partial u(x)}{\partial x_j},$$

$$L(\overset{1}{x}, \overset{2}{D}_x)u(x) = -i\frac{\partial}{\partial x_j}(a(x)u(x)).$$

Operators of the form (2.1) are called *pseudodifferential operators* (p.d.

operators); function $L(x, p)$ is called the *symbol* of a p.d. operator. The theory of p.d. operators was developed in works [41], [33].

Operators $L(\overset{2}{x}, \overset{1}{D}_x)$ and $L^*(\overset{1}{x}, \overset{2}{D}_x)$, where L^* is the operator with symbol $\overline{L(x, p)}$, are formally adjoint:

$${}^t L(\overset{2}{x}, \overset{1}{D}_x) = L^*(\overset{1}{x}, \overset{2}{D}_x).$$

This means that

$$(L(\overset{2}{x}, \overset{1}{D}_x)\varphi, \psi) = (\varphi, L^*(\overset{1}{x}, \overset{2}{D}_x)\psi)$$

for arbitrary functions $\psi(x)$, $\varphi(x) \in C_0^\infty(\mathbf{R}^n)$, and the corresponding conditions on the symbol (see below); here the scalar product is given by

$$(\varphi, \psi) = \int_{\mathbf{R}^n} \varphi(x)\, \overline{\psi(x)}\, \mathrm{d}x.$$

Thus, e.g., the operator

$$\mathscr{L} = L(\overset{1}{x}, \overset{2}{D}_x) + L^*(\overset{2}{x}, \overset{1}{D}_x)$$

is formally symmetric.

We shall consider p.d. operators which depend on a real parameter $\lambda \neq 0$. Formally, a λ-p.d. operator is introduced via the formula

$$L(\overset{2}{x}, \lambda^{-1}\overset{1}{D}_x ; (i\lambda)^{-1})\, u(x) = F_{\lambda, p \to x}^{-1} L(x, p; (i\lambda)^{-1}) F_{\lambda, x \to p}\, u(x). \tag{2.2}$$

For $\lambda = 1$ we obtain the usual p.d. operator. Function $L(x, p; (i\lambda)^{-1})$ is called the *symbol* of the λ-p.d. operator (2.2). Moreover,

$$L(x, p; (i\lambda)^{-1}) \to L(x, p; 0)$$

for $\lambda \to \infty$. The exact conditions on the symbol will be given later.

We consider operators which depend on parameter λ in the form (2.2), since many equations of mathematical physics involve parameter λ just in this way.

EXAMPLE 2.1.   The Schrödinger equation:

$$ih \frac{\partial \psi(x, t)}{\partial t} = -\frac{h^2}{2m} \Delta \psi(x, t) + V(x)\psi(x, t). \tag{2.3}$$

This equation describes the motion of a non-relativistic quantum particle of mass m in a field with the potential energy $V(x)$. Setting $h = \lambda^{-1}$, we obtain the λ-differential operator of form (2.2) with the symbol

$$L(x, t, p, E) = E + \frac{1}{2m} \langle p, p \rangle + V(x).$$

Here E (energy) is the conjugate variable to t.

EXAMPLE 2.2.   The Helmholtz equation:

$$(\Delta + k^2 n^2(x))u(x) = 0. \tag{2.4}$$

This equation describes propagation of light (and also of electromagnetic and acoustic waves) with frequency k in a medium with refraction index $n(x)$. Dividing both sides of the equation by k^2 and setting $\lambda = k$, we obtain the λ-differential operator of form (2.2) with the symbol

$$L(x, p) = - \langle p, p \rangle + n^2(x).$$

EXAMPLE 2.3.   Consider the difference scheme

$$\frac{u_m^{n+1} - 2u_m^n + u_m^{n-1'}}{h^2} = \alpha^2 \frac{u_{m+1}^n - 2u_m^n + u_{m-1}^n}{h^2} \tag{2.5}$$

for the wave equation $u_{tt} = u_{xx}$, where $u_m^n = u(hn/2; \alpha hm/2)$ and $\alpha > 0$, $h > 0$. By using the formula $e^{hd/dx} f(x) = f(x + h)$ and an analogous formula for $f(t + \alpha h)$ we obtain the equation

$$\left[\sin^2\left(\frac{h}{2} D_t\right) - \alpha^{-2} \sin^2\left(\frac{\alpha h}{2} D_x\right) \right] u(x, t) = 0$$

from which the difference scheme follows, whenever the values of u are taken in lattice points. This equation has the form

$$L(\lambda^{-1}D_t, \lambda^{-1}D_x)u = 0, \qquad h = \lambda^{-1},$$

where L is the λ-p.d. operator with the symbol

$$L(x, t, p_1, p_0) = \sin^2\left(\frac{p_0}{2}\right) - \alpha^{-2} \sin^2\left(\frac{\alpha p_1}{2}\right).$$

Now we shall describe the class of symbols.

DEFINITION 2.4.   Function $L(x, p)$ *belongs to the class T^m*, if the following conditions are satisfied:
(1) $L(x, p) \in C^\infty(\mathbf{R}_x^n \times \mathbf{R}_p^n)$.
(2) For any multi-indices α, β,

$$\left| D_x^\alpha D_p^\beta L(x, p) \right| \leq C_{\alpha\beta}(1 + |x|)^m (1 + |p|)^m, \quad (x, p) \in \mathbf{R}_x^n \times \mathbf{R}_p^n.$$
$$(2.6)$$

Thus at infinity, function $L(x, p)$ grows not faster than a polynomial
in x and p, and the differentiation of L with respect to x and p does not
change the growth. Examples of the symbols of class T^m are: (1) the polynomials in (x, p) of degree m; (2) the symbol in Example 2.3.
In the symbol of class T^m, variables x and p play equivalent roles.

REMARK. Class T^m differs from the Hörmander classes $S^m_{\rho, \delta}$ (since
differentiation with respect to p of the symbol $L \in S^m_{\rho, \delta}$ diminishes the
degree of growth of L in p). This difference is caused by the difference
in treated cases.

DEFINITION 2.5.   Function $L(x, p; (i\lambda)^{-1})$ *belongs to class T^m_+*, if
the following conditions are satisfied:
(1) $L(x, p; (i\lambda)^{-1}) \in C^\infty(\mathbf{R}_x^n \times \mathbf{R}_p^n \times \mathbf{R}_\lambda^+)$.
(2) For arbitrary integer $N \geq 0$,

$$L(x, \text{p}; (i\lambda)^{-1}) = \sum_{k=0}^N (i\lambda)^{-k} L_k(x, p) + \lambda^{-N-1} R_N(x, p; (i\lambda)^{-1}).$$
$$(2.7)$$

Here $L_k \in T^m$, and for arbitrary multi-indices α, β, and for any integer
$\gamma \geq 0$ we have

$$\left| D_x^\alpha D_p^\beta D_\varepsilon^\gamma R_N(x, p; \varepsilon) \right|$$
$$\leq C_{\alpha, \beta, \gamma}(1 + |x|)^m (1 + |p|)^m, \quad \varepsilon = (i\lambda)^{-1} \qquad (2.8)$$

where $(x, p, \lambda) \in \mathbf{R}_x^n \times \mathbf{R}_p^n \times \mathbf{R}_\lambda^+$.
The class T^m_- is defined similarly. As a typical example of the symbol
of class T^m_+ we can take the function $L = \sum_{k=0}^N (i\lambda)^{-k} L_k(x, p)$ where $L_k \in T^m_+$.
In particular, it follows from Definition 2.5 that

$$\left| D_x^\alpha D_p^\beta L(x, p; (i\lambda)^{-1}) \right| \leq C_{\alpha\beta}(1 + |x|)^m (1 + |p|)^m,$$
$$(x, p, \lambda) \in \mathbf{R}_x^n \times \mathbf{R}_p^n \times \mathbf{R}_\lambda^+. \qquad (2.8')$$

If

$$L(x, p;(i\lambda)^{-1}) \in T_+^m,$$

then

$$\frac{\partial L}{\partial x_j} \in T_+^m, \qquad \frac{\partial L}{\partial p_j} \in T_+^m, \qquad x_j L \in T_+^{m+1}, \qquad p_j L \in T_+^{m+1}.$$

2. The Action of λ-Pseudodifferential Operator on a Rapidly Oscillating Exponential

Our next task will be to derive formal asymptotic (particular) solutions for $\lambda \to \pm \infty$ of the equation

$$L(\overset{2}{x}, \lambda^{-1}\overset{1}{D}_x ; (i\lambda)^{-1}) u(x, \lambda) = 0.$$

If $Lu = \sum_{k=0}^m a_k (\lambda^{-1} D_x)^k u$, $x \in \mathbf{R}^1$, and the coefficients a_k are constant, then these solutions can be sought in the form $\exp(i\lambda S(x))$. Moreover, whenever the characteristic equation has simple roots, any solution will be a linear combination of the exponential solutions. Thus, in accordance with this example, we shall look for a solution in the form of a formal series,

$$\exp[i\lambda S(x)] \sum_{k=0}^{\infty} (i\lambda)^{-k} \varphi_k(x).$$

However, it is necessary to specify how a λ-p.d. operator acts on a rapidly oscillating exponential, i.e., on a function of the form $\varphi(x) \exp[i\lambda S(x)]$.

Let $x \in \mathbf{R}^1$; then

$$\left(\frac{1}{i\lambda}\frac{d}{dx}\right) \exp[i\lambda S(x)] = S'(x) \exp[i\lambda S(x)],$$

$$\left(\frac{1}{i\lambda}\frac{d}{dx}\right)^2 \exp(i\lambda S(x)) = \left[(S'(x))^2 + \frac{1}{i\lambda}S'(x)S''(x)\right] \exp[i\lambda S(x)].$$

It is not hard to show by induction that

$$\left(\frac{1}{i\lambda}\frac{d}{dx}\right)^m (\exp[i\lambda S(x)]) =$$

$$= \exp[i\lambda S(x)]\left[(S'(x))^m + \frac{m(m-1)}{2i\lambda}(S'(x))^{m-1}S''(x) + O(\lambda^{-2})\right].$$

By using the Leibniz formula we obtain

$$\left(\frac{1}{i\lambda}\frac{d}{dx}\right)^m [\varphi(x)\exp(i\lambda S(x))] =$$

$$= \exp[i\lambda S(x)]\left[(S'(x))^m\varphi(x) + \frac{m}{i\lambda}(S''(x))^{m-1}\varphi'(x) + \right.$$

$$\left. + \frac{m(m-1)}{2i\lambda}(S'(x))^{m-2}S''(x)\varphi(x) + \sum_{j=2}^m (i\lambda)^{-j}(R_j\varphi)(x)\right],$$

where R_j is a differential operator of order j. Operator $((1/i\lambda)(d/dx))^m$ has the symbol $L(p) = p^m$, and since $dL/dp = mp^{m-1}$, $d^2L/dp^2 = m(m-1)p^{m-2}$, we get for the operator in question the following expression:

$$L(\lambda^{-1}D_x)[\varphi(x)\exp(i\lambda S(x))] = \exp(i\lambda S(x))\left[L(S'(x))\varphi(x) + \right.$$

$$\left. + \frac{1}{i\lambda}\frac{dL(S'(x))}{dp}\varphi'(x) + \frac{1}{2i\lambda}\frac{d^2L(S'(x))}{dp^2}S''(x)\varphi(x) + \dots\right].$$

This formula is true also for differential operators $L(\overset{2}{x}, \lambda^{-1}\overset{1}{D}_x)$ with variable coefficients, since we differentiate first with respect to x, and then multiply the obtained expressions by functions of x. Finally, as the operators $\partial/\partial x_j, \partial/\partial x_k$ commute on smooth functions, we have

$$L(x, \lambda^{-1}D_x; (i\lambda)^{-1})[\varphi(x)\exp(i\lambda S(x))] =$$

$$= \exp(i\lambda S(x))\sum_{j=0}^m (i\lambda)^{-j}R_j(x, D_x)\varphi(x), \tag{2.9}$$

where R_j are linear differential operators of order j. If, in particular, L is a differential operator of order m, the coefficients of which are polynomials in $(i\lambda)^{-1}$, then we obtain

$$(R_0\varphi)(x) = L\left(x, \frac{\partial S(x)}{\partial x}; 0\right)\varphi(x), \tag{2.10}$$

$$(R_1\varphi)(x) = \left\langle \frac{\partial L\left(x, \frac{\partial S}{\partial x}; 0\right)}{\partial p}, \frac{\partial\varphi(x)}{\partial x}\right\rangle +$$

$$+ \left[\frac{1}{2}\operatorname{Sp}\left(\frac{\partial^2 L\left(x, \frac{\partial S(x)}{\partial x}; 0\right)}{\partial p^2}\frac{\partial^2 S(x)}{\partial x^2}\right) + \right.$$

$$+ \frac{\partial}{\partial \varepsilon} L\left(x, \frac{\partial S(x)}{\partial x}; \varepsilon\right)\Bigg|_{\varepsilon=0}\Bigg] \varphi(x).$$

THEOREM 2.6. *Let symbol* $L(x, p; (i\lambda)^{-1}) \in T_+^m$, $\varphi(x) \in C_0^\infty(\mathbf{R}^n)$, $S(x) \in C^\infty(\mathbf{R}^n)$, *and function* $S(x)$ *be real-valued. Then, for* $\lambda \geq 1$ *and for arbitrary integer* $N \geq 0$, *we have*

$$L(\overset{2}{x}, \lambda^{-1}\overset{1}{D}_x; (i\lambda)^{-1})[\varphi(x) \exp(i\lambda S(x))] =$$

$$= \exp(i\lambda S(x)) \sum_{j=0}^{N} (i\lambda)^{-j} R_j(x, D_x)\varphi(x) + O_{-N-1}(x, \lambda). \quad (2.12)$$

Here $R_j(x, D_x)$ *is a linear differential operator of order* $\leq j$ *with coefficients from class* $C^\infty(\mathbf{R}^n)$.

The estimate of the remainder is given by

$$|D_x^\alpha O_{-N-1}(x, \lambda)| \leq C_r \lambda^{-N-1+|\alpha|}(1 + |x|)^{-r} \quad (2.12')$$

with arbitrary $r > 0$, $x \in \mathbf{R}^n$. R_0 *and* R_1 *satisfy the formulae* (2.10) *and* (2.11).

Proof. Let $u(x, \lambda)$ be the left-hand side of (2.12). Then

$$u(x, \lambda) = \left(\frac{\lambda}{2\pi}\right)^n \int \exp[i\lambda \langle x, p \rangle] L(x, p; (i\lambda)^{-1}) I(p, \lambda) dp, \quad (2.13)$$

$$I(p, \lambda) = \int \varphi(x) \exp[i\lambda S(x) - \langle x, p \rangle] dx.$$

Let $M = \{p: p = \partial S/\partial x, x \in \text{supp } \varphi\}$ and $G(p) \subset \mathbf{R}_p^n$ be the exterior of a finite domain, $\overline{G(p)} \cap M = \varnothing$. We construct a C^∞-partition of unity: $\eta_1(p) + \eta_2(p) = 1$, $p \in \mathbf{R}^n$, where $\eta_1(p)$ has compact support, $\text{supp } \eta_2(p) \subset G(p)$, and we correspondingly set $u(x, \lambda) = u_1(x, \lambda) + u_2(x, \lambda)$. Further, by using formula (1.16), we obtain

$$u_2(x, \lambda) = -\frac{1}{i\lambda|x|^2} \int \exp[i\lambda \langle x, p \rangle] \sum_{j=1}^{n} \frac{\partial}{\partial p_j}(I(p, \lambda)L(x, p)) \, dp$$

for $x \neq 0$. Consequently, for $|x| \geq 1$ and arbitrary $N \geq 0$, we have

$$|u_2(x, \lambda)| \leq C\lambda^{-N}(1 + |x|)^{m-1}.$$

By applying (1.16) again, and taking into account the above estimate for

$|u_2|$, we derive

$$|u_2(x, \lambda)| \leq C_N \lambda^{-N}(1 + |x|)^{-N}, \quad x \in \mathbf{R}^n,$$

for arbitrary $N \geq 0$, and find that the same estimates are true for all derivatives of u_2 with respect to x. Further,

$$u_1(x, \lambda) = \left(\frac{\lambda}{2\pi}\right)^n \int L(x, p; (i\lambda)^{-1})\varphi(y)\eta_1(p) \times$$

$$\times \exp[i\lambda\psi(x, y, p)] \, dy \, dp,$$

$$\psi(x, y, p) = \langle x - y, p \rangle + S(y), \tag{2.14}$$

where integration is performed over a finite domain in $\mathbf{R}_y^n \times \mathbf{R}_p^n$. Let $x \in K$. We apply Theorem 1.4. The function ψ (as function of y and p) has a single stationary point $Q(x)$: $y = x$, $p = \partial S/\partial x$. Let $H(x)$ be the matrix composed of the second derivatives with respect to y and p of function ψ in the point $Q(x)$, i.e. $H = \|\partial^2\psi/\partial y_j \partial p_k\|$, $1 \leq j$, $k \leq n$. Then $\det H(x) = (-1)^n$, the signature of $H(x)$ is zero, and the eigenvalues of $H(x)$ are ± 1. Further,

$$\psi(x, Q(x)) = S(x).$$

If $|x| \leq R$, then from Theorem 1.4 we obtain (2.12), (2.12'). If $R > 0$ is large enough, then the integral for u_1 does not contain stationary points, and by virtue of Lemma 1.5

$$|u_1(x, \lambda)| \leq C_N \lambda^{-N}(1 + |x|)^{-N}, \quad |x| \geq R,$$

where $N \geq 0$ is arbitrary; an analogous estimate holds for all derivatives of u_1 with respect to x. Thus (2.12) is proved.

COROLLARY 2.7. *The asymptotic expansion* (2.12) *can be differentiated with respect to x and λ any number of times. For the case $L \in T^m$, Theorem 2.6 is valid without any change.*

If K is a compact subset of $\mathbf{R}_x^n$, formula (2.12) gives the asymptotic expansion of the function

$$L(\overset{2}{x}, \lambda^{-1}\overset{1}{D}_x; (i\lambda)^{-1})[\varphi(x) \exp(i\lambda S(x))]$$

into the asymptotic series in powers of $(i\lambda)^{-1}$ for $\lambda \to +\infty$ uniformly in $x \in K$. It follows from (2.12) that this expansion has the form

$$\exp[-i\lambda S(x)]L(\overset{2}{x}, \lambda^{-1}\overset{1}{D}_x; (i\lambda)^{-1})(\exp[i\lambda S(x)]\varphi(x)) \sim$$

$$\sim \sum_{|\alpha|=0}^{\infty} \frac{\lambda^{-|\alpha|}}{\alpha!} \partial_p^\alpha L\left(x, \frac{\partial S(x)}{\partial x}; (i\lambda)^{-1}\right) \times$$

$$\times D_y^\alpha(\varphi(x) \exp[i\lambda S(x, y)])|_{y=x}$$

where

$$S(x, y) = S(x) - S(y) - \left\langle \frac{\partial S(x)}{\partial x}, x - y \right\rangle.$$

For classical p.d. operators the asymptotic expansion of the function $u(x, \lambda) = L(\overset{2}{x}, \overset{1}{D}_x) [\exp(i\lambda S)\varphi(x)]$ was obtained in [33].

REMARK 2.8.   Formula (2.12) is also true under less restrictive assumptions on the symbol.

PROPOSITION 2.9.   *Let* $\Omega \subset \mathbf{R}^n$ *be a finite domain,* $L(x, p; (i\lambda)^{-1}) \in C^\infty(\Omega \times \mathbf{R}_p^n \times \mathbf{R}_\lambda^+)$, *function* L *have the expansion* (2.7), *and condition* (2.8) *be fulfilled for* $(x, p, \lambda) \in K \times \mathbf{R}_p^n \times \mathbf{R}_\lambda^+$ *with constant* C *depending on* α, β, γ, K, *for arbitrary compact subset* $K \subset \Omega$. *We shall denote this class by the symbol* $T_+^m(\Omega)$. *Then Theorem 2.6 holds provided* $S(x) \in C^\infty(\Omega)$, $\varphi(x) \in C_0^\infty(\Omega)$; *the remainder* $O_{-N-1}(x, \lambda) \in O_{-N-1}^+(\Omega)$.

The proposition follows from the proof of Theorem 2.6. The result for the class $T_-^m(\Omega)$ is formulated analogously.

Let domain Ω and functions $\varphi(x)$, $S(x)$ be the same as above, and the functions $\varphi(x)$ and $S(x)$ be fixed. We set

$$M(S, \varphi) = \left\{ p \colon p = \frac{\partial S(x)}{\partial x}, x \in \operatorname{supp} \varphi \right\}.$$

PROPOSITION 2.10.   *Let the assumptions of Definition 2.5 be satisfied for* $(x, p) \in \Omega \times \overline{U(p)}$ *and* $D_x^\alpha L(x, p; (i\lambda)^{-1}) \in \mathscr{L}_1(U(p))$ *for arbitrary multi-index* α *and for arbitrary* $x \in \Omega$. *(Here* $U(p)$ *is a domain containing* $M(S, \varphi)$.) *Then all conclusions of Theorem 2.6 remain true.*

Proof. We construct a partition of unity of the form: $\eta_1(p) + \eta_2(p) = 1$, $p \in \mathbf{R}^n$, where $\eta_1(p)$ is a function with compact support, $\operatorname{supp} \eta_1(p) \subset U(p)$ and $\eta_1(p) \equiv 1$ on $M(S, \varphi)$. Let Q_j be the λ-p.d. operator with symbol $L(x, p; (i\lambda)^{-1})\eta_j(p)$. It follows from the proof of Theorem 2.6 that formula (2.12) is valid for the function $Q_1(\varphi e^{i\lambda S})$. Further, $Q_2(\varphi e^{i\lambda S})$ has the form (2.13), and since $\partial S(x)/\partial x \neq p$ for $\lambda \geq 1, p \in \operatorname{supp} \eta_1$, we find $|D_p^\alpha(L(p, \lambda)| \leq \leq C_{N\alpha}\lambda^{-N}(1 + |p|)^{-N}$ for $\lambda \geq 1, p \in \operatorname{supp} \eta_1$. Consequently, for any x,

$$|Q_2(\varphi e^{i\lambda S})| \leq C_N \lambda^{-N} \int |L(x, p; (i\lambda)^{-1})| \eta_1(p)(1 + |p|)^{-N} \, dp \leq$$
$$\leq C_N'(x) \lambda^{-N}.$$

For instance, if $L = |p|$, it is sufficient to assume that $\partial S(x)/\partial x \neq 0$ for $x \in \text{supp}\, \varphi$.

The following theorem can be proved exactly in the same way as Theorem 2.6.

THEOREM 2.11. *All statements of Theorem 2.6 remain valid for operator* $L(\overset{1}{x}, \lambda^{-1}\overset{2}{D}_x; (i\lambda)^{-1})$ *except that*

$$(R_1 \varphi)(x) = \left\langle \frac{\partial L(x, \partial S(x)/\partial x; 0)}{\partial p}, \frac{\partial \varphi(x)}{\partial x} \right\rangle +$$

$$+ \left[\frac{1}{2} \text{Sp} \left(\frac{\partial^2 L\left(x, \dfrac{\partial S(x)}{\partial x}; 0\right)}{\partial p^2} \frac{\partial^2 S(x)}{\partial x^2} \right) + \right.$$

$$+ \left. \frac{\partial}{\partial \varepsilon} L\left(x, \frac{\partial S(x)}{\partial y}; \varepsilon\right)\Big|_{\varepsilon = 0} + \right.$$

$$\left. + \text{Sp}\, \frac{\partial^2 L\left(x, \dfrac{\partial S}{\partial x}; 0\right)}{\partial x\, \partial p} \right] \varphi(x). \qquad (2.15)$$

Notice that now

$$L\left(\overset{1}{x}, -\frac{1}{\lambda}\overset{2}{D}_x; (i\lambda)^{-1}\right)[\varphi(x)\exp(i\lambda S(x))] =$$

$$= \left(\frac{\lambda}{2\pi}\right)^n \int\int L(y, p; (i\lambda)^{-1})\varphi(y) \times$$

$$\times \exp\left[i\lambda(S(y) + \langle p, x - y\rangle)\right] dp\, dy. \qquad (2.16)$$

3. Composition of λ-Pseudodifferential Operators

Let $\mathscr{T}_+^m$ be the class of all λ-p.d. operators with the symbols from class T_+^m, and $\mathscr{T}$ be the union of all classes $\mathscr{T}_+^m$. It is obvious that $\mathscr{T}_+^m$ and $\mathscr{T}$ are linear spaces over the field C. We shall show that class $\mathscr{T}$ is an algebra, i.e. composition of two λ-p.d. operators is again a λ-p.d. operator. We take the space $C_0^\infty(\mathbf{R}^n)$ as a common domain of all λ-p.d. operators.

THEOREM 2.12. *Let* $\mathscr{L}_1, \mathscr{L}_2$ *be λ-pseudodifferential operators of the classes* $\mathscr{T}_+^{N_1}, \mathscr{T}_+^{N_2}$, *respectively. Then their product* $\mathscr{L}_2\mathscr{L}_1$ *is a λ-pseudo-differential operator of class* $\mathscr{T}_+^{N_1 + N_2}$.

REMARK. This theorem does not follow from known theorems on composition of p.d. operators [41], [33]. This is due to the fact there are other classes of symbols L in the cited works. For these classes differentiation of symbol $L(x, p)$ with respect to the variables p diminishes the growth of the symbol for $|p| \to \infty$. Namely, it is assumed that the estimates

$$\left| D_x^\alpha D_p^\beta L(x, p) \right| \leqq C_{\alpha, \beta, \mathcal{K}} \left(1 + |p| \right)^{m - |\beta|}, \qquad x \in \mathcal{K}, p \in \mathbf{R}^n,$$

hold for any compact subset $\mathcal{K} \subset \mathbf{R}_x^n$ and for arbitrary multi-indices α, β with some constant $C_{\alpha, \beta, \mathcal{K}}$. There are also other generalizations of the classes, however, they are not containing class T_+^m even in the case symbol L does not depend on λ.

For proving Theorem 2.12 it is useful to introduce a class of λ-p.d. operators larger than class T_+^m. More precisely, we shall consider λ-p.d. operators the symbols of which have different growths with respect to x and to p for $|x| \to \infty$ and for $|p| \to \infty$, respectively. Thus we introduce

DEFINITION 2.13. Function $L(x, p; (i\lambda)^{-1})$ *belongs to class* $T_+^{m_1, m_2}$, if the following conditions are satisfied:

(1) $L(x, p; (i\lambda)^{-1}) \in C^\infty(\mathbf{R}_x^n \times \mathbf{R}_p^n \times \mathbf{R}_\lambda^+)$.

(2) The estimates

$$\left| D_x^\alpha D_p^\beta D_\varepsilon^\gamma L(x, p; \varepsilon) \right|$$
$$\leqq C_{\alpha, \beta, \gamma} (1 + |p|)^{m_1} (1 + |p|)^{m_2}, \qquad (\varepsilon = (i\lambda)^{-1}), \qquad (2.17)$$

are true for arbitrary multi-indices α, β, any integer $\gamma \geqq 0$ and for $(x, p, \lambda) \in \mathbf{R}_x^n \times \mathbf{R}_p^n \times \mathbf{R}_\lambda^+$.

(3) The expansion

$$L(x, p; \varepsilon) = \sum_{k=0}^{N} \varepsilon^k L_k(x, p) + \varepsilon^{N+1} L_{N+1}(x, p; \varepsilon) \qquad (2.18)$$

holds for arbitrary integer $N \geqq 0$.

Assumptions (1) and (2) are fulfilled for functions L_k, L_{N+1} (for L_k, if $\lambda = 0$).

Let us denote by $\mathcal{T}_+^{m_1, m_2}$ the class of all λ-p.d. operators with the symbols belonging to class $T_+^{m_1, m_2}$.

As an example of the symbol of class $T_+^{m_1, m_2}$, one may consider the function

$$L(x, p) = (1 + |x|^2)^{m_1/2}(1 + |p|^2)^{m_2/2} \exp \left[i(\langle x, a \rangle + \langle p, b \rangle) \right], \tag{2.19}$$

where a, b are constant n-vectors. For $n = 1$ and for integers $m_1, m_2 \geqq 0$, the simplest example is the function

$$L(x, p) = x^{m_1} p^{m_2} \exp [i(x + p)].$$

Now we shall prove Theorem 2.12. For definiteness we restrict ourselves to the case when differential operators act first, i.e. operators $\mathscr{L}_j$ have the form

$$\mathscr{L}_j = L_j(\overset{2}{x}, \lambda^{-1} \overset{1}{D}_x ; \varepsilon)$$

Here and later on $\varepsilon = (i\lambda)^{-1}$.

LEMMA 2.14. *Let operators $\mathscr{L}_1, \mathscr{L}_2$ belong to the classes $\mathscr{T}_+^{m_1, k_1}$ and $\mathscr{T}_+^{m_2, k_2}$, respectively, and let $m_1 < -n, k_2 < -n$. Let function $u(x) \in C_0^\infty(\mathbf{R}^n)$. Then the formula*

$$(\mathscr{L}_2 \mathscr{L}_1 u)(x) = F_{\lambda, p \to x}^{-1} L(x, p; \varepsilon) F_{\lambda, x \to p} u(x) \qquad (2.20)$$

holds, where function L is equal to

$$L(x, p; \varepsilon) = \left(\frac{\lambda}{2\pi} \right)^n \int \exp [i\lambda \langle p - \tilde{p}, \tilde{x} - x \rangle] \times$$
$$\times L_2(x, \tilde{p}; \varepsilon) L_1(\tilde{x}, p; \varepsilon) \, d\tilde{x} \, d\tilde{p}. \qquad (2.21)$$

Proof. According to the definition of λ-p.d. operator (see (2.2)) we have

$$(\mathscr{L}_2 \mathscr{L}_1 u)(x) = \frac{\lambda^{3n/2}}{(-2\pi i)^{n/2} (2\pi)^n} \int \exp [i\lambda(\langle \tilde{p}, x - \tilde{x} \rangle +$$
$$+ \langle p, \tilde{x} \rangle)] L_2(x, \tilde{p}; \varepsilon) L_1(\tilde{x}, p; \varepsilon) \tilde{u}(p) \, dp \, d\tilde{x} \, d\tilde{p}, \qquad (2.22)$$

where the integral is taken over the whole space $\mathbf{R}_p^n \times \mathbf{R}_{\tilde{x}}^n \times \mathbf{R}_{\tilde{p}}^n$, and $\tilde{u}(p) = F_{\lambda, x \to p} u(x)$. The function $\tilde{u}(p)$ decreases faster than any power of $|p|$ for $|p| \to \infty$. If $m_1 < -n, k_2 < -n$, then the integral (2.22) is absolutely convergent. Exchanging the order of integrations we obtain formulae (2.20), (2.21).

LEMMA 2.15. *Let the assumptions of Lemma 2.14 be satisfied. Then the function $L(x, p; \varepsilon)$ given by formula (2.21) belongs to class $\mathscr{T}_+^{m_1 + m_2, k_1 + k_2}$.*
Proof. Substituting $\tilde{x} = y + x, \tilde{p} = q + p$ in the integral (2.21), we get

$$L(x, p; \varepsilon) = \left(\frac{\lambda}{2\pi} \right)^n \int \exp [-i\lambda \langle y, q \rangle] \times$$
$$\times L_1(y + x, p; \varepsilon) L_2(x, q + p; \varepsilon) \, dy \, dq. \qquad (2.23)$$

The phase function $S = -\langle y, q \rangle$ has the single stationary point $y = 0, q = 0$. We construct a C^∞-partition of unity: $1 = \eta_1(y, q) + \eta_2(y, q)$ in $\mathbf{R}^{2n}_{y, q}$. Here η_1 is a function with compact support equal to 1 in a neighbourhood of the origin. Correspondingly, integral (2.23) is decomposed into two integrals: $L = L^{(1)} + L^{(2)}$ (the integrands in $L^{(j)}$ differ from those in (2.23) by multipliers η_j).

The estimate of integral $L^{(2)}$ is obtained via integration by parts, whereas that of $L^{(1)}$ via the stationary phase method. Integrating by parts, and taking into account the identity $(S = -\langle y, q \rangle)$

$$\exp(i\lambda S) = i\lambda^{-1} M(\exp(i\lambda S)),$$

$$M = (|q|^2 + |y|^2)^{-1} \sum_{j=1}^{n} \left(q_j \frac{\partial}{\partial y_j} + y_j \frac{\partial}{\partial q_j} \right),$$

$$\tag{2.24}$$

we obtain

$$L^{(2)} = i\lambda^{n-1}(2\pi)^{-n} \int \exp(i\lambda S) \, {}^t M(L_1 L_2 \eta_2) \, dy \, dq.$$

Here ${}^t M$ is the operator formally adjoint to M and the arguments of all functions are as in (2.23).

By virtue of (2.24), we have

$$^t M = \sum_{j=1}^{n} \left(a_j \frac{\partial}{\partial y_j} + b_j \frac{\partial}{\partial q_j} \right) + c.$$

Here the functions a_j, b_j are of the order $O(r^{-2})$ for $r = \sqrt{|y|^2 + |q|^2} \to \infty$, $c = O(r^{-1})$, and the obtained asymptotics can be differentiated any number of times (i.e. $D^\alpha_y a_j = O(r^{-2-|\alpha|})$ for $r \to \infty$ etc.). As immediately follows from Definition 2.13, class $T^{m,k}_+$ is invariant with respect to the actions of the differential operators $D_{x_j}, D_{p_j}, D_\varepsilon$.

According to the assumptions

$$|L_1 L_2| \leq C(1 + |y + x|^2)^{m_1/2}(1 + |p|^2)^{k_1/2} \times$$
$$\times (1 + |x|^2)^{m_2/2}(1 + |q + p|^2)^{k_2/2} \equiv CQ. \tag{2.25}$$

This and all the following estimates are true for any real x, p, y, q and for $\lambda \geq 1$. All derivatives of the function $L_1 L_2$ satisfy estimates of the form (2.25), but with other constants C.

Due to the established properties of the coefficients of operator ${}^t M$ and because of estimate (2.25) we have

$$|{}^t M(L_1 L_2 \eta_2)| \leq CQ(1 + r^2)^{-1/2}$$

for $(q, y) \in \mathrm{supp}\, \eta_2$ (recall that $r \geq r_0 > 0$ on $\mathrm{supp}\, \eta_2$). The same estimate hold for all derivatives of the function ${}'M(L_1 L_2 \eta_2)$. One can prove that

$$|{}'M^j(L_1 L_2 \eta_2)| \leq CQ(1 + r^2)^{-j/2} \tag{2.26}$$

and all derivatives of the function ${}'M^j(L_1 L_2 \eta_2)$ admit similar estimates. For brevity, all constants appearing in the estimates will be denoted by the same letter C.

By using the Peetre inequality [41]:

$$2(1 + |\xi + \eta|^2) \geq \frac{1 + |\xi|^2}{1 + |\eta|^2},$$

where $\xi, \eta \in \mathbf{R}^n$, we obtain $(m_1 < 0)$

$$(1 + |x + y|^2)^{m_1/2} \leq C(1 + |x|^2)^{m_1/2} (1 + |y|^2)^{-m_1/2}.$$

Consequently,

$$Q \leq C(1 + |x|^2)^{(m_1 + m_2)/2} (1 + |p|^2)^{(k_1 + k_2)/2} (1 + |y|^2)^{-m_1/2} \times$$
$$\times (1 + |q|^2)^{-m_2/2}. \tag{2.27}$$

Recall that integration in formula (2.23) is performed with respect to the variables y and q. Integrating by parts, the estimate of the integrand takes the form (2.27) but with the additional multiplier $\lambda^{-j}(1 + |y|^2 + |q|^2)^{-j/2}$, for arbitrary $j \geq 0$. Thus, taking j sufficiently large and integrating this estimate with respect to $dy\, dq$, the estimate of function $L^{(2)}$ becomes

$$|L^{(2)}| \leq C\lambda^{-j}(1 + |x|^2)^{(m_1 + m_2)/2} (1 + |p|^2)^{(k_1 + k_2)/2}. \tag{2.28}$$

Here $j \geq 0$ is arbitrary. Exactly the same estimates hold for all derivatives of function L^2 due to the facts proved above. Hence it is evident that $L^{(2)} \in T_+^{m_2, k_2}$.

Consider now the integral $L^{(1)}$ taken over a bounded domain since function η_1 has compact support.

We put

$$f = L_1(y + x, p; \varepsilon) L_2(x, q + p; \varepsilon) \eta_1(y, q).$$

According to the Taylor formula we have

$$f = \sum_{|\alpha| + |\beta| \leq N} f_{\alpha\beta}(x, p) y^\alpha q^\beta + f_N. \tag{2.29}$$

The phase function $S = -\langle y, q \rangle$ in integral $L^{(1)}$ has one and, moreover, non-degenerate stationary point $y = 0, q = 0$. Applying the stationary phase method (Theorem 1.1) we see that the integrals

$$J_{\alpha\beta} = \left(\frac{\lambda}{2\pi}\right)^n \int \exp{(i\lambda S)} y^{\alpha} q^{\beta} \eta_1(y, q) \, dy \, dq$$

are of the form $J_{\alpha\beta} = \lambda^{-|\alpha|-|\beta|}$ for $\lambda \to +\infty$ (the asymptotic series of λ^{-1}). Further, functions $f_{\alpha\beta}$ are proportional to the derivatives $D_x^{\alpha} D_p^{\beta}(L_1(x, p; \varepsilon)L_2(x, p; \varepsilon))$, and, therefore, belong to class $T_+^{m_1+m_2, k_1+k_2}$. Consequently, all functions

$$f_{\alpha\beta} J_{\alpha\beta}^+ \in T_+^{m_1+m_2, k_1+k_2}.$$

By using the integral formula for remainder f_N in (2.29) we obtain

$$f_N = \sum_{|\alpha|+|\beta|=N+1} y^{\alpha} q^{\beta} g_{\alpha\beta}(x, p, y, q; \varepsilon),$$

where functions $g_{\alpha\beta}$ satisfy the estimates

$$|g_{\alpha\beta}| \leq C(1 + |x|^2)^{(m_1+m_2)/2} (1 + |p|^2)^{(k_1+k_2)/2}$$

for $(y, q) \in \operatorname{supp} \eta_1$; similar estimates (with other constants) are true for all derivatives of functions $g_{\alpha\beta}$. By using identity (2.24) and by integrating by parts we get

$$\tilde{J}_{\alpha\beta} = \int y^{\alpha} q^{\beta} \eta_1 \exp{(i\lambda S)} \, dy \, dq$$

$$= i\lambda^{-j} \int \exp{(i\lambda S)} \, {}^t M^j(y^{\alpha} q^{\beta} g_{\alpha\beta} \eta_1) \, dy \, dq.$$

The coefficients of operator ${}^t M^j$ have the form φr^{-2j}, where φ is a polynomial in the variables y, q, and the function $y^{\alpha} q^{\beta}$ has a zero of order $N + 1$ at the origin. Therefore, for fixed j and for $N \geq j$ large enough, the function ${}^t M^j(y^{\alpha} q^{\beta} g_{\alpha\beta} \eta_1)$ is a sum of derivatives of function $g_{\alpha\beta}$ with the coefficients being continuous functions of y and q in a neighbourhood of the origin. This yields the estimate (2.28) for integral $\tilde{J}_{\alpha\beta}$ and for all its derivatives with respect to the variables x and p provided $j \leq j(N)$, where $j(N) \to +\infty$ with $N \to \infty$.

In this way, for function $L^{(1)}$, the existence of representation (2.18) and estimate (2.17) for $\gamma = 0$ have been proved. We have

$$L^{(1)} = \sum_{|\alpha|+|\beta| \leq N} J_{\alpha\beta} + \left(\frac{\lambda}{2\pi i}\right)^n \sum_{|\alpha|+|\beta|=N+1} \tilde{J}_{\alpha\beta} \cdot$$

Integrals $J_{\alpha\beta}$ can be expanded into asymptotic series in powers of $\varepsilon = (i\lambda)^{-1}$ that can be differentiated term by term (Theorem 1.1). Further, $\partial/\partial\varepsilon = i\lambda^2(\partial/\partial\lambda)$, so that the derivative of integral $\tilde{J}_{\alpha\beta}$ with respect to ε yields an integral of the same form, multiplied by λ^2. Since, for sufficiently large N, the estimate (2.28) for integral $\tilde{J}_{\alpha\beta}$ holds with $j = -2 - n$, the inequality $|\partial\tilde{J}_{\alpha\beta}/\partial\varepsilon| \leq C$ is true for the same N. In this way the estimate (2.17) for function $L^{(1)}$ has been proved for $\gamma = 1$; the estimate for all $\gamma \geq 2$ can be found analogously.

Thus we have shown that $L^{(1)} \in T_+^{m_1+m_2,\,k_1+k_2}$ which, together with the estimate (2.28) for function $L^{(2)}$, proves that $L \in T_+^{m_1+m_2,\,k_1+k_2}$.

The proof of Theorem 2.12 will be reduced to the case considered in Lemma 2.15. For this purpose it is necessary to transform the composition $\mathscr{L}_2\mathscr{L}_1$ in such a way that the growths of symbol L_1 in x and of symbol L_2 in p are diminished. We shall establish still one important property of λ-p.d. operators in advance.

THEOREM 2.16 *Let operator $\mathscr{L} \in \mathscr{T}_+^{m,k}$, function $u(x) \in C_0^\infty(\mathbf{R}^n)$. Then the following estimates hold*

$$|D_x^\alpha(\mathscr{L}u)(x)| \leq C_{\alpha,N}(1 + |x|)^{-N}, \quad (x \in \mathbf{R}^n, \lambda \geq 1) \tag{2.30}$$

for arbitrary multi-indices α and for arbitrary integer $N \geq 0$.

Thus the function $\mathscr{L}u$ together with all its derivatives decreases faster than any power of $|x|$ for $|x| \to \infty$.

Proof. We construct a C^∞-partition of unity in $\mathbf{R}_x^n : \eta_1(x) + \eta_2(x) \equiv 1$ where function $\eta_1(x) \in C_0^\infty(\mathbf{R}^n)$ and $\eta_1(x) \equiv 1$ in a neighbourhood of the point $x = 0$. Then $\mathscr{L} = \mathscr{L}_1 + \mathscr{L}_2$, where $\mathscr{L}_j$ is the operator with symbol $\eta_j L$. We have

$$(\mathscr{L}_2 u)(x) = \left(\frac{\lambda}{-2\pi i}\right)^{n/2} \int \exp[i\lambda\langle x, p\rangle]\eta_2(x)L(x, p; \varepsilon)\tilde{u}(p)\, dp. \tag{2.31}$$

Using the identity

$$\exp[i\lambda\langle x, p\rangle] = \frac{1}{i\lambda} A(\exp[i\lambda\langle x, p\rangle]),$$

$$A = |x|^{-2} \sum_{k=1}^n x_k \frac{\partial}{\partial p_k},$$

and integrating by parts we obtain

$$(\mathcal{L}_2 u)(x) = -\left(\frac{\lambda}{-2\pi i}\right)^{n/2} \frac{(i\lambda)^{-1}}{|x|^2} \int \exp\left[i\lambda\langle x, p\rangle\right]\eta_2(x) \times$$

$$\times \left[\sum_{k=1}^{n}\left(x_k \frac{\partial}{\partial p_k} L(x, p)\tilde{u}(p) + x_k \frac{\partial \tilde{u}(p)}{\partial p_k}\right)\right] dp. \quad (2.32)$$

Since the functions $\partial L/\partial p_j \in T_+^{m,k}$ and function $\tilde{u}(p)$ together with all its derivatives decreases faster than any power of $|p|$ for $|p| \to \infty$ we get

$$|(\mathcal{L}_2 u)(x)| \leq \lambda^{(n/2)-1}(1 + |x|)^{m-1}.$$

By repeating the integration by parts we derive the estimate (2.30) for the function $\mathcal{L}_2 u$. Moreover, the right-hand side of the estimate can be multiplied by the factor λ^{-N} (for $\lambda \geq 1$) in addition.

The function $\mathcal{L}_1 u$ is of the form (2.31), in which $\eta_2(x)$ is replaced by $\eta_1(x)$. The integrand has compact support with respect to x, since $\eta_1(x) \in C_0^\infty(\mathbf{R}^n)$. Furthermore, there exist estimates, for function $\tilde{u}(p)$ and all its derivatives, of the form $|\tilde{u}(p)| \leq C_N |p|^{-N} \lambda^{-N}$ with $|p| \geq 1$, $\lambda \geq 1$, and arbitrary integer $N \geq 0$ (Lemma 1.5). One may, therefore, assume that symbol L has compact support with respect to p, i.e. $L \equiv 0$ for $|p| \geq 1$.

Consequently,

$$(\mathcal{L}_1 u)(x) = \eta_1(x)\left(\frac{\lambda}{2\pi}\right)^n \int\int \exp\left[i\lambda\langle x - y, p\rangle\right] \times$$

$$\times L(x, p; \varepsilon)u(y)dy\,dp,$$

where the integral is taken over a finite domain in $\mathbf{R}_y^n \times \mathbf{R}_p^n$. By using the stationary phase method (Theorem 1.4) one can see that the last integral is bounded function of x for $\lambda \geq 0$; the estimate (2.30) of the function $\mathcal{L}_1 u$ for $|\alpha| = 0$ follows from compactness of $\operatorname{supp}\eta_1$. In order to derive estimate (2.30) for $|\alpha| \geq 1$, we express the formula for $\mathcal{L}_1$ in the form

$$(\mathcal{L}_1 u)(x) = \eta_1(x)\left(\frac{\lambda}{2\pi}\right)^n \int\int \exp\left[-i\lambda\langle \tilde{y}, p\rangle\right] L(x, p; \varepsilon) \times$$

$$\times u(\tilde{y} + x)\,d\tilde{y}\,dp$$

and notice that differentiation of the integral with respect to x leads to an integral of the same form, q.e.d.

Proof of Theorem 2.12. By virtue of Theorem 2.16 the function

$(\mathscr{L}_1 u)(x)$ decreases faster than any power of $|x|$ when $|x| \to \infty, \lambda \geq 1$. The composition $(\mathscr{L}_2(\mathscr{L}_1 u))(x)$ is, therefore, well-defined. Actually,

$$(\mathscr{L}_2(\mathscr{L}_1 u))(x) =$$
$$= \left(\frac{\lambda}{-2\pi i}\right)^{n/2} \int \exp[i\lambda\langle x, p\rangle] L_2(x, p; \varepsilon) \widetilde{\mathscr{L}_1 u}(p)\, dp,$$

since the function $(\widetilde{\mathscr{L}_1 u})(p)$ for $\lambda \geq 1$ decreases faster than any power of $|p|$ for $|p| \to \infty$, the last integral is absolutely convergent. Let us write this integral in the form

$$(\mathscr{L}_2(\mathscr{L}_1 u))(x) =$$
$$= \left(\frac{\lambda}{-2\pi i}\right)^{n/2} \int \exp[i\lambda\langle x, p\rangle] L_2^{(N)}(x, p; \varepsilon) \times$$
$$\times \widetilde{(-\lambda^{-2}\Delta_x + 1)^N (\mathscr{L}_1 u)(\tilde{x})}\, dp$$

where $L_2^{(N)}(x, p; \varepsilon) = L_2(x, p; \varepsilon)(1 + |p|^2)^{-N}$, and $N > 0$ is chosen sufficiently large. Symbol $L_2^{(N)}$ belongs to class $T_+^{m_2, k_2 - N}$. The function $(\mathscr{L}_1 u)(x)$ has the form (2.31) where $L = L_1, \eta_2 \equiv 1$, so that

$$(-\lambda^{-2}\Delta_x + 1)^N (\mathscr{L}_1 u)(x) = (\mathscr{L}_1^{(-2N)} u)(x),$$

where $\mathscr{L}_1^{(-2N)}$ is a λ-p.d. operator with the symbol of class $T_+^{m_1 + 2N, k_1}$. Hence

$$\mathscr{L}_2 \mathscr{L}_1 u = \mathscr{L}_2^{(N)} \mathscr{L}_1^{(-2N)} u,$$

where operator $\mathscr{L}_2^{(N)}$ satisfies the assumptions of Lemma 2.15. This lemma can be applied to the composition $\mathscr{L}_2^{(N)} \mathscr{L}_1^{(-2N)}$, provided the growth of symbol $L_1^{(-2N)}$ with respect to x is decreased. For this purpose, like in the proof of Theorem 2.16, we represent operator $\mathscr{L}_1^{(-2N)}$ in the form $\mathscr{L}_1^{(-2N)} = \mathscr{L}_{11} + \mathscr{L}_{12}$, where $\mathscr{L}_{11}$ is an operator with the compactly supported symbol with respect to x, and symbol L_{12} of the operator $\mathscr{L}_{12}$ is identically equal to zero for $|x| \leq 1$. By applying, to the function $(\mathscr{L}_{12} u)(x)$, the procedure previously used for the function $(\mathscr{L}_{12} u)(x)$ in the proof of Theorem 2.16, we obtain (see (2.32))

$$(\mathscr{L}_{12} u)(x) = `$$
$$= \left(\frac{\lambda}{-2\pi i}\right)^{n/2} \int \exp[i\lambda\langle x, p\rangle] \times$$
$$\times \eta_2(x) \sum_{|\alpha|=0}^{M} L_{12}^{(\alpha)}(x, p; \varepsilon)(i\lambda)^{-|\alpha|} D_p^\alpha \tilde{u}(p)\, dp.$$

Here $M \geq 1$ is a sufficiently large integer and symbols $L_{12}^{(\alpha)}$ belong to class T^{m_1-q,k_1}, in which $q \geq 1$ can be made arbitrarily large on account of increase of M.

The Theorem follows, for composition $\mathscr{L}_2^{(N)} \mathscr{L}_{11}$, due to Lemma 2.15. In just the same way this Lemma can be applied to the compositions $\mathscr{L}_2^{(N)} \mathscr{L}_{12}^{(\alpha)}$, where $\mathscr{L}_{12}^{(\alpha)}$ is the λ-p.d. operator with symbol $L_{12}^{(\alpha)}$. However, the operators $\mathscr{L}_2^{(N)} \mathscr{L}_{12}^{(\alpha)}$ do not act now on function $u(x)$ but on function $x^{\alpha}u(x)$, where $x^{\alpha} = x_1^{\alpha_1} \dots x_n^{\alpha_n}$. If $\mathscr{L}$ is a λ-p.d. operator with symbol L of class $T_+^{m,k}$, then, because of compactness of $\operatorname{supp} u$, we have

$$\mathscr{L}(x^{\alpha}u(x)) = (\tilde{\mathscr{L}}u)(x),$$

where $\tilde{\mathscr{L}}$ is a λ-p.d. operator with the symbol from class $T_+^{m+|\alpha|,k}$. To prove this fact, it suffices to integrate by parts with respect to dp in formula (2.31) for the λ-p.d. operator.

Thus it was proved that the composition $\mathscr{L}_2 \mathscr{L}_1$ of two λ-p.d. operators is again a λ-p.d. operator with the symbol from class $T^{m,k}$ for some m, k. By computing the orders of the considered operators $\mathscr{L}_2^{(N)}, \mathscr{L}_{12}^{(\alpha)}$, and by taking Lemma 2.15 into account, it is not hard to show that $m = m_1 + m_2, k = k_1 + k_2$.

There is a more general result which follows from the proof of Theorem 2.12.

THEOREM 2.17. *Let* $\mathscr{L}_j \in \mathscr{T}_+^{m_j,k_j}, j = 1, 2.$ *Then the composition* $\mathscr{L}_2 \mathscr{L}_1 \in \mathscr{T}_+^{m,k}, m = m_1 + m_2, k = k_1 + k_2.$

Now we shall introduce formulae for composition of two λ-p.d. operators.

THEOREM 2.18. *Let the* λ-*pseudodifferential operators* $\mathscr{L}_j, j = 1, 2,$ *belong to classes* $\mathscr{T}_+^{m_j,k_j}$, *with their differential operators acting first. Then the symbol L of the operator* $\mathscr{L} = \mathscr{L}_2 \mathscr{L}_1$ *has the form*

$$L_2(\overset{2}{x}, \lambda^{-1}\overset{1}{D}_x; \varepsilon)[L_1(\overset{2}{x}, \lambda^{-1}\overset{1}{D}_x; \varepsilon)u(x)] =$$
$$= [L^0(\overset{2}{x}, \lambda^{-1}\overset{1}{D}_x) + \varepsilon L^1(\overset{2}{x}, \lambda^{-1}\overset{1}{D}_x)]u(x) =$$
$$+ \varepsilon^2 L^3(\overset{2}{x}, \lambda^{-1}\overset{1}{D}_x; \varepsilon)u(x). \tag{2.33}$$

Here $\varepsilon = (i\lambda)^{-1}$, *operator* $\mathscr{L}^3 \in \mathscr{T}_+^{m_1+m_2,k_1+k_2}$, *and*

$$L^0(x, p) = L_1(x, p; 0)L_2(x, p, 0), \tag{2.34}$$

$$L^1(x, p) = \sum_{j=1}^{n} \frac{\partial L_2(x, p; 0)}{\partial p_j} \frac{\partial L_1(x, p; 0)}{\partial x_j} + \frac{\partial L_1(x, p; \varepsilon)}{\partial \varepsilon}\bigg|_{\varepsilon=0}. \tag{2.35}$$

Proof. Representation (2.33) follows from Theorem 2.17 and Definition 2.13. It remains to prove the formulae (2.34) and (2.35). By acting with operator $\mathscr{L}$ on a rapidly oscillating exponential $u = \varphi(x) \exp[i\lambda S(x)]$ where $\varphi \in C_0^\infty(\mathbf{R}^n), S \in C^\infty(\mathbf{R}^n)$ and phase $S(x)$ is real-valued, and by virtue of Theorem 2.6, we obtain

$$(\mathscr{L}u)(x) = \exp[i\lambda S(x)][R_0 \varphi + \varepsilon R_1 \varphi + O(\varepsilon^2)], \quad (\varepsilon \to 0).$$
(2.36)

The estimate $O(\varepsilon^2)$ is uniform with respect to x on an arbitrary compact set. The functions $R_0 \varphi, R_1 \varphi$ are of the form (2.10), (2.11) where L is the symbol of operator $\mathscr{L}$.

Now we shall successively apply the operators $\mathscr{L}_1$ and $\mathscr{L}_2$ to function u. We get

$$(\mathscr{L}_1 u)(x) = \exp[i\lambda S(x)][R_0^1 \varphi + \varepsilon R_1^1 \varphi] + O(\varepsilon^2), \quad (\varepsilon \to 0),$$

where functions $R_j^1 \varphi$ are determined by symbol L_1 with the help of the formulae (2.10) and (2.11). The remainder in the last formula belongs to class $O_{-2}^+(x, \lambda)$ (Theorem 2.6) so that

$$|\mathscr{L}_2 O(\varepsilon^2)| \leq C|\varepsilon|^2$$

for $x \in \mathbf{R}^n$, $\lambda \geq 1$. Consequently,

$$(\mathscr{L}_2 \mathscr{L}_1 u)(x) = \exp[i\lambda S(x)][\mathscr{L}_2(R_0^1 \varphi + \varepsilon R_1^1 \varphi)] + O(\varepsilon^2) =$$
$$= \exp[i\lambda S(x)][R_0^2(R_0^1 \varphi) + \varepsilon R_0^2(R_1^1 \varphi) +$$
$$+ \varepsilon R_1^2(R_0^1 \varphi)] + O(\varepsilon^3), \quad (\varepsilon \to 0).$$

The comparison of this formula with formula (2.36) yields

$$R_0 \varphi = R_0^2(R_0^1 \varphi), \qquad R_1 \varphi = R_0^2(R_1^1 \varphi) + R_1^2(R_0^1 \varphi)$$
(2.37)

for any function $\varphi \in C_0^\infty(\mathbf{R}^n)$. Consequently,

$$L^0(x, p) = L_1(x, p; 0)L_2(x, p; 0),$$

where $p = \partial S(x)/\partial x$. Since S may be an arbitrary linear function, formula (2.34) is proved.

Notice that the identities (2.37) have the same form for both λ-differential and λ-pseudodifferential operators. Consequently, it suffices to prove (2.37) for λ-differential operators $\mathscr{L}_1, \mathscr{L}_2$; in this case the expressions (2.33)–(2.35) follow from the Leibniz formula.

When the order of the differentiations and multiplications in operators

$\mathcal{L}_j$ differs from that in (2.33), the composition formula can be proved exactly in the same manner.

§ 3. The Hamilton–Jacobi Equation. The Hamilton System

We look for a particular formal asymptotic solution of the equation $L(\overset{2}{x}, \lambda^{-1}\overset{1}{D}_x)u = 0$ in the form

$$u(x, \lambda) = \varphi(x) \exp\left[i\lambda S(x)\right].$$

Then function $S(x)$ satisfies the *Hamilton–Jacobi equation* which is characteristic equation for the equation $Lu = 0$. Characteristic equation for the Schrödinger equation turns out to be the Hamilton–Jacobi equation of classical dynamics; and for the Helmholtz equation – the eikonal equation.

The Cauchy problem for the Hamilton–Jacobi equation is equivalent to the Cauchy problem for the Hamilton system

$$\frac{dx}{dt} = \frac{\partial L(x, p)}{\partial p}, \qquad \frac{dp}{dt} = -\frac{\partial L(x, p)}{\partial x}.$$

Initial data form an $(n-1)$-dimensional manifold: $x = x^0(\alpha), p = p^0(\alpha)$, $\alpha = (\alpha_1, \ldots, \alpha_{n-1})$, in the phase space (x, p). Let us denote the solution of the Cauchy problem for this system by $(x(t, \alpha), p(t, \alpha))$.

The amplitude $\varphi(x)$ satisfies the transport equation, i.e. an ordinary linear first order differential equation along the ray $x = x(t, \alpha)$. By applying the Liouville formula we obtain the leading term of the asymptotics in the small:

$$u(x) = \sqrt{\frac{J(0, \alpha)}{J(t, \alpha)}} \exp\left[i\lambda S(x) + \frac{1}{2}\int_0^t \sum_{j=1}^n \frac{\partial^2 L(x, p)}{\partial x_j \partial p_j} dt' \right].$$

Here, the Jacobian $J(t, \alpha) = \det\left[\partial x(t, \alpha)/\partial(t, \alpha)\right]$ characterizes the change of the cross-section area of the ray tube. In general, this formula is applicable only for small $|t|$.

1. *Characteristic Equation (The Hamilton–Jacobi Equation)*

Consider the equation

$$L(\overset{1}{x}, \lambda^{-1}\overset{2}{D}_x; (i\lambda)^{-1})u(x) = 0. \tag{3.1}$$

We are interested in its formal asymptotic solutions for $\lambda \to \pm \infty$. Let G be a finite domain in $\mathbf{R}^n_x$.

DEFINITION 3.1. Function $u(x, \lambda) \in O_0^+(G)$ is called a *solution of equation* (3.1) *modulo* $O^+_{-m}(G)$ (mod $O^+_{-m}(G)$), if $Lu(x, \lambda) \in O^+_{-m}(G)$.

In this case one says that "$u(x, \lambda)$ is an asymptotic solution of Equation (3.1) with accuracy to $O(\lambda^{-m})$ for $\lambda \to +\infty$" in accordance with the common phraseology in works on asymptotic methods in the theory of differential equations.

We shall seek a particular solution mod $O^+_{-1}(G)$ in the form

$$u(x, \lambda) = \varphi(x) \exp[i\lambda S(x)]. \tag{3.2}$$

Theorem 2.6 together with formula (2.10) yields

PROPOSITION 3.2. *Let symbol* $L(x, p; (i\lambda^{-1})) \in T_+^m$, *function* $S(x) \in C^\infty(G)$ *be real-valued, and function* $\varphi(x) \in C_0^\infty(G)$. *Let* $S(x)$ *satisfy the equation*

$$L(x, \partial S(x)/\partial x; 0) = 0, \quad x \in G. \tag{3.3}$$

Then the function (3.2) *is a solution of Equation* (3.1) *mod* $O^+_{-1}(G)$.

Equation (3.3) is called the *characteristic equation* or the *Hamilton–Jacobi equation* for Equation (3.1).

DEFINITION 3.3. Real-valued function $S(x)$ is a *λ-characteristic* of the λ-p.d. operator $L(\overset{1}{x}, \lambda^{-1} \overset{2}{D}_x; (i\lambda)^{-1})$, if, for any function $\varphi(x) \in C_0^\infty(G)$,

$$L(\overset{2}{x}, \lambda^{-1} \overset{1}{D}_x; (i\lambda)^{-1})[\varphi(x) \exp(i\lambda S(x))] = O(\lambda^{-1}),$$
$$(\lambda \to +\infty) \tag{3.4}$$

for each fixed $x \in G$. The λ-characteristic $S(x)$ satisfies the characteristic equation.

REMARK 3.4. In the general theory of partial differential equations another definition of characteristic is used (see e.g. [36]). Namely, function $S(x) \in C^\infty(G)$ is said to be a *characteristic* of differential operator L, if it satisfies the equation

$$L^0(x, \partial S(x)/\partial x) = 0, \tag{3.5}$$

where L^0 is the principal part of operator L (i.e. the part of polynomial

$L(x, p)$ which is homogeneous and of the highest degree with respect to the variables p).

This definition is equivalent to the following one: function $S(x)$ is a characteristic of operator L, if for any function $\varphi(x) \in C_0^\infty(G)$ the relation

$$L(\overset{2}{x}, \overset{1}{D}_x)[\varphi(x) \exp(i\lambda S(x))] = \lambda^m O(\lambda^{-1})$$

holds for $x \in G$, $\lambda \to +\infty$, where m is the order of operator L.

EXAMPLE 3.5. Characteristic equations for Schrödinger's equation (2.3) and Helmholtz's equation (2.4) have the forms

$$\frac{\partial S}{\partial t} + \frac{1}{2m}\left(\frac{\partial S}{\partial x}\right)^2 + V(x) = 0, \tag{3.6}$$

$$\left(\frac{\partial S}{\partial x}\right)^2 = n^2(x), \tag{3.7}$$

respectively.

The former is the Hamilton–Jacobi equation which describes the motion of a classical particle of mass m in a field with potential energy $V(x)$. The latter is the equation of geometrical optics (the eikonal equation) which describes propagation of light in a medium with refraction index $n(x)$. In classical mechanics a solution of Equation (3.6) is called an *action*; a solution of Equation (3.7) in optics is called an *eikonal*.

According to the definition of the characteristic via (3.5), the characteristics of Equations (3.6) and (3.7) fulfil the equations

$$\left(\frac{\partial S(t, x)}{\partial x}\right)^2 = 0, \qquad \left(\frac{\partial S(x)}{\partial x}\right)^2 = 0,$$

respectively. In particular, the Helmholtz equation has no real characteristics different from constants. On the other hand, the characteristics of the Schrödinger equation are of the form $S = S(t)$. They coincide according to (3.5) with the characteristics of the heat transfer equation

$$\frac{\partial u}{\partial t} = a^2 \Delta u$$

the properties of which immensely differ from those of the Schrödinger equation.

The physicists actually take as characteristics of the Schrödinger and the Helmholtz equations just the solutions of the Hamilton–Jacobi equation (3.6) and of the eikonal equation (3.7), since just these equations describe the correspondence between quantum and classical mechanics, and between wave and geometrical optics, respectively.

The notion of a characteristic of a differential or pseudo-differential operator is ambiguous and is determined by the class of problems in question. For the class of problems considered here, the definition (3.4) is more suitable than (3.5).

PROPOSITION 3.6. *The* λ-*characteristics for* λ-*pseudodifferential operators*

$$L(\overset{2}{x}, \lambda^{-1}\overset{1}{D}_x; (i\lambda)^{-1}), \qquad L(\overset{1}{x}, \lambda^{-1}\overset{2}{D}_x; (i\lambda)^{-1})$$

coincide.

2. Bicharacteristics

Equation (3.3) is a non-linear first order partial differential equation. As known from classical analysis, integration of this equation can be reduced to integration of the system of ordinary differential equations

$$\frac{dx}{dt} = \frac{\partial L(x, p; 0)}{\partial p}, \qquad \frac{dp}{dt} = -\frac{\partial L(x, p; 0)}{\partial x}. \tag{3.8}$$

This system will be called the *bicharacteristic system* or the *Hamilton system* for operator $L(x, D_x; (i\lambda)^{-1})$.

We shall use the following terminology. The space $\mathbf{R}^{2n} = \mathbf{R}^n_x \times \mathbf{R}^n_p$ will be called a *phase space*. Let $x = x(t)$, $p = p(t)$, $t \in I = (t_1, t_2)$, be a solution of the Hamilton system. The curve $\{(x, p): x = x(t), p = p(t), t \in I\}$ in the phase space will be called a *bicharacteristic*, and its projection on $\mathbf{R}^n_x$, i.e. the curve $\{x: x = x(t), t \in I\}$, a *ray* or a *trajectory*. This terminology is from geometrical optics and classical mechanics (see Example 3.5) and has the following relation to that usually accepted:

a bicharacteristic $\leftrightarrow$ a bicharacteristic strip,

a ray (a trajectory) $\leftrightarrow$ a bicharacteristic.

It is assumed throughout § 3 that the following assumption holds.

L3.1. Function $L(x, p; 0)$ belongs to $C^\infty(\mathbf{R}^n_x \times \mathbf{R}^n_p)$ and is real-valued.

We shall summarize a number of well-known properties of the Hamilton system [82], [31], [11], [77]. First recall that the *Poisson bracket* (f, g) of scalar functions $f(x, p), g(x, p)$ is the function

$$(f, g)(x, p) = \left\langle \frac{\partial f}{\partial x}, \frac{\partial g}{\partial p} \right\rangle - \left\langle \frac{\partial f}{\partial p}, \frac{\partial g}{\partial x} \right\rangle.$$

PROPOSITION 3.7. *The time derivative of function $f(x, p)$ w.r.t. the Hamilton system is equal to the Poisson bracket (f, L). The function $L(x, p; 0)$ is a first integral of the Hamilton system.*

Proof. The derivative $\dot{f}(x, p)$ of function $f(x, p)$ w.r.t. the Hamilton system is equal to

$$\dot{f}(x, p) = \left\langle \frac{\partial f}{\partial x}, \frac{\partial L}{\partial p} \right\rangle - \left\langle \frac{\partial f}{\partial p}, \frac{\partial L}{\partial x} \right\rangle = (f, L)(x, p).$$

$L(x, p; 0)$ is a first integral, since $\dot{L} = (L, L) = 0$.

PROPOSITION 3.8. *Let $S(x)$ be a solution of the characteristic equation in a domain $G \ni x^0$, and $\Gamma = \{(x, p): x = x(t), p = p(t), t \in I \ni 0\}$ is a bicharacteristic corresponding to the initial data*

$$x\big|_{t=0} = x^0, \qquad p\big|_{t=0} = \frac{\partial S(x^0)}{\partial x}. \tag{3.9}$$

Then, for $(x, p) \in \Gamma$, we have

$$p = \frac{\partial S(x)}{\partial x}, \tag{3.10}$$

$$\frac{dS(x)}{dt} = \left\langle p(t), \frac{dx(t)}{dt} \right\rangle. \tag{3.11}$$

Proof. We suppose here that $t \in I = (-\delta, \delta)$, where $\delta > 0$ is sufficiently small. Consider the Cauchy problem

$$\frac{dy}{dt} = \frac{\partial L(y, \partial S(y)/\partial y; 0)}{\partial p}, \qquad y\big|_{t=0} = x^0 \tag{3.12}$$

and set

$$q(t) = (S \circ y)(t). \tag{3.12'}$$

A solution of the problem (3.12) exists and is unique for $t \in I$. We shall show that $(y(t), q(t))$ is a solution of the Hamilton system (3.8) for $t \in I$; then (3.12), (3.12′) and the uniqueness of the solution of the Cauchy problem lead to (3.10). By differentiating both (3.12′) with respect to t and the identity $L(x, \partial S(x)/\partial x; 0) = 0$ with respect to x we get

$$\frac{dq(t)}{dt} = \frac{\partial^2 S(y(t))}{\partial x^2} \frac{\partial L(y(t), q(t); 0)}{\partial p},$$

$$\frac{\partial L\left(x, \frac{\partial S(x)}{\partial x};0\right)}{\partial x} + \frac{\partial^2 S(x)}{\partial x^2} \frac{\partial L\left(x, \frac{\partial S(x)}{\partial x};0\right)}{\partial p} = 0.$$

Substituting $x = y(t)$ in the last relation we obtain

$$\frac{dq(t)}{dt} = -\frac{\partial L(y(t), q(t); 0)}{\partial p},$$

the second equation in (3.8).

Further, we have

$$\frac{dS(x, t)}{dt} = \left\langle \frac{\partial S(x(t))}{\partial x}, \frac{dx(t)}{dt} \right\rangle = \left\langle p(t), \frac{dx(t)}{dt} \right\rangle$$

along Γ that yields (3.10).

3. The Cauchy Problem in a plane

We shall formulate the Cauchy problem for equation (3.3) in the plane $x_n = 0$. Since this equation is non-linear, it is, in general, not sufficient to specify $S(x)$ for $x_n = 0$. We set $x' = (x_1, \ldots, x_{n-1})$, $p' = (p_1, \ldots, p_{n-1})$. The Cauchy problem for the Hamilton–Jacobi equation should mean: find a solution of this equation satisfying the Cauchy data

$$S\big|_{x_n=0} = S_0(x'), \quad x' \in U', \tag{3.13}$$

$$\frac{\partial S}{\partial x}\bigg|_{x_n=0} = p^0(x'), \quad x' \in U'. \tag{3.14}$$

Here U' is a neighbourhood of the point $x' = 0'$, and $S_0(x'), p^0(x')$ are given real-valued functions of class $C^\infty(U')$.

Functions $S_0(x'), p^0(x')$ must satisfy the compatibility conditions: for $x' \in U'$

$$L(x', 0, p^0(x'); 0) = 0, \tag{3.15}$$

$$\langle p^0(x'), dx' \rangle = dS_0(x'). \tag{3.16}$$

The last condition is called a *strip condition* [11], [77].

The problem given by (3.3), (3.13), (3.14) together with the compatibility conditions (3.15), (3.16) will be called a *Lagrangian Cauchy problem* (see § 4).

LEMMA 3.9. *Let*

$$\frac{\partial L(0, p_n(0');0)}{\partial p_n} \neq 0. \tag{3.17}$$

Then a solution of the Lagrangian Cauchy problem exists and is unique in some neighbourhood of the point $x = 0$.

Proof. Set $y = (y_1, \ldots, y_{n-1})$, and let $\{x(t, y), p(t, y)\}$ be a solution of the Cauchy problem

$$x|_{t=0} = (y, 0), \qquad p|_{t=0} = p^0(y) \tag{3.18}$$

for the Hamilton system (3.8). This solution exists and is unique for $t \in I(y) \ni 0$, if the interval $I(y)$ is sufficiently small. We can choose (y, t), as local coordinates in a neighbourhood of the point $x = 0$, since the Jacobian $J(t, y) = \det \left[\partial x(t, y)/\partial(t, y) \right]$ in the point $t = 0, y = 0$ is equal to $\partial L(0, p^0(0'); 0)/\partial p_n \neq 0$ (see (3.17), (3.18)).

If $S(x)$ is a solution of the Lagrangian–Cauchy problem, then we have

$$\frac{dS}{dt} = \left\langle p, \frac{dx}{dt} \right\rangle$$

along a trajectory of the Hamilton system (Proposition 3.8). We set therefore

$$S(x(t, y)) = S_0(y) + \int_0^t \left\langle p(y, \tau), \frac{dx(y, \tau)}{d\tau} \right\rangle d\tau. \tag{3.19}$$

Function $S(x)$ is infinitely differentiable in a neighbourhood of the point $x = 0$ and satisfies the conditions (3.13) and (3.14). Further, if t and y are small,

$$L(x(t, y), p(t, y);0) = 0. \tag{3.20}$$

Thus L appears to be a first integral of system (3.8), and (3.20) holds for $t = 0$ due to (3.13). We shall show that

$$p(t, y) = \frac{\partial S(x(t, y))}{\partial x}. \tag{3.21}$$

Then (3.20) implies that $S(x)$ is a solution of (3.3). Set $S_1(t, y) = S(x(t, y))$;

then equality (3.21) is equivalent to the system

$$\frac{\partial S_1}{\partial t} - \left\langle p, \frac{\partial x}{\partial t} \right\rangle = 0,$$

$$M_j \equiv \frac{\partial S_1}{\partial y_j} - \left\langle p, \frac{\partial x}{\partial y_j} \right\rangle = 0, \qquad 1 \leq j \leq n - 1, \qquad (3.21')$$

where $x = x(t, y)$, $p = p(t, y)$, $S_1 = S_1(x, y)$. The first equality in (3.21') follows from (3.19). Differentiating it with respect to y, and the other equations in (3.21') with respect to t, we obtain

$$\frac{\partial^2 S_1}{\partial t \, \partial y_j} - \sum_{k=1}^{n} \frac{\partial p_k}{\partial y_j} \frac{\partial x_k}{\partial t} - \sum_{k=1}^{n} p_k \frac{\partial^2 x_k}{\partial t \, \partial y_j} = 0,$$

$$\frac{\partial M_j}{\partial t} = \frac{\partial^2 S_1}{\partial t \, \partial y_j} - \sum_{k=1}^{n} \frac{\partial p_k}{\partial t} \frac{\partial x_k}{\partial y_j} - \sum_{k=1}^{n} p_k \frac{\partial^2 x_k}{\partial t \, \partial y_j} = 0.$$

Consequently,

$$\frac{\partial M_j}{\partial t} = \left\langle \frac{\partial p}{\partial y_j}, \frac{\partial x}{\partial t} \right\rangle - \left\langle \frac{\partial p}{\partial t}, \frac{\partial x}{\partial y_j} \right\rangle = \frac{\partial}{\partial y_j} L(x(t, y), p(t, y); 0) = 0$$

by virtue of (3.20). Since

$$M_j(0, y) = \frac{\partial S_1(t, y)}{\partial y_j} - \left\langle p(t, y), \frac{\partial x(t, y)}{\partial y_j} \right\rangle \bigg|_{t=0} =$$

$$= \frac{\partial S_0(y)}{\partial y_j} - p_j^0(y) = 0$$

due to (3.13), we have $M_j(t, y) \equiv 0$, and (3.21) is proved.

Thus we have shown the existence of a solution of the Cauchy problem. Now we shall prove its uniqueness. Let $S(x)$ be the solution previously constructed and $S_1(x)$ another solution of the Cauchy problem. Since $\partial S_1(0)/\partial x = \partial S(0)/\partial x$, and (3.14) holds, we get $\partial S_1(x', 0)/\partial x = p^0(x')$. If a bicharacteristic starts in the point $(x', 0, p^0(x'))$, then $p(x) = \partial S(x)/\partial x$, $p(x) = \partial S_1(x)/\partial x$ will hold along it due to Proposition 3.8. Consequently $S_1(x)$ coincides with the right-hand side of (3.19), i.e. with $S(x)$, q.e.d.

4. The Transport Equation

Consider Equation (3.1) where $L \in T_+^m$ (or T_-^m). Our goal will be to construct formally asymptotic solutions (f.a. solutions) of this equation with accuracy to $O(\lambda^N)(\lambda \to +\infty)$ for arbitrary N. Later we will exploit the known Liouville formula [4], [77], [78]; we shall prove it right now. Consider the autonomous real system of n equations

$$\frac{dx}{dt} = f(x), \tag{3.22}$$

where, for simplicity, $f(x) \in C^\infty(\mathbf{R}_x^n)$. Let $x(t, \alpha) \in C^\infty(I \times V)$ be an $(n-1)$-parameter family of its solutions, where $\alpha = (\alpha_1, \ldots, \alpha_{n-1}) \in \mathbf{R}_\alpha^{n-1}$, V is a domain in $\mathbf{R}_\alpha^{n-1}$, and I is the interval $|t| < \delta$. Set

$$J(t, \alpha) = \det \frac{\partial x(t, \alpha)}{\partial(t, \alpha)}. \tag{3.23}$$

THEOREM 3.10.   *Let $J(t, \alpha) \neq 0$ for $(t, \alpha) \in I \times V$. Then the Liouville formula*

$$\frac{d}{dt} \ln J(t, \alpha) = \operatorname{Sp} \frac{\partial f}{\partial x}(x(t, \alpha)) \tag{3.24}$$

is valid.

   Proof. If matrix $A(t)$ belongs to $C^1(I)$ and is non-degenerate for $t \in I$, then

$$\frac{d}{dt} \ln J(t, \alpha) = \operatorname{Sp}\left(A^{-1}(t) \frac{dA(t)}{dt} \right). \tag{3.25}$$

From (3.22) we have

$$\frac{\partial}{\partial t}\left(\frac{\partial x(t, \alpha)}{\partial(t, \alpha)} \right) = \frac{\partial f(x(t, \alpha))}{\partial x} \frac{\partial x(t, \alpha)}{\partial(t, \alpha)}.$$

Inserting this expression into (3.25) and taking into account that $\operatorname{Sp}(AB) = \operatorname{Sp}(BA)$, we obtain the Liouville formula (3.24).

   Consider now the non-autonomous real system of n equations

$$\frac{dx}{dt} = f(t, x), \tag{3.26}$$

where, for simplicity, $f(t, x) \in C^\infty(\mathbf{R}_t \times \mathbf{R}_x^n)$. Let $x(t, \alpha) \in C^\infty(I \times V)$ be an n-parameter family of solutions of system (3.26), where I is the interval $|t| < \delta, \alpha \in \mathbf{R}_\alpha^n$, and V is a domain in $\mathbf{R}_\alpha^n$. Set

$$J_1(t, \alpha) = \det \frac{\partial x(t, \alpha)}{\partial \alpha}. \tag{3.27}$$

Then the Liouville formula

$$\frac{d}{dt} \ln J_1(t, \alpha) = \operatorname{Sp} f'_x(x(t, \alpha)) \tag{3.28}$$

holds if $J_1(t, \alpha) \neq 0$ for $(t, \alpha) \in I \times V$. The proof is the same as before.

We shall seek now a solution of Equation (3.1) mod $O_2^{\pm}(\Omega)$ in the form

$$u(x, \lambda) = \varphi(x) \exp(i\lambda S(x)). \tag{3.29}$$

Inserting (3.29) into (3.1) and applying (2.10)–(2.12) we see that the functions φ and S should satisfy the equations

$$L\left(x, \frac{\partial S}{\partial x}; 0\right) = 0, \tag{3.30}$$

$$\left\langle \frac{\partial \varphi}{\partial x}, \frac{\partial L\left(x, \frac{\partial S}{\partial x}; 0\right)}{\partial p} \right\rangle + \frac{1}{2} \operatorname{Sp}\left(\frac{\partial^2 L\left(x, \frac{\partial S}{\partial x}; 0\right)}{\partial p^2} \frac{\partial^2 S}{\partial x^2}\right)\varphi(x) +$$

$$+ \frac{\partial L\left(x, \frac{\partial S}{\partial x}; \varepsilon\right)}{\partial \varepsilon}\Bigg|_{\varepsilon = 0} \varphi(x) = 0. \tag{3.31}$$

Equation (3.31) is called a *transport equation*.

Let us formulate the assumptions on symbol L. We recall condition
L3.1: function $L(x, p; 0)$ belongs to C^{∞} and is real-valued for all real x, p.
L3.2. $L(x, p; (i\lambda)^{-1}) \in T^m_+$.
L3.3. Function $L(x, p; (i\lambda)^{-1})$ is real-valued for $x \in \mathbf{R}^n$, $p \in \mathbf{R}^n$, $\lambda \geq 1$.
Consider the Cauchy problem for the Hamilton system (3.8):

$$\begin{aligned} x|_{t=0} &= x^0(\alpha), \\ p|_{t=0} &= p^0(\alpha), \quad \alpha \in U. \end{aligned} \tag{3.32}$$

Here $\alpha = (\alpha_1, \ldots, \alpha_{n-1}) \in \mathbf{R}^{n-1}_\alpha$, U is a domain in $\mathbf{R}^{n-1}_\alpha$, $\{x : x = x^0(\alpha)\}$ is a C^{∞}-manifold of dimension $(n-1)$ in $\mathbf{R}^n_x$, and the compatibility condition

$$\langle p^0(\alpha), dx^0(\alpha) \rangle = 0, \quad \alpha \in U, \tag{3.33}$$

is satisfied. The next assumption on symbol L and on the Cauchy data will be formulated in terms of bicharacteristics.
L3.4. The Cauchy problem (3.8), (3.31) has the unique solution

$\{x(t, \alpha), p(t, \alpha)\}$ for $(t, \alpha) \in I \times U$, where $I = (-\delta, +\delta)$ and the mapping

$$(t, \alpha) \rightarrow x(t, \alpha), \quad (t, \alpha) \in I \times U, \tag{3.34}$$

is a diffeomorphism.

We set

$$S(x(t, \alpha)) = S_0(x^0(\alpha)) + \int_0^t \left\langle p(\tau, \alpha), \frac{dx(\tau, \alpha)}{d\tau} \right\rangle d\tau, \tag{3.35}$$

where $S_0(x^0(\alpha)) \in C^\infty(U)$. Then function $S(x)$ belongs to $C^\infty(\Omega)$, where $\Omega = \{x : x = x(t, \alpha); (t, \alpha) \in I \times U\}$ and is a solution of the characteristic equation for $x \in \Omega$.

THEOREM 3.11.   *Let assumptions L3.1–L3.4 be satisfied. Then the function*

$$u(x, \lambda) = \frac{\varphi_0(\alpha)\sqrt{|J(0, \alpha)|}}{\sqrt{|J(t, \alpha)|}} \exp\left[i\lambda S(x(t, \alpha))\right] \times$$

$$\times \exp\left(\int_0^t \left[\frac{1}{2} \mathrm{Sp} \frac{\partial^2 L(x(\tau, \alpha), p(\tau, \alpha); 0)}{\partial x\, \partial p} - \right.\right.$$

$$\left.\left. - \frac{\partial L(x(\tau, \alpha), p(\tau, \alpha); \varepsilon)}{\partial \varepsilon}\right|_{\varepsilon = 0}\right] d\tau \right) \tag{3.36}$$

represents a solution mod $O_{-2}^{\pm}(\Omega)$ *of Equation* (3.1).

Here S and J are determined by (3.35) and (3.23), $\varphi_0(\alpha)$ is an arbitrary function of class $C^\infty(U)$, and $x = x(t, \alpha)$.

Proof. Insert (3.36) into (3.1). The Equation (3.29) will be fulfilled. Denoting by $L^0, \partial L^0/\partial x, \partial L^0/\partial(i\lambda)^{-1}$ etc. the values of these functions in the point $(x, \partial S/\partial x; 0)$, we have

$$\frac{d\varphi}{dt} = \left\langle \frac{\partial \varphi(x)}{\partial x}, \frac{\partial L^0}{\partial p} \right\rangle, \quad x = x(t, \alpha).$$

Hence Equation (3.31) along the ray $x = x(t, \alpha)$ (for fixed α) takes the form

$$\frac{d\varphi}{dt} + \left[\frac{1}{2} \mathrm{Sp} \left(\frac{\partial^2 L^0}{\partial p^2} \frac{\partial^2 S}{\partial x^2} \right) \varphi + \frac{\partial L^0}{\partial (i\lambda)^{-1}} \varphi \right] = 0, \tag{3.37}$$

By applying the Liouville formula (3.24) to the system $dx/dt = \partial L^0/\partial p$, we obtain

$$\frac{d}{dt} \ln J = \text{Sp}\left(\frac{\partial}{\partial x}\left(\frac{\partial L^0\left(x, \frac{\partial S}{\partial x}; 0 \right)}{\partial p} \right) \right) =$$

$$= \text{Sp}\left(\frac{\partial^2 L^0}{\partial p^2} \frac{\partial^2 S}{\partial x^2} \right) + \text{Sp}\left(\frac{\partial^2 L^0}{\partial x\, \partial p} \right).$$

Consequently, Equation (3.37) can be written in the form

$$\frac{1}{\sqrt{J}} \frac{d}{dt}(\sqrt{J}\, \varphi) = \left(\left(\frac{1}{2} \text{Sp}\, \frac{\partial^2 L^0}{\partial x\, \partial p} \right) - \frac{\partial L^0}{\partial (i\lambda)^{-1}} \right)\varphi. \tag{3.38}$$

Equation (3.36) is then obtained by integrating the last equation.

Consider an operator with a symmetric symbol.

THEOREM 3.12. *Let the assumptions of Theorem 3.11 be fulfilled. Then the function*

$$u(x, \lambda) = \frac{\varphi_0(\alpha)\sqrt{|J(0, \alpha)|}}{\sqrt{|J(t, \alpha)|}} \exp\left(i\lambda S(x) \right) \tag{3.39}$$

represents a solution of the equation

$$[L(\overset{2}{x}, \lambda^{-1}\overset{1}{D}_x) + L(\overset{1}{x}, \lambda^{-1}\overset{2}{D}_x)]\, u(x, \lambda) = 0, \qquad (\text{mod } O^{+}_{-2}(\Omega)). \tag{3.40}$$

Proof. Because of (2.15), the transport equation (3.37) in the considered case takes the form

$$\frac{d\varphi}{dt} + \frac{1}{2} \text{Sp}\left(\frac{\partial^2 L^0}{\partial p^2} \frac{\partial^2 S}{\partial x^2} \right)\varphi + \frac{1}{2} \text{Sp}\left(\frac{\partial^2 L^0}{\partial x\, \partial p} \right)\varphi = 0. \tag{3.41}$$

Repeating the proof of Theorem 3.11 we obtain the equation for φ:

$$\frac{1}{\sqrt{J}} \frac{d}{dt}(\sqrt{J}\, \varphi) = 0.$$

REMARK. Consider an n-dimensional manifold $\Lambda^n = \{(x, p); x = x(t, \alpha),\ p = p(t, \alpha), \alpha \in V, t \in I\}$ in the phase space. The volume element on Λ^n has the form $d\sigma^n(x) = a(x)dx$. We demand this volume to be invariant with respect to the displacements along phase trajectories; then $(a \circ x)(t, \alpha) = b(\alpha)$. Consequently,

$$\frac{d\sigma''(x)}{dx} = \frac{J(0, \alpha)}{J(t, \alpha)}, \qquad (x = x(t, \alpha)),$$

so that formula (3.39) takes the form (for $\varphi_0(\alpha) \equiv 1$)

$$u(x, \lambda) = \sqrt{\left| \frac{d\sigma''(x)}{dx} \right|} \exp(i\lambda S(x)). \tag{3.39'}$$

5. The Cauchy Problem

We shall consider a λ-pseudodifferential operator with separated derivative with respect to t:

$$L = \lambda^{-1} D_t + H(t, \overset{2}{x}, \lambda^{-1} \overset{1}{D}_x ; (i\lambda)^{-1}) \tag{3.42}$$

where $t \in [0, +\infty)$, $x \in \mathbf{R}^n$, and H is a λ-p.d. operator. The variable t plays the role of a parameter in operator H. Consider the Cauchy problem with rapidly oscillating initial data:

$$Lu(t, x) = 0, \tag{3.43}$$

$$u|_{t=0} = u_0(x) \exp(i\lambda S_0(x)), \tag{3.44}$$

where $u_0(x) \in C^\infty(\mathbf{R}_x^n)$, $S_0(x) \in C^\infty(\mathbf{R}_x^n)$ and function $S_0(x)$ is real-valued.

The Hamilton–Jacobi equation for L is of the form

$$\frac{\partial S(t, x)}{\partial t} + H\left(t, x, \frac{\partial S(t, x)}{\partial x}; 0\right) = 0. \tag{3.45}$$

It follows from the Hamilton system that $dt/d\tau = 1$, where τ is a parameter along a bicharacteristic so that τ can be identified with t and the Hamilton system takes the form

$$\frac{dx}{dt} = \frac{\partial H(t, x, p; 0)}{\partial p}, \qquad \frac{dp}{dt} = -\frac{\partial H(t, x, p; 0)}{\partial x}, \tag{3.46}$$

$$\frac{dS}{dt} = \left\langle p, \frac{\partial H(t, x, p; 0)}{\partial p} \right\rangle - H(t, x, p; 0), \tag{3.47}$$

$$\frac{dt}{d\tau} = 1, \qquad \frac{dE}{d\tau} = -\frac{\partial H(t, x, p; 0)}{\partial t}, \tag{3.48}$$

where E is the variable conjugate to t. System (3.46) is closed, and, moreover, function S can be found from (3.47) if solution of (3.46) is known.

We drop therefore Equations (3.48), and call (3.46) a *truncated Hamilton system* for operator L. Thus we shall investigate bicharacteristics not in the $(2n + 2)$-dimensional phase space (x, t, p, E) but in the $2n$-dimensional phase space (x, p).

Let us formulate the assumptions on operator H.

H3.1. Function $H(t, x, p; (i\lambda)^{-1}) \in C^\infty$ for $t \geq 0, (x, p) \in \mathbf{R}^{2n}, \lambda \geq 1$, and is real-valued.

H3.2. For each fixed $t \geq 0$, function $H(x, p, t; (i\lambda)^{-1}) \in T_+^m(\mathbf{R}_x^n)$, where m does not depend on t. Estimates (2.6) hold with constants $C_{\alpha\beta}(t)$ which can be chosen that $\sup_{t \in [0, T]} C_{\alpha\beta}(t) < \infty$ for any $T > 0$ and for arbitrary fixed α, β.

H3.3. The Cauchy problem.

$$x\big|_{t=0} = y, \quad p\big|_{t=0} = \frac{\partial S_0(y)}{\partial y}, \quad y \in \mathbf{R}_x^n, \tag{3.49}$$

for system (3.46) has a unique solution

$$\{x(t, y), p(t, y)\} \in C^\infty([0, T] \times \mathbf{R}_y^n).$$

We shall seek a solution of the Cauchy problem (3.43), (3.44) in the form of the formal series

$$u(t, x) = \exp(i\lambda S(t, x)) \sum_{k=0}^\infty \varphi_k(t, x)(i\lambda)^{-k}. \tag{3.50}$$

Then we obtain Equation (3.45) and the Cauchy data

$$S\big|_{t=0} = S_0(x) \tag{3.51}$$

for $S(t, x)$, and the Cauchy data

$$\varphi_0\big|_{t=0} = u_0(x), \quad \varphi_j\big|_{t=0} = 0, \quad j \geq 1, \tag{3.52}$$

for φ_k.

LEMMA 3.13. *Let $\varphi(x, t)$ and $S(x, t) \in C^\infty([0, T] \times \mathbf{R}_x^n)$, function S be real-valued, function $\varphi(0, t) = 0$ for $x \geq a, t \in [0, T]$. Suppose that the symbol of operator L fulfils assumptions H3.1–H3.3. Then, for arbitary integer $N \geq 1$,*

$$L(t, \overset{2}{x}, \lambda^{-1} \overset{1}{D}_x ; (i\lambda)^{-1})(\varphi(t, x) \exp(i\lambda S(t, x))) =$$

$$= \exp(i\lambda S(t, x)) \sum_{j=0}^N (i\lambda)^{-j}(R_j \varphi)(t, x) + \varphi_{N+1}(t, x, \lambda). \tag{3.53}$$

The estimate of the remainder is

$$\left| D_x^\alpha D_t^\beta \varphi_{N+1}(t, x, \lambda) \right| \leq C_{\alpha, \beta, K} \lambda^{-N-1+|\alpha|+\beta} \tag{3.54}$$

for arbitrary multi-indices α, β with $|\lambda| \geq 1, t \in [0, T]$, and $x \in K \subset \mathbf{R}_x^n$ (K a compact set). R_j are differential operators with respect to x of the order $\leq 2j$ and with the coefficients of class $C^\infty([0, T] \times \mathbf{R}_x^n)$.

The proof follows from Theorem 2.6.

Now we put

$$S(t, x(t, y)) = \int_0^t \left[\left\langle p(\tau, y), \frac{dx(\tau, y)}{dt} \right\rangle - \right.$$

$$\left. - H(\tau, x(\tau, y), p(\tau, y); 0) \right] d\tau + S_0(y). \tag{3.55}$$

$$J(t, y) = \det \frac{\partial x(t, y)}{\partial y}. \tag{3.56}$$

We introduce the notation: $d\varphi/dt$ is the derivative of function $\varphi(x, t)$ with respect to system (3.42), i.e.

$$\frac{d\varphi}{dt} = \frac{\partial \varphi}{\partial t} + \left\langle \frac{\partial \varphi}{\partial x}, \frac{\partial H}{\partial p} \right\rangle. \tag{3.57}$$

LEMMA 3.14. *Let the mapping*

$$(t, y) \to x(t, y), \quad t \in [0, T], y \in \mathbf{R}_y^n, \tag{3.58}$$

be a diffeomorphism and function S have the form (3.55). Moreover, let assumptions H3.1–H3.3 be satisfied. Then

$$R_0 \varphi \equiv 0, \tag{3.59}$$

$$(R_1 \varphi)(t, x(t, y)) = \frac{1}{\sqrt{J}} \frac{d}{dt}(\sqrt{J} \varphi) - \tfrac{1}{2} \mathrm{Sp} \left(\frac{\partial^2 H^0}{\partial x \, \partial p} \right) \varphi +$$

$$+ \left. \frac{\partial H(t, x, p; \varepsilon)}{\partial \varepsilon} \right|_{\varepsilon = 0} \varphi. \tag{3.60}$$

Here $\varphi = \varphi(t, x)$, $x = x(t, y)$, $p = p(t, y)$.

Proof. It follows from (2.10) that

$$R_0 \varphi \equiv \left(\frac{\partial S}{\partial t} + H^0 \right) \varphi \equiv 0,$$

since S satisfies the Hamilton–Jacobi equation. Further, (2.11) implies

$$(R_1 \varphi)(t,x) = \frac{\partial \varphi}{\partial t} + \left\langle \frac{\partial \varphi}{\partial x}, \frac{\partial H^0}{\partial p} \right\rangle +$$

$$+ \left[\tfrac{1}{2} \operatorname{Sp} \left(\frac{\partial^2 H^0}{\partial p^2} \frac{\partial^2 S}{\partial x^2} \right) + \frac{\partial H^0}{\partial (i\lambda)^{-1}} \Big|_{(i\lambda)^{-1}=0} \right] \varphi.$$

Applying the Liouville formula (3.28) to the non-autonomous system

$$\frac{dx}{dt} = \frac{\partial H^0}{\partial p},$$

as in Theorem 3.11, we obtain (3.60).

Let function $S(t,x)$ have the form (3.55):

$$\varphi_0(t, x(t,y)) = u_0(y) \frac{\sqrt{J(0,y)}}{\sqrt{J(t,y)}} \times$$

$$\times \exp\left[\int_0^t \left(\tfrac{1}{2} \operatorname{Sp}\left(\frac{\partial^2 H^0}{\partial x \, \partial p} \right) - \frac{\partial H^0}{\partial (i\lambda)^{-1}} \Big|_{(i\lambda)^{-1}=0} \right) dt \right]. \qquad (3.61)$$

Functions $\varphi_j(x,t)$ for $j \geq 1$ are solutions of the Cauchy problem

$$R_1 \varphi_1 = -R_2 \varphi_0, \qquad \varphi_1|_{t=0} = 0,$$

$$R_1 \varphi_2^{\bullet} = -R_2 \varphi_1 - R_3 \varphi_0, \qquad \varphi_2|_{t=0} = 0,$$

$$\dots\dots\dots\dots\dots\dots \qquad (3.62)$$

$$R_1 \varphi_j = -R_2 \varphi_{j-1} - \dots - R_{j+1} \varphi_0, \qquad \varphi_j|_{t=0} = 0.$$

THEOREM 3.15. *Suppose assumptions H3.1–H3.3 are fulfilled and map (3.58) is a diffeomorphism. Then the function*

$$u_N(t,x,\lambda) = \exp(i\lambda S(t,x)) \sum_{j=0}^{N} (i\lambda)^{-j} \varphi_j(t,x) \qquad (3.63)$$

satisfies Cauchy data (3.44) and the equation

$$L u_N(t,x,\lambda) = \psi_{N+1}(t,x,\lambda) \qquad (3.64)$$

for $t \in [0,T]$ and $x \in \mathbf{R}_x^n$, where ψ_{N+1} satisfies estimates (3.54).

Proof. For functions $S(t,x)$, $\varphi(t,x)$, the assumptions of Lemma 3.14 are true so that formula (3.53) holds for function u_N. By virtue of (3.62)

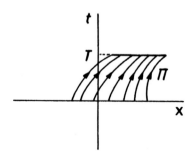

Fig. 4.

and the choice of function S we obtain

$$Lu_N = \exp(i\lambda S(t,x))(i\lambda)^{-N} R(\varphi_0, \ldots, \varphi_N) + \chi_{N+1}(t,x,\lambda) \equiv$$
$$\equiv \chi^0_{N+1}(t,x,\lambda) + \chi_{N+1}(t,x,\lambda),$$

where χ_{N+1} satisfies estimate (3.54). Further, $R(\varphi_0, \ldots, \varphi_N)$ is a sum of linear differential operators of a finite order acting on functions $\varphi_0, \ldots, \varphi_N$, which, according to (3.62), belong to $C^\infty([0,\tau] \times \mathbf{R}^n_x)$. Function $u_0(x)$ has compact support; because of (3.61), we have $\varphi_0(t, x(t,y)) \equiv 0$ for $t \in [0, T]$, $y \notin \mathrm{supp}\, u_0(y)$ (i.e. $\varphi_0 \equiv 0$ outside the strip Π (Figure 4) which is filled by the rays $x = x(t,y)$, $t \in [0, T]$, $y \in \mathrm{supp}\, u_0(y)$). Therefore $R_j \varphi_0 \equiv 0$ outside Π for all j. Since $\varphi_j|_{t=0} = 0, j \geq 1$, we obtain from (3.62) that $\varphi_j(t,x) \equiv 0$ outside Π for all j. Consequently, estimates (3.54) are true for function $\chi^0_{N+1}(t,x,\lambda)$, q.e.d.

This construction is not applicable whenever $J(t,y)$ vanishes.

6. *The Examples*

We shall consider the Helmholtz equation

$$(\Delta + k^2 n^2(x))u(x) = 0.$$

In this case, the characteristic equation appears to be *eikonal equation* (3.7) – the equation of geometrical optics in an isotropic medium. The surfaces of the levels $S(x) = \text{const.}$ are called *wave fronts* or surfaces of constant phase.

The Hamilton system has the form

$$\frac{dx}{dt} = 2p, \qquad \frac{dp}{dt} = -\nabla n^2(x), \tag{3.65}$$

and function S is determined by the equation

$$\frac{dS}{dt} = 2n^2(x), \qquad\qquad (3.65')$$

where t is a parameter along the trajectory. The projection of the phase trajectory $(x(t), p(t))$ on x-space is a *light ray*. In particular, $x(t)$ satisfies the Newton system

$$\ddot{x} + \nabla n^2(x) = 0.$$

The light rays are perpendicular to the wave fronts; this follows from the relations $dx/dt = p, p = \partial S/\partial x$ and from the fact that gradient is orthogonal to level surfaces.

By virtue of (3.36) the leading term of an asymptotic solution of the Helmholtz equation is of the form

$$u_0(x, k) = \varphi_0(\alpha)\sqrt{\frac{J(0, \alpha)}{J(t, \alpha)}}\exp(ikS(x)).$$

The amplitude in the last formula can be evaluated with the help of the light energy-flux conservation law. Take an arbitrary part of the wave front $S(x) = c_0$, say Ω_0, set $t = 0$ on it and send off a family of rays. We get the so-called *tube of rays* (Figure 5). Let $u(x, k) = A(x, k)\exp(ikS(x))$; then the light energy passing through infinitesimal part Ω_0 equals $|u|^2 S(\Omega_0) = |A|^2 S(\Omega_0)$, where $S(\Omega_0)$ is the area of Ω_0. This energy is equal to $|A|^2 S(\Omega_t)$ for any t because of the conservation law; here, area Ω_t is obtained from Ω_0 via the displacement g^t along the rays in the tube. Consequently,

$$\frac{|A(t, \alpha)|^2}{|A(0, \alpha)|^2} = \frac{S(\Omega_0)}{S(\Omega_t)} \approx \frac{J(0, \alpha)}{J(t, \alpha)},$$

which gives the desired expression for the amplitude.

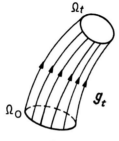

Fig. 5.

Let us consider the case of constant refraction index $n(x) = 1$ in more detail. Then

$$(\Delta + k^2) u(x) = 0,$$

and the eikonal equation and the Hamilton system are of the form

$$(\nabla S(x))^2 = 1, \tag{3.66}$$

$$\frac{dx}{dt} = 2p, \qquad \frac{dp}{dt} = 0, \qquad \frac{dS}{dt} = 2.$$

The solution of the Hamilton system with the Cauchy data for $t = 0$ is given by

$$x(t) = 2tp(0) + x(0), \qquad p(t) = p(0),$$
$$S(t) = 2t + S(0),$$

so that both the phase trajectories and the rays are straight lines. Let us consider a smooth orientable manifold M^{n-1} of dimension $n-1$ in $\mathbf{R}^n$, given by the equations $x = x^0(\alpha), \alpha = (\alpha_1, \ldots, \alpha_{n-1}) \in U$, where U is a domain in $\mathbf{R}_\alpha^{n-1}$. The manifold M^{n-1} may be compact or non-compact. We put

$$S(x) \equiv 0, \qquad x \in M^{n-1}.$$

In order to formulate the Cauchy problem for the eikonal equation, it is still necessary to specify $\nabla S(x)$ on M^{n-1}. Since $S(x) \equiv 0$ on M^{n-1}, all derivatives of function S with respect to the tangent variables are equal to zero, and the eikonal equation on M^{n-1} takes the form $(\partial S/\partial v)^2 = 1$, where $\partial/\partial v$ is the derivative in the normal direction to M^{n-1}. At each point $x \in M^{n-1}$ we fix the normal unit vector v_x depending continuously on x, and put

$$\frac{\partial S}{\partial v_x} = 1, \qquad x \in M^{n-1}.$$

Thus the Cauchy problem for the eikonal equation is fully specified. The corresponding Lagrangian Cauchy problem for the Hamilton system has the form

$$x|_{t=0} = x^0(\alpha), \qquad p|_{t=0} = p^0(\alpha), \qquad \alpha \in U.$$

Vector $p^0(\alpha)$ coincides, by its construction, with the normal unit vector v_x at the point $x = x^0(\alpha)$. The solution of the Cauchy problem leads to

$$x(t, \alpha) = 2tp^0(\alpha) + x^0(\alpha), \qquad p = p^0(\alpha), \qquad (3.67)$$
$$S(x(t, \alpha)) = 2t.$$

This implies that the wave fronts are *equidistant surfaces*. Indeed, taking the point $x = x^0(\alpha)$ on the wave front $S(x) \equiv 0$ and shifting it along the normal through distance h, we reach the point on the wave front $S(x) = h$.

Equations (3.67) express eikonal $S(x)$ in the parametric form: namely, expressing (t, α) in terms of x from the first equation, we get $t = t(x)$, $\alpha = \alpha(x)$, so that $S(x) = 2t(x)$. Function $S(x)$ is smooth and single-valued provided the Jacobian $J(t, \alpha) \neq 0$, where

$$J(t, \alpha) = \det \frac{\partial x(t, \alpha)}{\partial(t, \alpha)}.$$

Let there be an $(n - 1)$-parameter family of curves in $\mathbf{R}_x^n : x = x(t, \alpha)$, $t \in [T_1, T_2]$, $\alpha = (\alpha_1, \dots, \alpha_{n-1})$. The set of points in which $J(t, \alpha) = 0$ is called a *caustic* of the family. The point $x^0 = x(t_0, \alpha^0)$ in which $J(t_0, \alpha^0) = 0$, is called a *focal point* of the curve $x = x(t, \alpha^0)$. If the family appears to be a family of light rays, then, on the caustic, the leading term of the short-wave approximation tends to infinity; hence, near caustics, the approximation of geometrical optics is inappropriate [47].

It follows from the geometrical interpretation of wave fronts that, in the case the wave front $S(x) = 0$ is a strictly convex smooth compact manifold, the wave fronts for small $|t|$ have the same property and eikonal $S(x)$ is a C^∞-function. However, when t is large, this property is not preserved.

EXAMPLE 3.16. Consider the Cauchy problem

$$S(x) \equiv 0, \qquad \frac{\partial S}{\partial v} = 1, \qquad x \in S^{n-1},$$

for the eikonal equation

$$(\nabla S(x))^2 = 1.$$

Here S^{n-1} is the unit sphere $|x| = 1$, $\partial/\partial v$ is the derivative along the inward normal to S^{n-1}. The Cauchy problem for the Hamilton system has the form $x|_{t=0} = \omega, p|_{t=0} = -\omega$, where $\sum_{j=1}^{n} \omega_j^2 = 1$; from here

$$x(t, \omega) = (1 - 2t)\omega, \qquad S = 2t,$$

so that

$$S(x) = 1 \pm |x|,$$

and the eikonal appears to be a multi-valued function. In the ray picture: each point moves along the ray with the velocity 2, and, for $t = \frac{1}{2}$, all rays reach the point $x = 0$, so that the focusation of rays takes place. The wave front $S(x) = 2t$ for $0 < t < \frac{1}{2}$ has the form of a sphere of radius $1 - 2t$ with the centre in the point $x = 0$; for $t = \frac{1}{2}$, the wave front shrinks into the point $x = 0$. It follows from the geometrical meaning of the Jacobian that the ratio $J(0, \alpha)/J(t, \alpha)$ is equal to the ratio of areas of the spheres with the radii 1 and $1 - 2t$, i.e. to $|x|^{-n+1}$. Consequently, the Helmholtz equation has a f.a. solution u which is a wave converging to the centre, and the leading term u_0 of which is equal to

$$u_0(x, k) = |x|^{-(n-1)/2} \exp(-ik|x|).$$

This f.a. solution is uniform with respect to x in any finite domain not containing the point $x = 0$.

Let us remark that the point $x = 0$ – the focus of the ray family – is a singular point of eikonal $S(x)$.

Now we shall discuss the phase trajectories of the considered problem. We have

$$x(t) = (1 - 2t)\omega, \qquad p(t) = -\omega, \qquad \sum_{j=1}^{n} \omega_j^2 = 1.$$

Let g^t be the displacement along the trajectories through time t, $\Lambda_t = g^t \Lambda_0$, where Λ_0 is the initial manifold $x = \omega$, $p = -\omega$, $\omega \in S^{n-1}$. Manifold Λ_t is an $(n-1)$-sphere embedded in the n-plane $x = (2t - 1)p$; it can be diffeomorphically projected on x-space for $t \neq \frac{1}{2}$, and on p-space for all t. The projection of manifold $\Lambda_{1/2}$ on $\mathbf{R}_x^n$ is the point $x = 0$ (the focus).

The union $\Lambda^n = \bigcup_{-\infty < t < \infty} \Lambda_t$ is an n-dimensional C^∞-manifold, too; this manifold is Lagrangian (see § 4). It lies in the 'cylinder' $|p| = 1$ and is given by the equation

$$p = \frac{\partial S(\alpha)}{\partial(x)} \, (S(x) = 1 - |x|), \qquad x \in \mathbf{R}^n \setminus \{0\}$$

for $x \neq 0$. Further, $\Lambda^n \setminus \Lambda_{1/2}$ is diffeomorphically projected on $\mathbf{R}_x^n$. The manifold $\Lambda_{1/2} = \{(x, p): x = 0, |p| = 1\}$ is called *a cycle of singularities* (with respect to the projection on $\mathbf{R}^n_x$) of the Lagrangian manifold Λ^n.

This example has the following characteristic features:

(1) The family of trajectories in phase space $\mathbf{R}^{2n}_{x,p}$ generated by the Lagrangian Cauchy problem has no singularities (i.e., it forms an n-dimensional C^∞-manifold Λ^n, and its cross-section Λ_t at time t is an $(n-1)$-dimensional C^∞-manifold).

(2) The family of rays (the projections of phase trajectories on $\mathbf{R}^n_x$) has singularities.

This situation is typical for majority of our problems.

EXAMPLE 3.17. Consider the family of trajectories (3.67); the Cauchy data for the eikonal equation $(\nabla S(x))^2 = 1$ are given as previously on an $(n-1)$-dimensional C^∞-manifold M^{n-1} so that $S(x) = 0$, $x \in M^{n-1}$. Let

$$M^{n-1} = \{x : x = x^0(\alpha),\ \alpha \in U\}, \qquad \alpha = (\alpha_1, \ldots, \alpha_{n-1}),$$

U be a domain in $\mathbf{R}^{n-1}_\alpha$; the vector $p^0(\alpha)$ coincides with the unit normal to M^{n-1} at the point $x = x^0(\alpha)$. Further,

$$\operatorname{rank} \frac{\partial x^0(\alpha)}{\partial \alpha} = n - 1, \qquad \alpha \in U.$$

We shall consider the sets

$$\Lambda_t = \{(x, p): x = x(t, \alpha),\ p = p(t, \alpha),\ \alpha \in U\},$$

in phase space $\mathbf{R}^{2n}_{x,p}$, where $x(t, \alpha)$, and $p(t, \alpha)$ are determined by (3.67), and their union $\Lambda^n = \bigcup_{-\infty < t < \infty} \Lambda_t$. Set Λ_t is obtained from initial manifold Λ_0 by the displacement through time t along phase trajectories. We shall show that Λ_t and Λ^n are C^∞-manifolds of dimensions $n-1$ and n, respectively. From (3.67) we have

$$\frac{\partial(x, p)}{\partial \alpha} = \left(2t \frac{\partial p^0}{\partial \alpha} + \frac{\partial x^0}{\partial \alpha}, \frac{\partial p^0}{\partial \alpha} \right),$$

and since

$$\operatorname{rank}\left(\frac{\partial x^0}{\partial \alpha} \right) = n - 1$$

by assumption, the rank of the matrix is given by

$$\operatorname{rank}\left(\frac{\partial x^0}{\partial \alpha}, \frac{\partial p^0}{\partial \alpha} \right) = n - 1, \qquad \alpha \in U.$$

Consequently, Λ_t is an $(n-1)$-dimensional C^∞-manifold. Further, the matrix

$$\frac{\partial(x,p)}{\partial(t,\alpha)} = \left\| \begin{array}{cc} p^0 & 2t\dfrac{\partial p^0}{\partial \alpha_0} + \dfrac{\partial x^0}{\partial \alpha} \\ 0 & \dfrac{\partial p}{\partial \alpha} \end{array} \right\|,$$

and one of its minors of order n equals the determinant

$$\Delta(\alpha) = \det \left\| p^0(\alpha), \frac{\partial x^0}{\partial \alpha_1}, \dots, \frac{\partial x^0}{\partial \alpha_{n-1}} \right\|$$

(all vectors are columns). Since the vectors $\partial x^0/\partial \alpha_j$ are linearly independent and belong to a plane tangent to M^{n-1}, and vector $p^0(\alpha)$ is perpendicular to M^{n-1}, the determinant $\Delta(\alpha) \neq 0$, $\alpha \in U$. Consequently,

$$\mathrm{rank}\, \frac{\partial(x,p)}{\partial(t,\alpha)} = n,$$

so Λ^n is an n-dimensional C^∞-manifold. In this example, manifold Λ^n is the normal vector bundle NM^{n-1}.

However, the family of rays $x = x(t, \alpha)$, $\alpha \in U$, $-\infty < t < \infty$, has singularities; there is only one exception, namely, when M^{n-1} is an $(n-1)$-plane, i.e. $S(x)$ is a linear function.

As known from differential geometry [66], the set of points of family (3.67) in which $J(t, \alpha) = 0$, coincides with the geometrical locus of the centres of curvature of manifold M^{n-1}. Suppose $x = x(t, \alpha^0)$ is one of the rays; then the point $x(t_0, \alpha^0)$ lies on a caustic, if and only if the distance between the points $x(0, \alpha^0)$ and $x(t_0, \alpha^0)$ is equal to k_l^{-1}, where k_l is one of the principal curvatures of manifold M^{n-1} at the point $x = x^0(\alpha^0)$. In the case of a general position, i.e. when the caustic consists of $n-1$ smooth manifolds (which may mutually intersect), the ray $x = x(t, \alpha^0)$ touches the caustic at the focal points.

This example leads to the following conclusion: the eikonal equation $(\nabla S(x))^2 = 1$ has no solutions of class $C^\infty(\mathbf{R}_x^n)$ except the linear function

$$S(x) = C + \langle x, \xi \rangle, \quad \sum_{j=1}^{n} \xi_j^2 = 1.$$

EXAMPLE 3.18. Let us consider a medium with the refraction index

$n(x) = \sqrt{x_1}$. Setting $x' = (x_2, \ldots, x_n)$ and $p' = (p_2, \ldots, p_n)$, we obtain the Hamilton system:

$$\frac{dx_1}{dt} = 2p_1, \qquad \frac{dx'}{dt} = 2p', \qquad \frac{dp_1}{dt} = -1, \qquad \frac{dp'}{dt} = 0.$$

The solution of the Lagrangian Cauchy problem

$$x|_{t=0} = x^0(\alpha), \qquad p|_{t=0} = p^0(\alpha)_1, \qquad \alpha = (\alpha_1, \ldots, \alpha_{n-1}) \in U$$

has the form

$$x_1(t, \alpha) = -2t^2 + 2p_1^0(\alpha)t + x_1^0(\alpha),$$
$$p_1(t, \alpha) = -2t + p_1^0(\alpha),$$
$$x'(t, \alpha) = 2tp^{0'}(\alpha) + x^{0'}(\alpha),$$
$$p'(t, \alpha) = p^{0'}(\alpha).$$

Consequently, the rays appear to be quadratic parabolas, and the eikonal is given by

$$S(x(t, \alpha)) = S_0(\alpha) - \tfrac{4}{3}t^3 + 2p_1^0(\alpha)t^2 + tx_1^0(\alpha).$$

EXAMPLE 3.19.   Consider the equation

$$L(\lambda^{-1}D_x)u(x) = 0$$

which does not explicitly depend on x.
The Hamilton system has the form

$$\frac{dx}{d\tau} = \frac{\partial L(p)}{\partial p}, \qquad \frac{dp}{d\tau} = 0,$$

so that the corresponding phase trajectories and rays are the straight lines

$$x(\tau) = \tau \frac{\partial L(p)}{\partial p}\bigg|_{p=p(0)} + x(0), \qquad p(\tau) = p(0).$$

EXAMPLE 3.20.   Consider the Hamilton–Jacobi equation

$$\frac{\partial S}{\partial t} + H\left(t, x, \frac{\partial S}{\partial x}\right) = 0. \tag{3.45'}$$

We introduce a *two-point characteristic function* $S(t, x, t', x')$ [4], [46], where $x \in \mathbf{R}^n$, $x' \in \mathbf{R}^n$. This function is by definition equal to the action along the ray connecting the points x and x'. More precisely, let $(x(\tau), p(\tau))$ be the solution of the truncated Hamilton system

$$\frac{dx}{d\tau} = \frac{\partial H}{\partial p}, \qquad \frac{dp}{d\tau} = -\frac{\partial H}{\partial x},$$

satisfying the boundary conditions

$$x(t) = x, \qquad x(t') = x'. \tag{3.68}$$

Then, by definition,

$$S(t, x; t', x') = \int_t^{t'} [\langle p, dx \rangle - H\, d\tau], \tag{3.69}$$

where $p = p(\tau)$, $x = x(\tau)$, and $H = H(\tau, x(\tau), p(\tau))$.

For instance, the function

$$S(t, x; t', x') = \int_t^{t'} \left[\frac{m}{2} \dot{x}^2(\tau) - V(x(\tau)) \right] d\tau, \tag{3.70}$$

corresponds to Equation (3.6), where $x(\tau)$ is the solution of the following boundary problem for the Newton equation:

$$m\frac{d^2 x}{d\tau^2} = -\nabla V(x), \qquad x(t) = x, \qquad x(t') = x'. \tag{3.70'}$$

In classical mechanics, action S is defined as the integral

$$S = \int_{t_1}^{t_2} L\, dt$$

over a classical trajectory (L is the Lagrange function). Thus the action is a function of trajectory (a functional). In the previously introduced Hamilton's definition of action (Section 1), the action was regarded as a function of the coordinates (including time). If both the initial time t' and the initial point x' are fixed, the two-point characteristic function will be a function of (t, x): $S = S(t, x)$. It can be evaluated in the following way: select among all rays launched (with different initial momenta)

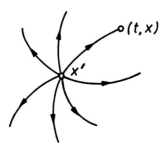

Fig. 6.

from x' at time t' one, which reaches point x through time $t - t'$ (Figure 6); then evaluate the action $S = \int_{t'}^{t} L dt$ over this ray.

The question of existence of a two-point function in a global setting is non-trivial, since the boundary-value problem (3.68) might not be uniquely solvable. The existence and smoothness of this function are guaranteed, whenever the points x and x' are sufficiently close to each other. Below we shall assume that function $S(t, x; t', x')$ is single-valued and infinitely differentiable with respect to all its variables in a certain domain of the space $(t, x; t', x')$.

We shall derive the most important properties of function S. By variating expression (3.69) with respect to the variables t, x, t', x' we get a vanishing integral, since it is taken along the ray. Consequently,

$$\delta S = \langle p, \delta x \rangle - H \delta t |_{(t, x)}^{(t', x')}.$$

From here we get the relations

$$\frac{\partial S}{\partial x} = - p(t), \qquad \frac{\partial S}{\partial x'} = p(t'),$$

$$\frac{\partial S}{\partial t} = H(\tau, x(\tau), p(\tau))|_{\tau = t}, \tag{3.71}$$

$$\frac{\partial S}{\partial t'} = - H(\tau, x(\tau), p(\tau))|_{\tau = t'}.$$

Thus function S satisfies the Hamilton–Jacobi equation

$$\frac{\partial S}{\partial t'} + H\left(t', x', \frac{\partial S}{\partial x'}\right) = 0$$

(with respect to the variables (t', x') of the end-point of the ray), and the conjugate equation

$$\frac{\partial S}{\partial t} - H\left(t, x, \frac{\partial S}{\partial x}\right) = 0$$

(with respect to the variables (t, x) of the initial point).

For the one-dimensional Newton equation we have [21]

$$V(x) = 0, \qquad S(t, x; t', x') = \frac{m(x' - x)^2}{2(t' - t)};$$

$$V(x) = -Fx, \qquad S(t, x; t', x') =$$
$$= \frac{m(x' - x)^2}{2(t' - t)} + \tfrac{1}{2}F(x' + x)(t' - t) - \frac{F(t' - t)^3}{24}; \qquad (3.72)$$

$$V(x) = \tfrac{1}{2}m\omega^2 x^2, \qquad S(t, x; t', x') =$$
$$= m\omega[(x^2 + x'^2)\cos \omega(t' - t) - 2xx'](2\sin \omega(t' - t))^{-1}.$$

The last formula holds only for $0 < (t' - t) < \pi$.

Consider still one example: a particle of charge e and mass m moving in a constant external magnetic field B directed along the z-axis. The corresponding Lagrangian is

$$L = \frac{m}{2}(\dot{x}^2 + \dot{y}^2 + \dot{z}^2) + \frac{Be}{2c}(x\dot{y} - y\dot{x}),$$

where c is the velocity of light in the vacuum, and S given by the formula [21]

$$S(t, r; t', r') = \frac{m\omega}{2}\left[\frac{(z' - z)^2}{t' - t} + \frac{\omega}{2}\cot\frac{\omega(t' - t)}{2} \times \right.$$
$$\left. \times ((x' - x)^2 + (y' - y)^2 + \omega(xy' - x'y))\right], \qquad (3.72')$$

where $\omega = Be/mc, r = (x, y, z)$.

There is yet one important feature of the two-point function to mention. Consider the determinant

$$Y(t', x') = \det\frac{\partial^2 S}{\partial x\, \partial x'}. \qquad (3.73)$$

LEMMA 3.21. *Determinant (3.73) satisfies the continuity equation*

$$\frac{\partial Y}{\partial t'} + \sum_{j=1}^{n}\frac{\partial}{\partial x_j'}\left(Y\frac{\partial H}{\partial p_j'}\right) = 0. \qquad (3.74)$$

Proof. From (3.71) we have $\partial S/\partial x = -p(t)$ so that

$$\frac{\partial^2 S}{\partial x\,\partial x'} = -\frac{\partial p(t)}{\partial x'} = -\left(\frac{\partial x'}{\partial p(t)}\right)^{-1}.$$

If initial point x and initial time t are fixed, while initial momentum $p(t)$ varies, the solution of the Hamilton system depends on the parameters $p(t) = \alpha = (\alpha_1, \ldots, \alpha_n)$, so that

$$\frac{\partial^2 S}{\partial x\,\partial x'} = -\left(\frac{\partial x'}{\partial \alpha}\right)^{-1}.$$

By virtue of the Liouville theorem we have

$$\frac{d}{dt}\ln J(t,\alpha) = \operatorname{div}_{x'}\frac{\partial H}{\partial p'},$$

where $J(t,\alpha) = \det \|\partial \alpha'/\partial \alpha\|$; hence

$$\frac{dJ}{dt} = J\sum_{j=1}^{n}\frac{\partial}{\partial x'_j}\left(\frac{\partial H}{\partial p'_j}\right).$$

Since $Y = J^{-1}$, we obtain

$$\frac{dY}{dt} + Y\sum_{j=1}^{n}\frac{\partial}{\partial x'_j}\left(\frac{\partial H}{\partial p'_j}\right) = 0. \tag{3.74'}$$

Moreover, d/dt is the total derivative (the derivative with respect to the system), so that

$$\frac{dY}{dt} = \frac{\partial Y}{\partial t} + \left\langle \frac{\partial Y}{\partial x'}, \frac{dx'}{dt}\right\rangle.$$

Since $dx'/dt = \partial H/\partial p'$, then (3.74') implies (3.74).

EXAMPLE 3.22. The equation of the quantum mechanical oscillator in $\mathbf{R}^n$ has the form

$$ih\frac{\partial \psi}{\partial t} = -\frac{h^2}{2m}\Delta\psi + \frac{m\omega^2}{2}\langle x, x\rangle. \tag{3.75}$$

The associated truncated Hamilton system is linear:

$$m\frac{dx}{dt} = p. \qquad \frac{dp}{dt} = -m\omega^2 x. \tag{3.76}$$

The solution of this system with the Cauchy data

$$x|_{t=0} = x^0, \qquad p|_{t=0} = p^0$$

has the form

$$x = -\frac{p^0}{m\omega} \sin \omega t + x^0 \cos \omega t,$$

(3.77)

$$p = p^0 \cos \omega t + m\omega x^0 \sin \omega t.$$

If we choose the system of units in such a way that $m\omega = 1$, then

$$\begin{pmatrix} x(t) \\ p(t) \end{pmatrix} = \begin{pmatrix} I \cos \omega t, & I \sin \omega t \\ -I \sin \omega t, & I \cos \omega t \end{pmatrix} \begin{pmatrix} x^0 \\ p^0 \end{pmatrix} = g' \begin{pmatrix} x^0 \\ p^0 \end{pmatrix},$$

i.e. g' is a rotation in the phase space (here I is the unit $(n \times n)$-matrix). Since the functions $L_j(x, p) = x_j^2 + (p_j^2/m^2\omega^2)$ are first integrals of the system (3.76), the projection of the phase point on the 2-plane (x_j, p_j) moves along the ellipse

$$x_j^2 + \frac{p_j^2}{m^2\omega^2} = x_j^{0^2} + \frac{p_j^{0^2}}{m^2\omega^2},$$

(3.78)

and the phase trajectory itself lies in an n-dimensional surface T^n given by Equation (3.78), $1 \leq j \leq n$. For $m\omega = 1$, surface T^n appears to be an n-torus, and the phase trajectory a circle on it, since the motion of the point is periodic with the period $T = 2\pi/\omega$. The whole phase space $\mathbf{R}_{x,p}^{2n}$ splits into n-tori invariant with respect to displacements along trajectories of the dynamical system (3.76).

The origin of the coordinates $(x = 0, p = 0)$ is a stable equilibrium position of the 'n-dimensional centre' type.

It follows from (3.77) that

$$x\left(\frac{\pi}{\omega}\right) = -x^0, \qquad p\left(\frac{\pi}{\omega}\right) = -p^0.$$

This means that all rays in $\mathbf{R}_x^n$ originating from the point x_0 meet at point $-x^0$ (Figure 7), i.e. focusation of the rays takes place. The solution $x(\tau)$ of the boundary-value problem

$$x(t) = x, \qquad x(t') = x',$$

has the form (if $\sin \omega(t' - t) \neq 0$)

$$x(\tau) = \frac{1}{\sin \omega(t' - t)} [x' \sin \omega(\tau - t) - x \sin \omega(\tau - t')],$$

(3.79)

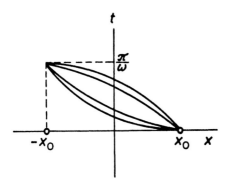

Fig. 7.

and the two-point function $S(t, x; t', x')$ is given in (3.72). Thus

$$\det\left(-\frac{\partial^2 S}{\partial x\, \partial x'}\right) = \left(\frac{m\omega}{\sin \omega(t' - t)}\right)^n. \tag{3.80}$$

We notice that function S has singularities at $\omega(t - t') = k\pi, k \in \mathbf{Z}$, i.e. in the focal points.

EXAMPLE 3.23.   Consider the Schrödinger equation

$$ih\frac{\partial \psi}{\partial t} = -\frac{h^2}{2}\Delta\psi + \frac{1}{2}\sum_{j=1}^{n}\varepsilon_j x_j^2\psi = 0, \tag{3.81}$$

where $\varepsilon_j = \pm 1$. The linear oscillator approximates the motion of a particle near a stable equilibrium configuration; similarly, the potential $\frac{1}{2}\sum_{j=1}^{n}\varepsilon_j x_j^2$ approximates the potential function near the non-degenerate extremum $x = 0$. Since in this example the Hamilton system splits into n subsystems like in the previous one, it suffices to investigate the system in a plane of the form

$$\frac{\mathrm{d}x}{\mathrm{d}t} = p, \qquad \frac{\mathrm{d}p}{\mathrm{d}t} = x.$$

The solution of this system with the Cauchy data $x(0) = x^0, p(0) = p^0$ can be written as

$$x(t) = p^0 \operatorname{sh} t + x^0 \operatorname{ch} t,$$
$$p(t) = p^0 \operatorname{ch} t + x^0 \operatorname{sh} t,$$

and the function $p^2 - x^2$ is a first integral of the system. The phase trajectory is the hyperbola $p^2 - x^2 = p^{02} - x^{02}$, and transformation g^t (the displacement along trajectory through time t) is a hyperbolic rotation. The equilibrium position $x = 0, p = 0$ is a saddle. Solution of the boundary-value problem

$$x(t) = x, \qquad x(t') = x'$$

has the form

$$x(\tau) = \frac{1}{\text{sh}\,(t' - t)}[x'\,\text{sh}\,(\tau - t) - x\,\text{sh}\,(\tau - t')],$$

and the two-point function is given by

$$S(t, x; t', x') = \frac{1}{2\,\text{sh}\,(t' - t)}[(x^2 + x'^2)\,\text{ch}\,(t' - t) - 2xx'].$$

In this example the two-point function S is well-defined and is infinitely differentiable for all t, x, t', x', provided $t \neq t'$.

Thus the Hamilton system associated with Equation (3.81) has the first integrals $p_j^2 + \varepsilon_j x_j^2$, $1 \leq j \leq n$, and its phase space splits into invariant (with respect to g^t) n-dimensional manifolds Λ^n, given by the equations

$$p_j^2 + \varepsilon_j x_j^2 = C_j, \qquad 1 \leq j \leq n,$$

where C_j are constants. If negative numbers occur among ε_j, manifolds Λ^n are non-compact.

The previous examples are simple to serve as illustrations. Note that any integrable problem of classical mechanics could be taken as an example as well.

§ 4. The Lagrangian Manifolds and Canonical Transformations

The semi-classical approximation of solutions of Equation (3.1) leads to the Hamilton system (3.8) in phase space $\mathbf{R}_{x,p}^{2n}$. As known from classical mechanics, phase flow g^t (i.e. the displacement along trajectory of the Hamilton system through time t) conserves the differential form

$$\omega^2 = \sum_{j=1}^{n} dp_j \wedge dx_j.$$

Such transformations of the phase space are called *canonical*. Linear canonical transformations are symplectic.

There exists a remarkable class of manifolds (closely related with classical mechanics) in the phase space, on which the integral $\int \langle p, dx \rangle$ is (locally) path-independent. Here

$$\langle p, dx \rangle = \sum_{j=1}^{n} p_j \, dx_j.$$

These manifolds are called *Lagrangian*. The dimension of Lagrangian manifold does not exceed n. The class of Lagrangian manifolds is invariant with respect to classical dynamics; in other words, if manifold Λ is Lagrangian, the displaced manifold $g^t \Lambda$ is Lagrangian, too.

The most interesting Lagrangian manifolds are those with maximal dimension n. If such (simply connected) manifold Λ^n admits injective projection on x-space, then it is given by the equation

$$p = \frac{\partial S(x)}{\partial x};$$

it is clear that

$$S = \int \langle p, dx \rangle.$$

If, in addition, Λ^n is injectively projected on p-space, then it is given by the equation

$$x = \frac{\partial \tilde{S}(p)}{\partial p}.$$

In this case the mapping

$$(x, S(x)) \rightarrow (p, \tilde{S}(p))$$

is a Legendre transformation, so that functions $S(x)$, $\tilde{S}(p)$ are Young-dual. Thus, in classical mechanics, the transition from x-representation to p-representation (i.e. the passage from the coordinate space to the momentum space) is a contact transformation.

Lagrangian manifolds possess still one very important property: as their local coordinates it is always possible to select the set

$$x_{i_1}, \ldots, x_{i_k}, p_{j_1}, \ldots, p_{j_{n-k}},$$

which does not contain any conjugate coordinates and momenta (i.e. the pairs like (x_j, p_j)).

Let us describe now one of the most important constructions of n-dimensional Lagrangian manifolds. Consider the Cauchy problem for the Hamilton–Jacobi equation (3.3); it induces the Cauchy problem for the Hamilton system, i.e. an $(n-1)$-dimensional (Lagrangian) manifold Λ^{n-1} in the phase space. Its displacements through time $t, t_1 \leq t \leq t_2$, fill up an n-dimensional manifold Λ^n (a tube of trajectories) – the desired Lagrangian manifold. Such construction naturally arises when deriving the semi-classical approximation.

The rest of the paragraph will be devoted to the detailed exposition of the above mentioned facts.

1. Symplectic Geometry

Let $\mathbf{R}^{2n}$ be a phase space, i.e. a $2n$-dimensional real space

$$\mathbf{R}^{2n} = \{r\}, r = (x, p), x = (x_1, \ldots, x_n), p = (p_1, \ldots p_n).$$

We introduce the Euclidean structure in $\mathbf{R}^{2n}$ by means of the quadratic form

$$\langle r, r \rangle = \langle x, x \rangle + \langle p, p \rangle,$$

and the symplectic structure by the *skew-scalar product*

$$[r^1, r^2] = \langle p^1, x^2 \rangle - \langle x^1, p^2 \rangle = \sum_{i=1}^{N} (p_i^1 x_i^2 - p_i^2 x_i^1) \tag{4.1}$$

with the properties

$$[r^1, r^2] = -[r^2, r^1],$$
$$[r^1, r^2] = \langle Jr^1, r^2 \rangle,$$
$$J = \begin{pmatrix} 0 & I \\ -I & 0 \end{pmatrix} \tag{4.2}$$

where 0 and I are the zero and the unit $(n \times n)$-matrices, respectively.

The variables x_j, p_j are called *conjugate*.

DEFINITION 4.1. A hyperplane $\lambda \subset \mathbf{R}^{2n}$ is called *Lagrangian*, if $[r^1, r^2] = 0$ for arbitrary $r^1, r^2 \in \lambda$. A smooth manifold Λ in the phase space is called *Lagrangian*, if all its tangent planes are Lagrangian.

PROPOSITION 4.2. *The dimension of a Lagrangian plane does not exceed n.*

Proof. Since $J^2 = -I_{2n}$, the necessary and sufficient condition for plane λ to be Lagrangian is the orthogonality of the planes λ and $J\lambda$. As dim $\lambda = \dim J\lambda$, we see that dim $\lambda \leq n$.

It is also evident from this that the maximal number of linearly independent mutually skew-orthogonal vectors in the phase space is equal to n.

EXAMPLE 4.3. Any straight line in $\mathbf{R}^2$ going through the origin is Lagrangian. Any smooth curve in $\mathbf{R}^2$ is a Lagrangian manifold.

EXAMPLE 4.4. Decompose the set $(1, 2, \ldots, n)$ into two disjoint subsets $(\alpha) = \{\alpha_1, \ldots, \alpha_k\}$, $(\beta) = \{\beta_1, \ldots, \beta_l\}$, $k + l = n$.

Let $\lambda^n(p_{(\alpha)}, x_{(\beta)})$ be the plane $(p_{(\alpha)}, x_{(\beta)})$, i.e. $x_{(\alpha)} = 0, p_{(\beta)} = 0$. The plane $\lambda^n(p_{(\alpha)}, x_{(\beta)})$ is Lagrangian. In particular, the planes $x = 0$ and $p = 0$ are Lagrangian.

There are no conjugate variables among $x_i, p_j, i \in (\alpha), j \in (\beta)$.

Planes $\lambda^n(p_{(\alpha)}, x_{(\beta)})$ will be called *coordinate Lagrangian planes*. There exist altogether 2^n of them. One fundamental property of Lagrangian n-planes is stated in

LEMMA 4.5. *Every Lagrangian n-plane is transversal to one of the coordinate Lagrangian planes.*

Proof. [2]. Let λ be the given plane, σ the plane $x = 0$, and $\lambda_0 = \lambda \cap \sigma$, dim $\lambda_0 = k < n$. Then λ_0 in σ is transversal to one of $\binom{n}{k}$ coordinate planes $\tau = \lambda(p_{(\alpha)}, x_{(\beta)}) \cap \sigma$, where (α) consists of k elements, so that $\lambda_0 \cap \sigma \cap \lambda(p_{(\alpha)}, x_{(\beta)}) = 0$. We shall prove that plane λ is transversal to $\lambda(p_{(\alpha)}, x_{(\beta)}) = \lambda'$.

According to the assumption $\lambda_0 + \tau = \sigma$. Since λ and λ' are Lagrangian, then $[\lambda, \lambda_0] = 0$ (as $\lambda_0 \subset \lambda$) and $[\lambda', \tau] = 0$ (as $\tau \subset \lambda'$). Consequently $[\lambda \cap \lambda', \sigma] = 0$. However, the maximal number of mutually skew-orthogonal independent vectors in $\mathbf{R}^{2n}$ is n. Therefore n-plane σ is the maximal plane skew-orthogonal to itself, so that $\lambda \cap \lambda' \subset \sigma$. Thus $(\lambda \cap \lambda') \subset (\lambda \cap \sigma) \cap (\lambda' \cap \sigma) = \lambda_0 \cap \tau = 0$, q.e.d.

This lemma yields

PROPOSITION 4.6. *Let Λ^n be an n-dimensional Lagrangian manifold in $\mathbf{R}^{2n}$. Then a sufficiently small neighbourhood of each point $r \in \Lambda^n$ is diffeomorphically projected on one of the coordinate Lagrangian planes. The Lagrangian n-plane $p = x$ (the 'diagonal') is transversal to all coordinate Lagrangian planes.*

DEFINITION 4.7. A linear map $g: \mathbf{R}^{2n} \to \mathbf{R}^{2n}$ is called *symplectic*, if it preserves the skew-scalar product

$$[gr^1, gr^2] = [r^1, r^2] \tag{4.3}$$

for any $r^1, r^2 \in \mathbf{R}^{2n}$.

As follows from this definition, symplectic transformations map Lagrangian planes into Lagrangian planes. Properties (4.2) imply

PROPOSITION 4.8.   *Let $\mathscr{G}$ be the matrix of a linear map $g: \mathbf{R}^{2n} \to \mathbf{R}^{2n}$ in an orthonormal basis in $\mathbf{R}^{2n}$. The necessary and sufficient condition for g to be symplectic is*

$${}^t\mathscr{G} J \mathscr{G} = J. \tag{4.4}$$

The $(n \times n)$-matrix $\mathscr{G}$ satisfying condition (4.4) is called *symplectic*. It follows from (4.4) that $\det \mathscr{G} = \pm 1$, so that $\mathscr{G}$ is non-degenerate. The symplectic transformations form a group; it is called *symplectic* and denoted by Sp(n).

Now we shall summarize some results of symplectic geometry without proof (see e.g. [4], [28]). Let $\{e^j, f^k\}$, $1 \leq j, k \leq n$ be a basis of $2n$ vectors e^j, f^k in $\mathbf{R}^{2n}$. This basis is called *symplectic*, if

$$[e^i, e^j] = [f^i, f^j] = 0, \qquad [e^j, f^k] = \delta_{jk}. \tag{4.5}$$

For instance, the basis formed by the vectors $e^j = \{x_i = \delta_{ij}, p_i = 0\}$, $f^k = \{x_i = 0, p_i = \delta_{ik}\}$ is symplectic. As in Euclidean geometry, the following statements are true:

(1) Symplectic transformations (and only these) map any symplectic basis into a symplectic one.

(2) Any symplectic basis can be transformed into any other symplectic basis by means of a symplectic transformation.

(3) Any set of m vectors e^j and m vectors f^k, $1 \leq m < n$, satisfying condition (4.5) can be completed into a symplectic basis in $\mathbf{R}^{2n}$.

(4) In any symplectic basis the matrix of a symplectic transformation is symplectic.

2. *The Canonical Transformation*

We shall introduce the differential forms

$$\omega^1 = \langle p, \mathrm{d}x \rangle \equiv \sum_{j=1}^{n} p_j \, \mathrm{d}x_j,$$

$$\omega^2 = dp \wedge dx \equiv \sum_{j=1}^{n} dp_j \wedge dx_j \qquad (4.5')$$

in $\mathbf{R}^{2n}$. It is clear that $d\omega^1 = \omega^2$, $d\omega^2 = 0$.

Let U, V be domains in $\mathbf{R}^{2n}$. All mappings considered below are assumed to be infinitely differentiable.

DEFINITION 4.9.   A diffeomorphism $g: U \to V$ is said to be a *canonical transformation*, if it preserves the form ω^2, i.e.

$$g^*\omega^2 = \omega^2 \qquad (4.6)$$

With the notation $(x', p') = g(x, p)$, (4.6) takes the form

$$dp' \wedge dx' = dp \wedge dx. \qquad (4.6')$$

The set of all canonical transformations $g: U \to U$ forms a group.

The forms $\omega^2, \omega^4 = \omega^2 \wedge \omega^2, \ldots, \omega^{2n} = \omega^2 \wedge \ldots \wedge \omega^2$ (n-times) are *invariants* of the canonical transformations. Since ω^{2n}, up to a constant multiplier, coincides with the phase volume form $dx_1 \wedge dx_2 \wedge \ldots \wedge dx_n \wedge dp_1 \wedge \ldots \wedge dp_n$, the canonical transformations preserve the phase volume (Liouville's theorem).

Further, since

$$[r^1, r^2] = \omega^2(r^1, r^2), \qquad (4.7)$$

Definition 4.9 yields

PROPOSITION 4.10.   *Transformation g is canonical (in the small) if and only if its Jacobi matrix*

$$g' = \begin{pmatrix} \dfrac{\partial x'}{\partial x} & \dfrac{\partial x'}{\partial p} \\[2mm] \dfrac{\partial p'}{\partial x} & \dfrac{\partial p'}{\partial p} \end{pmatrix} \qquad (4.8)$$

is symplectic.

The right-hand side of formula (4.7) is nothing else than the value of form ω^2 on the pair of vectors r^1, r^2. By virtue of the definition of a differential form we have (with $r = (x, p)$)

$$dp_i(r) = p_i, \qquad dx_i(r) = x_i,$$

$$(dp_i \wedge dx_i)(r^1, r^2) = \begin{vmatrix} dp_i(r^1) & dp_i(r^2) \\ dx_i(r^1) & dx_i(r^2) \end{vmatrix} = p_i^1 x_i^2 - p_i^2 x_i^1.$$

Consequently,

$$\omega^2(r^1, r^2) = \sum_{i=1}^{n} (p_i^1 x_i^2 - p_i^2 x_i^1) = [r^1, r^2].$$

EXAMPLE 4.11. The identity transformation $x' = x, p' = p$ is canonical.

EXAMPLE 4.12. (1) The transformation $x_1' = p_1, p_1' = -x_1, x_j' = x_j,$ $p_j' = p_j, j \geq 2$, is canonical.
   (2) Let $\{(\alpha), (\beta)\}$ be a decomposition of the set $\{1, 2, \ldots, n\}$ into two disjoint subsets (α), (β). The transformation

$$x_{(\alpha)}' = p_{(\alpha)}, \qquad p_{(\beta)}' = -x_{(\beta)}$$

is a product of transformations specified in (1), and it is therefore canonical.

EXAMPLE 4.13. Let the map $g: W \to W'$ be given by the formulae

$$x = \varphi(x'), \qquad p' = {}^t\left(\frac{\partial \varphi(x')}{\partial x'}\right) p \tag{4.9}$$

where a diffeomorphism $\varphi: \mathbf{R}^n \to \mathbf{R}^n$ is determined by specifying n functions $\varphi_1(x'), \ldots, \varphi_n(x')$. It is a canonical transformation since $\langle p', dx' \rangle = \langle p, dx \rangle$. Here W and W' are small neighbourhoods of the points (x^0, p^0) and (x'^0, p'^0), respectively,

$$x^0 = \varphi(x'^0), p'^0 = {}^t\left(\frac{\partial \varphi(x'^0)}{\partial x}\right) p^0, \quad \text{and} \quad \det \frac{\partial \varphi(x'^0)}{\partial x'} \neq 0.$$

EXAMPLE 4.14. The transformation

$$x' = Sx, \qquad p' = Sp, \tag{4.10}$$

where S is a real orthogonal matrix is canonical. This transformation is called a *rotation of the phase space*.
   The Poisson bracket is one of the invariants of canonical transformations. The *Poisson bracket* of two scalar functions $f(x, p)$ and $h(x, p)$ is defined by

$$(f, h)(x, p) = -[\nabla_{x, p} f, \nabla_{x, p} h] =$$

$$= \left\langle \frac{\partial f}{\partial x}, \frac{\partial h}{\partial p} \right\rangle - \left\langle \frac{\partial f}{\partial p}, \frac{\partial h}{\partial x} \right\rangle. \tag{4.11}$$

Canonical transformation g preserves the Poisson bracket, i.e.

$$(f, h)(x, p) = (f', h')(x, p). \tag{4.12}$$

Here $f'(x', p') = f(x, p)$, where $(x', p') = g(x, p)$, i.e. (4.12) has the form

$$\left\langle \frac{\partial f}{\partial x}, \frac{\partial h}{\partial p} \right\rangle - \left\langle \frac{\partial f}{\partial p}, \frac{\partial h}{\partial x} \right\rangle =$$

$$= \left\langle \frac{\partial f'}{\partial x}, \frac{\partial h'}{\partial p} \right\rangle - \left\langle \frac{\partial f'}{\partial p}, \frac{\partial h'}{\partial x} \right\rangle. \tag{4.12'}$$

Conversely, if a transformation preserves the Poisson bracket, then it is (locally) canonical. Thus the necessary and sufficient condition for transformation $(x', p') = g(x, p)$ to be canonical can be expressed in the form

$$(x_i', x_j') = (p_i', p_j') = 0,$$
$$(x_i', p_j') = \delta_{ij}, \tag{4.13}$$

or, in more detail,

$$\sum_{\alpha=1}^{n} \left(\frac{\partial x_i'}{\partial x_\alpha} \frac{\partial x_j'}{\partial p_\alpha} - \frac{\partial x_i'}{\partial p_\alpha} \frac{\partial x_j'}{\partial x_\alpha} \right) = 0,$$

$$\sum_{\alpha=1}^{n} \left(\frac{\partial p_i'}{\partial x_\alpha} \frac{\partial p_j'}{\partial p_\alpha} - \frac{\partial p_i'}{\partial p_\alpha} \frac{\partial p_j'}{\partial x_\alpha} \right) = 0, \tag{4.13'}$$

$$\sum_{\alpha=1}^{n} \left(\frac{\partial x_i'}{\partial x_\alpha} \frac{\partial p_j'}{\partial p_\alpha} - \frac{\partial x_i'}{\partial p_\alpha} \frac{\partial p_j'}{\partial x_\alpha} \right) = \delta_{ij}.$$

The canonical transformations are remarkable by being completely determined by a single function (the so-called *generating function*).

PROPOSITION 4.15. (1) *Let U be a simply connected domain in the phase space, $g : U \to V$ a canonical transformation. Then there exists a function $S(x, p) \in C^\infty(U)$ such that*

$$\langle p', dx' \rangle - \langle p, dx \rangle = - dS. \tag{4.14}$$

(2) *If $g : U \to V$ is a diffeomorphism and (4.14) holds, then g is a canonical transformation (simple connectedness is not required).*

Proof. The 1-form $\eta = \langle p', dx' \rangle - \langle p, dx \rangle$ is closed; by virtue of (4.14) and according to the Poincaré lemma, $\eta = - dS$. The proof of (2) is obvious.

By using function S, it is difficult to construct a canonical transformation explicitly. If (x, x') can be chosen as independent variables (i.e. p, p' are functions of (x, x')), then, from (4.14), we find

$$p = \frac{\partial S(x, x')}{\partial x}, \qquad p' = -\frac{\partial S(x, x')}{\partial x'}. \tag{4.15}$$

All next considerations in this section have local character. If

$$\det \frac{\partial^2 S(x, x')}{\partial x \, \partial x'} \neq 0, \tag{4.16}$$

then, according to the implicit function theorem, it is possible to express x' from the first relation (4.15) in terms of (x, p): $x' = x'(x, p)$. Inserting x' into the second relation, p' is expressed in terms of (x, p):

$$p' = \left. \frac{\partial S(x, x')}{\partial x} \right|_{x' = x'(x, p)}.$$

Function $S(x, x')$ which satisfies relation (4.14) is called a *generating function of the canonical transformation*. For example, for the transformation $x' = p, p' = -x$ we have $S = \langle x, x' \rangle$.

We shall introduce generating functions which depend on other sets of variables. Let $2n$ independent variables $x_{(\alpha)}, p_{(\beta)}, x'_{(\alpha')}, p'_{(\beta')}$ be chosen from $4n$ variables x, p, x', p', related by $(x', p') = g(x, p)$. Here $(\alpha) \cup (\beta) = \{1, 2, \ldots, n\}$, $(\alpha) \cap (\beta) = \varnothing$; $(\alpha'), (\beta')$ are defined analogously. There are no conjugate variables in this set. Let S in (4.14) be a function of the proposed set of variables. Then

$$\langle p'_{(\alpha')}, dx'_{(\alpha')} \rangle - \langle x'_{(\beta')}, dp'_{(\beta')} \rangle - \langle p_{(\alpha)}, dx_{(\alpha)} \rangle + \langle x_{(\beta)}, dp_{(\beta)} \rangle =$$
$$= -dS_1(x_{(\alpha)}, p_{(\beta)}, x'_{(\alpha')}, p'_{(\beta')}), \tag{4.17}$$
$$S_1 = S - \langle x'_{(\beta')}, p'_{(\beta')} \rangle + \langle x_{(\beta)}, p_{(\beta)} \rangle.$$

In this case function S_1 is called a *generating function* of the canonical transformation which itself is recovered from S_1 by means of the relations

$$p'_{(\alpha')} = -\frac{\partial S_1}{\partial x'_{(\alpha')}}, \qquad x'_{(\beta')} = \frac{\partial S_1}{\partial p'_{(\beta')}},$$
$$p_{(\alpha)} = \frac{\partial S_1}{\partial x_{(\alpha)}}, \qquad x_{(\beta)} = -\frac{\partial S_1}{\partial p_{(\beta)}}. \tag{4.18}$$

Now we shall summarize a number of known results from analytical mechanics [4], [23], [28], [29].

THEOREM 4.16. *Every canonical transformation maps a Hamilton system into another one.*

Namely, if $(x', p') = g(x, p)$, then the Hamilton system

$$\frac{dx}{dt} = \frac{\partial L(x, p)}{\partial p}, \qquad \frac{dp}{dt} = -\frac{\partial L(x, p)}{\partial x} \tag{4.19}$$

has in the new variables the form

$$\frac{dx'}{dt} = \frac{\partial L'(x', p')}{\partial p'}, \qquad \frac{dp'}{dt} = -\frac{\partial L'(x', p')}{\partial x'}, \tag{4.20}$$

where $L'(x', p') = L(x, p)$ (i.e. L' is the original Hamilton function in the new variables).

Proof. Let $r = (x, p)$ (all vectors are columns). Then system (4.19) has the form

$$\frac{dr}{dt} = J \frac{\partial L}{\partial r} \tag{4.19'}$$

Further,

$$\frac{dr'}{dt} = g' \frac{dr}{dt}, \qquad \frac{\partial L}{\partial r} = {}^t g' \frac{\partial L'}{\partial r'},$$

so that

$$\frac{dr'}{dt} = g' J \, {}^t g' \frac{\partial L'}{\partial r'}.$$

Since $J^{-1} = -J$, it follows from (4.4) that $g' J \, {}^t g' = J$, and we obtain $dr'/dt = J (\partial L'/\partial r')$.

Let g^t be a displacement along the trajectory of the Hamilton system (4.19) through a fixed time t, i.e.

$$g^t(x^0, p^0) = \{x(t; x^0, p^0), \quad p(t; x^0, p^0)\}, \tag{4.21}$$

where $\{x(t; x^0, p^0), p(t; x^0, p^0)\}$ is the solution of system (4.19) with the Cauchy data

$$x\big|_{t=0} = x^0, \qquad p\big|_{t=0} = p^0. \tag{4.22}$$

The map $g^t : \mathbf{R}^{2n} \to \mathbf{R}^{2n}$ is called a *phase flow*. We suppose that $L(x, p) \in C^\infty(\mathbf{R}^{2n})$ and, for simplicity, that every solution of the system is defined for all t. In the opposite case the result stated below will be true in the small (i.e. for small $|t|$ and in a small domain in $\mathbf{R}^{2n}$).

THEOREM 4.17. *Map g^t is a canonical transformation (for any fixed t).*
   Proof. Transformation g^t is defined by formula (4.21). We set $r^0 = (x^0, p^0)$ and

$$A = \frac{\partial g^t}{\partial r^0} = \begin{Vmatrix} \dfrac{\partial x}{\partial x^0} & \dfrac{\partial x}{\partial p^0} \\[2mm] \dfrac{\partial p}{\partial x^0} & \dfrac{\partial p}{\partial p^0} \end{Vmatrix}.$$

It is necessary to prove that

$$^T AJA = J \qquad\qquad (4.4')$$

for all t, where $^T A$ is the transposed matrix to A.

   For $t = 0$, we have $A = I_n$, so that relation (4.4') is fulfilled. We shall show that the left-hand side of this relation does not depend on t, and thus prove the theorem. By differentiating the Hamilton equation (4.19') with respect to the variables r^0 we get the equation

$$\frac{dA}{dt} = JL''_{rr} A.$$

As a consequence

$$\frac{d^T A}{dt} = -{}^T AL''_{rr} J,$$

since $^T J = -J$ and matrix L''_{rr} is symmetric. By differentiating the left-hand side of identity (4.4') with respect to t we obtain

$$\frac{d}{dt}(^T AJA) = -{}^T AL''_{rr} J^2 A + {}^T AJ^2 L''_{rr} A = 0,$$

since $J^2 = I_{2n}$, q.e.d.
   This fact can be also proved with the help of the Poincaré–Cartan integral invariant [4], [10]. Let Γ *be a simple smooth loop in* $\mathbf{R}^{2n}$. Then

$$\oint_{\Gamma} \langle p, dx \rangle = \oint_{g^t \Gamma} \langle p, dx \rangle. \qquad\qquad (4.22')$$

Let $(x', p') = g^t(x, p)$; then

$$\oint_{g^t \Gamma} \langle p, dx \rangle = \oint_{\Gamma} \langle p', dx' \rangle,$$

and, it follows from (4.22') that $dp' \wedge dx' = dp \wedge dx$.

3. Canonical Transformations and Lagrangian Manifolds

As follows from (4.1) and Definition 4.7, a manifold is Lagrangian if and only if the form $\omega^1 = \langle p, dx \rangle$ is closed on it. This yields

THEOREM 4.18. *Every canonical transformation maps a Lagrangian manifold into another one.*

We shall use also

PROPOSITION 4.19. *Let Λ^{n-1} be a Lagrangian C^∞-manifold such that $L(x, p) \equiv \text{const}$ on Λ^{n-1}. Let the map*

$$g^t : \Lambda^{n-1} \to g^t \Lambda^{n-1}$$

be a diffeomorphism for each $t \in [0, T]$, and

$$\Lambda^n = \bigcup_{0 < t < T} g^t \Lambda^{n-1}$$

be a C^∞-manifold of dimension n. Then the manifold Λ^n is Lagrangian.

Proof. The manifolds $g^t \Lambda^{n-1}$ are Lagrangian. Let $r^0 = \{x(t^0; x^0, p^0), p(t^0; x^0, p^0)\}$, where $t^0 \in (0, T), (x^0, p^0) \in \Lambda^{n-1}$ and λ^n, λ^{n-1} be tangent planes to $\Lambda^n, g^t \Lambda^{n-1}$ at the point r^0, respectively. Then $\lambda^{n-1} \subset \lambda^n$ and the vector $\eta^0 = (\partial x/\partial t, \partial p/\partial t)$ at the point (t^0, x^0, p^0) is transversal to λ^{n-1} in λ^n. Consequently, any vector $\eta \in \lambda^n$ has the form $\eta = \xi + C\eta^0$, where $\xi \in \lambda^{n-1}$, $C = \text{const}$. We shall show that $[\eta^0, \xi] = 0$ for each $\xi \in \lambda^{n-1}$; then λ^n is a Lagrangian plane and hence the proposition will be proved. Let $(\alpha_1, \ldots, \alpha_n)$ be local coordinates on $g^{t_0} \Lambda^{n-1}$ in a neighbourhood of the point r^0; then the vectors $\xi^j = (\partial x/\partial \alpha_j, \partial p/\partial \alpha_j)$ form a basis in $\mathbf{R}^{2n}$. We have

$$[\xi^j, \eta^0] = -\frac{\partial}{\partial \alpha_j} L(x(t, \alpha), p(t, \alpha)) = 0,$$

since $\{x(t, \alpha), p(t, \alpha)\}$ is a solution of the Hamilton system and $L(x, p)$ is its first integral.

This proposition provides the basic construction of n-dimensional Lagrangian manifolds. Namely, the Cauchy problem for the Hamilton–Jacobi equation induces an $(n - 1)$-dimensional initial manifold Λ_0^{n-1} which is Lagrangian. Its displacements along the trajectories of the Hamilton system fill up an n-dimensional Lagrangian manifold. If g^t fulfils the assumptions of Proposition 4.19 for all $t \in \mathbf{R}$, then

$$\Lambda^n = \bigcup_{t \in \mathbf{R}} g^t \Lambda_0^{n-1}$$

is a Lagrangian manifold invariant with respect to g^t.

For instance, the examples considered at the end of § 3 demonstrate various Lagrangian manifolds.

The following facts specify the local structure of n-dimensional Lagrangian manifold Λ^n. We shall consider the manifold which admits a diffeomorphic projection on $\mathbf{R}_x^n$.

THEOREM 4.20   *Let manifold Λ^n be given by the equation*

$$p = p(x), \quad x \in G,$$

where G is a simply connected domain in $\mathbf{R}_x^n$. The necessary and sufficient condition for Λ^n to be Lagrangian is the existence of a scalar function $S(x)$ such that

$$p = \frac{\partial S(x)}{\partial x}, \quad x \in G. \tag{4.23}$$

Proof. Let

$$S(x) = \int_{x^0}^{x} \langle p(x), dx \rangle.$$

In order that this integral is path-independent, it is necessary and sufficient that

$$dp(x) \wedge dx = 0.$$

If $p(x)$ has the form (4.23), then this condition is fulfilled and Λ^n is Lagrangian; if Λ^n is Lagrangian, then the integral is path-independent.

Notice that if manifold Λ^n has the form (4.23), then it is Lagrangian even if it is not simply connected.

Let us clarify now the local structure of an arbitrary Lagrangian manifold Λ^n.

Let, as usual, $(\alpha), (\beta)$ be a decomposition of the set $(1, \dots, n)$ into disjoint subsets.

THEOREM 4.21.   *Let manifold Λ^n be given by the equations*

$$x_{(\alpha)} = x_{(\alpha)}(p_{(\alpha)}, x_{(\beta)}), \quad p_{(\beta)} = p_{(\beta)}(p_{(\alpha)}, x_{(\beta)}), \quad (p_{(\alpha)}, x_{(\beta)}) \in G,$$

where G is a simply connected domain in $\mathbf{R}_{x_{(\alpha)}}^{k} \times \mathbf{R}_{p_{(\beta)}}^{n-k}$. The necessary and *sufficient condition for* Λ^n *to be Lagrangian is the existence of such scalar function* $S(p_{(\alpha)}, x_{(\beta)})$ *that*

$$x_{(\alpha)} = -\frac{\partial S}{\partial p_{(\alpha)}}, \qquad p_{(\beta)} = \frac{\partial S}{\partial x_{(\beta)}}. \tag{4.24}$$

Proof. Consider the function $S = \int_{(x^0, p^0)}^{(x, p)} \langle p, dx \rangle - \langle x_{(\alpha)}, p_{(\alpha)} \rangle$ on Λ^n. Let Λ^n be Lagrangian; then the integral $\int \langle p, dx \rangle$ is path-independent so that

$$dS = \langle p_{(\beta)}, dx_{(\beta)} \rangle - \langle x_{(\alpha)}, dp_{(\alpha)} \rangle, \tag{4.25}$$

and S is, according to the assumption, a function of $p_{(\alpha)}, x_{(\beta)}$. This yields (4.24). Conversely, if (4.24) holds, then (4.25) takes place so that

$$d[S + \langle x_{(\alpha)}, p_{(\alpha)} \rangle] = \langle p, dx \rangle$$

and form ω^1 is closed on Λ^n.

Function S which determines the Lagrangian manifold Λ^n (see (4.23), (4.24)) is also called a *generating function* of this manifold; it is defined by Λ^n up to an additive constant, provided Λ^n is simply connected.

PROPOSITION 4.22. *Let* Λ^n *be a Lagrangian manifold and let a sufficiently small neighbourhood* U *of the point* $r^0 = (x^0, p^0) \in \Lambda^n$ *admit diffeomorphic projections on both* $\mathbf{R}_x^n$ *and the plane* $\lambda^n(p_{(\alpha)}, x_{(\beta)})$. *If* $S(x)$ *and* $\tilde{S}(p_{(\alpha)}, x_{(\beta)})$ *are the associated generating functions, then the map*

$$(x, S(x)) \to ((p_{(\alpha)}, x_{(\beta)}), -\tilde{S}(p_{(\alpha)}, x_{(\beta)})) \tag{4.26}$$

is the Legendre transformation with respect to the variables $x_{(\alpha)}$.

Proof. Up to an inessential additive constant we have

$$S(x) = \int_{r^0}^{r(x, p)} \langle p, dx \rangle,$$

$$\tilde{S}(p_{(\alpha)}, x_{(\beta)}) = \int_{r^0}^{r(x, p)} \langle p, dx \rangle - \langle x_{(\alpha)}, p_{(\alpha)} \rangle,$$

so that

$$S(x_{(\alpha)}, x_{(\beta)}) - \tilde{S}(p_{(\alpha)}, x_{(\beta)}) = \langle x_{(\alpha)}, p_{(\alpha)} \rangle.$$

It follows from the proof of Theorem 4.21 that

$$x_{(\alpha)} = -\frac{\partial \tilde{S}(p_{(\alpha)}, x_{(\beta)})}{\partial p_{(\alpha)}},$$

and thus our proposition is proved. Moreover,

$$\det \frac{\partial^2 S(r^0)}{\partial x^2_{(\alpha)}} \neq 0, \qquad \det \frac{\partial^2 \tilde{S}(r^0)}{\partial p^2_{(\alpha)}} \neq 0 \qquad (4.27)$$

at the point (x^0, p^0).

4. *The Lagrangian Cauchy Problem*

Consider the Hamilton–Jacobi equation

$$L(x, \partial S(x)/\partial x) = 0 \qquad (4.28)$$

with a real-valued symbol $L(x, p) \in C^\infty(\mathbf{R}^{2n})$. The Lagrangian Cauchy problem for this equation with the initial data in a plane was discussed in §3; now, the initial data will be specified on a manifold. The Lagrangian Cauchy problem for Equation (4.28) is formulated in the following way.

(1) Introduce Lagrangian manifold Λ^{n-1} of dimension $n-1$ by

$$L(x, p) = 0, \qquad (x, p) \in \Lambda^{n-1} \qquad (4.29)$$

such that Λ^{n-1} admits a diffeomorphic projection on $\mathbf{R}^n_x$.

(2) Specify the initial data so that for $x \in \pi_x \Lambda^{n-1}$ (projection of Λ^{n-1} on $\mathbf{R}^n_x$) we have

$$S(x) = S_0(x), \qquad (4.30)$$

$$\frac{\partial S(x)}{\partial x} = p(x), \qquad (x, p(x)) \in \Lambda^{n-1}. \qquad (4.31)$$

(3) Introduce the compatibility condition

$$dS_0(x) = \langle p(x), dx \rangle, \qquad x \in \pi_x \Lambda^{n-1}. \qquad (4.32)$$

THEOREM 4.23.   *Let* $r^0 = (x^0, p^0) \in \Lambda^{n-1}$ *and let the vector* $\partial L(r^0)/\partial p$ *be transversal to* $\pi_x(T\Lambda^{n-1})(r^0)$. *Then there exists a solution of the Lagrangian Cauchy problem and is unique in a small neighbourhood of the point* x^0.

Proof. We shall reduce the theorem to Proposition 3.8. Let $r^0 = 0$ and Λ^{n-1} be given by the equations

$$x = \varphi(x_1, \ldots, x_{n-1}), \qquad p = \psi(x_1, \ldots, x_{n-1})$$

in a neighbourhood of the point 0. Then $S_0 = S_0(x_1, \ldots, x_{n-1})$. We shall transform the variables

$$x' = h(x), \qquad p_j = \sum_{k=1}^{n} \frac{\partial h_j(x)}{\partial x_k} p'_k, \qquad 1 \leq j \leq n. \tag{4.33}$$

where $x'_j = x_j$, $1 \leq j \leq n-1$, $x'_n = x_n - \varphi(x_1, \ldots, x_{n-1})$. This transformation is a diffeomorphism in a small neighbourhood of the origin of coordinates; it is canonical (Example 4.13) and transforms $(S, L, S_0, \Lambda^{n-1})$ into $(S', L', S'_0, \Lambda'^{n-1})$. Here

$$S'(x') = S(x), \quad L'(x', p') = L(x, p), \quad S'_0(x') = S_0(x),$$

so that, if $S(x)$ satisfies Equation (4.28), then

$$L'\left(x', \frac{\partial S'(x')}{\partial x'}\right) = 0. \tag{4.34}$$

Put $y = (x'_1, \ldots, x'_{n-1})$; then Lagrangian manifold Λ'^{n-1} has the form

$$\Lambda'^{n-1} = \left\{ (x', p') = \left(y, 0; \frac{\partial \psi(y)}{\partial y}, p'_n(y) \right), \quad y \in V \right\},$$

which can be proved in the same manner as Theorem 4.20. Here V is a neighbourhood of the point $y = 0$. Moreover,

$$L'\left(y, 0, \frac{\partial \psi(y)}{\partial y}, p'_n(y) \right) \equiv 0, \qquad y \in V. \tag{4.35}$$

It follows from the compatibility condition (4.32) that

$$\frac{\partial \psi(y)}{\partial y} = \frac{\partial S'_0(y)}{\partial y},$$

$y \in V$, and our problem is of the same kind as in Proposition 3.8:

$$S'(y, 0) = S'_0(y), \qquad \frac{\partial S'(y, 0)}{\partial x'_n} = p'_n(y), \qquad y \in V.$$

We shall show that

$$\frac{\partial L'}{\partial p'_n} \neq 0 \qquad \left(x' = 0, \quad p' = \left(\frac{\partial S'_0(0)}{\partial y}, \ p'_n(0) \right) \right). \tag{4.36}$$

Then all assumptions of Proposition 3.8 will be satisfied. We have

$$\frac{\partial L'(x', p')}{\partial p'_n} = \sum_{k=1}^{n} \frac{\partial L(x, p)}{\partial p_k} \frac{\partial p_k}{\partial p'_n} = \sum_{k=1}^{n} \frac{\partial L(x, p)}{\partial p_k} \frac{\partial x'_n}{\partial x_k}$$

by virtue of (4.33.) For $x' = 0, p' = (0, p'_n(0))$ we have

$$\frac{\partial L'}{\partial p'_n} = -\sum_{j=1}^{n-1} \frac{\partial L(r^0)}{\partial p_k} \frac{\partial \varphi(x_1^0, \ldots, x_{n-1}^0)}{\partial x_k} + \frac{\partial L(r^0)}{\partial p_n} \neq 0;$$

due to the assumption of the theorem $\partial L/\partial p'_n \neq 0$, since the vector $(-(\partial\varphi/\partial x_1), \ldots, -(\partial\varphi/\partial x_{n-1}), 1)$ is orthogonal to $\pi_x(T\Lambda^{n-1})(r^0)$.

5. Generalization of the Notion of Lagrangian Manifold

Let M^n be an n-dimensional C^v-manifold and T^*M^n its cotangent bundle. Let $x = (x_1, \ldots, x_n)$ be local coordinates on M^n; then $T^*M^n = \bigcup_{x\in M^n} T^*M^n_x$, and the elements of $T^*M^n_x$ are the 1-forms $\eta_x = \sum_{j=1}^n p_j \, dx_j$ on M^n, i.e. $\eta_x : TM^n_x \to R$, where $T^*M^n_x$ is the tangent space to M^n at the point x. We shall introduce the local coordinates $(p, x), p = (p_1, \ldots, p_n)$ on T^*M^n and consider the 1-*form* on T^*M^n

$$\omega^1 = \langle p, dx \rangle. \tag{4.37}$$

Notice that a 1-form on T^*M^n is, in general, given by

$$\theta^1 = \sum_{j=1}^n a_j(p, x) \, dx_j + \sum_{j=1}^n b_j(p, x) \, dp_j.$$

The form ω^1 is well defined. Indeed, if x' are new local coordinates on M^n, $x = \varphi(x')$, then

$$(p, x) \to (p', x'), \qquad p'_j = \sum_{k=1}^n \frac{\partial \varphi_k}{\partial x_j}(x') \, p_k.$$

As a consequence,

$$\langle p', dx' \rangle = \sum_{k=1}^n p_k \left(\sum_{j=1}^n \frac{\partial \varphi_k(x')}{\partial x'_j} \partial x'_j \right) = \langle p, dx \rangle.$$

We introduce the 2-*form* on M^n

$$\omega^2 = d\omega^1. \tag{4.38}$$

In local coordinates,

$$\omega^2 = \langle dp, dx \rangle.$$

DEFINITION 4.24.   The vectors $\xi, \eta \in T_x(T^*M)$ are said to be *skew-orthogonal*, if $\omega^2(\xi, \eta) = 0$.

DEFINITION 4.25.   A C^∞-submanifold $\Lambda^k \subset T^*M^n$ (of dimension k) is said to be *Lagrangian*, if any two tangent vectors to Λ^k are skew-orthogonal.

It follows from Lemma 4.5 that $(p_{(\alpha)}, x_{(\beta)})$ can be chosen as local coordinates in a neighbourhood of the point on the Lagrangian manifold.

Let $H: T^*M \to \mathbf{R}$ be a real-valued C^∞-function on the cotangent bundle and let X, Y be C^∞-vector fields on T^*M^n.

DEFINITION 4.26.   The vector field X_H on T^*M^n such that

$$\omega^2(X_H \quad Y) = -dH(Y) \tag{4.39}$$

for any $Y \in T(T^*M^n)$ is said to be a *Hamiltonian system* on T^*M^n with the Hamiltonian H.

In the local coordinates we have

$$X_H = \left\langle \frac{\partial H}{\partial p}, \frac{\partial}{\partial x} \right\rangle - \left\langle \frac{\partial H}{\partial x}, \frac{\partial}{\partial p} \right\rangle.$$

The trajectories of field X_H are the solutions of the Hamilton system (4.19).

Let $g_H^t: T^*M^n \to T^*M^n$ be the phase flow induced by the field X_H. Then g_H^t preserves the form ω^2.

§ 5. Fourier Transformation of a λ-Pseudodifferential Operator
(the Transition to p-Representation)

Our task is to construct a global f.a. solution of Equation (3.1), i.e. a solution in the whole space $\mathbf{R}_x^n$. The first step of the construction of global classical approximation consists in the derivation of solution $S(x)$ of the Hamilton–Jacobi equation (3.3) which is smooth everywhere in $\mathbf{R}_x^n$. However, this equation is non-linear and therefore may have no smooth solution in the whole space except the trivial one; examples of this kind were given at the end of § 3. As a consequence, solutions of the form (3.2) do not suffice for constructing global f.a. solutions.

The following observation appears to be basic [59]: the fundamental object associated with Hamilton–Jacobi equation is not function $S(x)$ but the Lagrangian manifold Λ^n. The specification of function S in some domain $\Omega \subset \mathbf{R}_x^n$ is equivalent to the specification of a 'part' of the Lagrangian manifold in the phase space,

$$p = \frac{\partial S(x)}{\partial x}, \qquad x \in \Omega.$$

We shall extend this manifold, i.e., we shall displace its points along the trajectories of the Hamilton system; we obtain a smooth Lagrangian manifold Λ^n (with the corresponding assumptions).

Thus suppose Lagrangian manifold Λ^n has no singularities. This Lagrangian manifold forms a kind of specific 'Riemann surface' of function S.

Though we assume Lagrangian manifold Λ^n without singularities, the function $S = \int \langle p, dx \rangle$ associated with it can, nevertheless, have singularities. Namely, the generating function S of Lagrangian manifold Λ^n has singularities in the points $(x, p) \in \Lambda^n$, the neighbourhoods of which are badly projected on x-space. However, in these points it is possible to choose other local coordinates on Λ^n, e.g. by going to p-representation. Then it is natural to consider the operator L in p-representation, too. This transition is realized by means of the Fourier transformation. We pass to p-representation of the equation

$$L(\overset{2}{x}, \lambda^{-1}\overset{1}{D}_x)u(x) = f(x),$$

by setting

$$u(x) = F^{-1}_{\lambda, p \to x}\hat{u}(p), \qquad f(x) = F^{-1}_{\lambda, p \to x}\hat{f}(p).$$

Then the equation takes the form

$$(F_{\lambda, x \to p}L(\overset{2}{x}, \lambda^{-1}\overset{1}{D}_x)F^{-1}_{\lambda, p \to x})\hat{u}(p) = \hat{f}(p).$$

Thus we obtained the original operator in p-representation. It has the form

$$L(-\lambda^{-1}\overset{2}{D}_p, \overset{1}{p}),$$

and is therefore a λ-p.d. operator as before.

Particular f.a. solution of the equation

$$L(-\lambda^{-1}\overset{2}{D}_p, \overset{1}{p})\hat{u}(p) = 0$$

can again be looked for in the form (3.2) with the replacement of x by p. It will be shown in the next paragraphs that by combining such solutions it is possible to obtain a global f.a. solution of Equation (3.1).

1. Partial Fourier Transformations and λ-p.d. Operators

We decompose the set $(1, 2, ..., n)$ into two disjoint subsets (α), (β): $(\alpha) = (\alpha_1, ..., \alpha_k)$, $(\beta) = (\beta_1, ..., \beta_l)$, where $k + l = n, \alpha_i \neq \beta_j$, for all i, j (one of the sets (α), (β) can be void). We set $x = (x_{(\alpha)}, x_{(\beta)})$, $x_{(\alpha)} =$

$(x_{\alpha_1}, \ldots, x_{\alpha_k})$, and analogously $p = (p_{(\alpha)}, p_{(\beta)})$. We denote

$$\langle x_{(\alpha)}, p_{(\alpha)} \rangle = \sum_{j=1}^{k} x_{\alpha_j} p_{\alpha_j}, \qquad dx_{(\alpha)} = dx_{\alpha_1} \ldots dx_{\alpha_k},$$

and analogously

$$\langle x_{(\beta)}, p_{(\beta)} \rangle, \quad dx_{(\beta)}, dp_{(\alpha)}, dp_{(\beta)}.$$

We introduce the λ-Fourier transformation over a part of the variables by

$$(F_{\lambda, x_{(\alpha)} \to p_{(\alpha)}} u(x))(p_{(\alpha)}, x_{(\beta)}) =$$

$$= \left(\frac{\lambda}{2\pi i}\right)^{k/2} \int \exp[-i\lambda \langle x_{(\alpha)}, p_{(\alpha)} \rangle] u(x) dx_{(\alpha)}.$$

where, as usual, $\sqrt{i} = e^{i\pi/4}$. Then the inverse transformation has the form

$$(F^{-1}_{\lambda, p_{(\alpha)} \to x_{(\alpha)}} v(p_{(\alpha)}, x_{(\beta)}))(x) =$$

$$= \left(\frac{\lambda}{-2\pi i}\right)^{k/2} \int \exp[i\lambda \langle x_{(\alpha)}, p_{(\alpha)} \rangle] v(p_{(\alpha)}, x_{(\beta)}) dp_{(\alpha)}. \qquad (5.2)$$

Let $L(x, p; (i\lambda)^{-1})$ be a polynomial in (x, p) with the coefficients smooth with respect to λ^{-1} for $\lambda \geq 1$. Then well-known properties of the Fourier transformation yield

$$F_{\lambda, x_{(\alpha)} \to p_{(\alpha)}} L(\overset{2}{x}, \lambda^{-1} \overset{1}{D}_x; (i\lambda)^{-1}) F^{-1}_{\lambda, p_{(\alpha)} \to x_{(\alpha)}} u(p_{(\alpha)}, x_{(\beta)}) =$$

$$= L(-\lambda^{-1} \overset{2}{D}_{p_{(\alpha)}}, \overset{2}{x}_{(\beta)}, \overset{1}{p}_{(\alpha)}, \lambda^{-1} \overset{1}{D}_{x_{(\beta)}}; (i\lambda)^{-1}) u(p_{(\alpha)}, x_{(\beta)}). \qquad (5.3)$$

PROPOSITION 5.1. *Let symbol* $L(x, p) \in T^m_{\pm}(\mathbf{R}^n_x)$. *Then the left-hand side of formula* (5.3) *defines a linear operator from* $C_0^\infty(\mathbf{R}^k_{p_{(\alpha)}} \times \mathbf{R}^{n-k}_{x_{(\beta)}})$ *into* $C^\infty(\mathbf{R}^k_{p_{(\alpha)}} \times \mathbf{R}^{n-k}_{x_{(\beta)}})$.

The proof is obvious. Formula (5.3) will be regarded as a definition of the operator

$$L(-\lambda^{-1} \overset{2}{D}_{p_{(\alpha)}}, \overset{2}{x}_{(\beta)}, \overset{1}{p}_{(\alpha)}, \lambda^{-1} \overset{1}{D}_{x_{(\beta)}}; (i\lambda)^{-1});$$

the λ-p.d. operator $L(\overset{2}{x}, \lambda^{-1} \overset{1}{D}_x; (i\lambda)^{-1})$ will be called its λ-Fourier transform with respect to the variables $x_{(\alpha)}$.

2. *The New Class of Partial Asymptotic Solutions*

Consider the equation

$$L(\overset{2}{x}, \lambda^{-1}D_x ; (i\lambda)^{-1}) u(x) = 0. \tag{5.4}$$

We shall assume throughout this paragraph that $L(x, p ; (i\lambda)^{-1}) \in T_+^m(\mathbf{R}^n)$ for some m.

In §3 we have constructed the class of f.a. solutions of Equation (5.4) of the form

$$u(x, \lambda) = \varphi(x) \exp (i\lambda S(x)).$$

Formula (5.3) allows us to construct a class of f.a. solutions of the form

$$u(x, \lambda) = F_{\lambda, \, p_{(\alpha)} \to x_{(\alpha)}}^{-1} (\varphi(p_{(\alpha)}, x_{(\beta)}) \exp [i\lambda S(p_{(\alpha)}, x_{(\beta)})]) \tag{5.5}$$

First we present a formal exposition and then we state the rigorous results. We shall seek a solution of Equation (5.4) in the form

$$u(x, \lambda) = F_{\lambda, \, p_{(\alpha)} \to x_{(\alpha)}}^{-1} v(p_{(\alpha)}, x_{(\beta)}, \lambda). \tag{5.6}$$

Inserting (5.6) in (5.4), applying operator $F_{\lambda, \, x_{(\alpha)} \to p_{(\alpha)}}$ from the left and using formula (5.3), we obtain the equation

$$L(- \lambda^{-1}\overset{2}{D}_{p_{(\alpha)}}, \overset{2}{x}_{(\beta)}, \overset{1}{p}_{(\alpha)}, \lambda^{-1}\overset{1}{D}_{x_{(\beta)}} ; (i\lambda)^{-1})v(p_{(\alpha)}, x_{(\beta)}, \lambda) = 0. \tag{5.7}$$

But this equation has the same form as (5.4) (i.e. a λ-p.d. operator stands again on the left-hand side) and, as in §3, we may look for a f.a. solution of the form

$$v(p_{(\alpha)}, x_{(\beta)}, \lambda) = \varphi(p_{(\alpha)}, x_{(\beta)}) \exp [i\lambda S(p_{(\alpha)}, x_{(\beta)})]. \tag{5.8}$$

Inserting it into (5.6) we get (5.7). This class of f.a. solutions was first constructed in [59], [60].

THEOREM 5.2. *Let symbol* $L(x, p; (i\lambda)^{-1}) \in T_+^m(\mathbf{R}_x^n)$, *function* $S(p_{(\alpha)}, x_{(\beta)})$ *be real-valued,*

$$\varphi(p_{(\alpha)}, x_{(\beta)}) \in C_0^\infty(\mathbf{R}_{p_{(\alpha)}}^k \times \mathbf{R}_{x_{(\beta)}}^{n-k}),$$
$$S(p_{(\alpha)}, x_{(\beta)}) \in C^\infty(\mathbf{R}_{p_{(\alpha)}}^k \times \mathbf{R}_{x_{(\beta)}}^{n-k}).$$

Then

$$L(\overset{2}{x}, \lambda^{-1}\overset{1}{D}_x; (-i\lambda)^{-1})F^{-1}_{\lambda, \, p_{(\alpha)} \to x_{(\alpha)}}(\varphi(p_{(\alpha)}, x_{(\beta)}) \times$$

$$\times \exp[i\lambda S(p_{(\alpha)}, x_{(\beta)})]) =$$

$$= F^{-1}_{\lambda, \, p_{(\alpha)} \to x_{(\alpha)}} [\exp(i\lambda S(p_{(\alpha)}, x_{(\beta)})) \times$$

$$\times \sum_{j=0}^{N-1} (i\lambda)^{-j} R_j \varphi(p_{(\alpha)}, x_{(\beta)}) + O_{-N}] \tag{5.9}$$

holds for $\lambda \geq 1$ and for an arbitrary integer $N \geq 0$.

Here R_j is a linear differential operator of the order $\leq j$ with C^∞-coefficients (depending on S). The remainder $O^+_{-N} \in O^+_{-N}(\mathbf{R}^k_{p_{(\alpha)}} \times \mathbf{R}^{n-k}_{x_{(\beta)}})$.

Proof. Let function $v(p_{(\alpha)}, x_{(\beta)})$ be infinitely differentiable and with compact support with respect to all its arguments. Then because of (5.3) we have

$$L(\overset{2}{x}, \lambda^{-1}\overset{1}{D}_x; (i\lambda)^{-1})F^{-1}_{\lambda, \, p_{(\alpha)} \to x_{(\alpha)}} v(p_{(\alpha)}, x_{(\beta)}) =$$

$$= F^{-1}_{\lambda, \, p_{(\alpha)} \to x_{(\alpha)}} \mathscr{P}v(p_{(\alpha)}, x_{(\beta)}), \tag{5.10}$$

$$\mathscr{P} = L(-\lambda^{-1}\overset{1}{D}_{p_{(\alpha)}}, \overset{2}{x}_{(\beta)}, \lambda^{-1}\overset{1}{D}_{x_{(\beta)}}, \overset{2}{p}_{(\alpha)}).$$

Let v be of the form (5.8). Applying Theorems 2.6 and 2.11 we obtain the expansions

$$\mathscr{P}v = e^{i\lambda S} \sum_{j=0}^{N-1} (i\lambda)^{-j} R_j \varphi + M_N(p_{(\alpha)}, x_{(\beta)}; (i\lambda)^{-1}).$$

The statement of the theorem concerning the remainder follows from (2.12')

We write the first two terms of expansion (5.9):

$$R_0 \varphi = L^0 \varphi, \tag{5.11}$$

$$R_1 \varphi = \left\langle \frac{\partial L^0}{\partial x_{(\alpha)}}, \frac{\partial \varphi}{\partial p_{(\alpha)}} \right\rangle - \left\langle \frac{\partial L^0}{\partial p_{(\beta)}}, \frac{\partial \varphi}{\partial x_{(\beta)}} \right\rangle -$$

$$- \frac{1}{2}\left[\mathrm{Sp}\left(\frac{\partial^2 L^0}{\partial x^2_{(\alpha)}} \frac{\partial^2 S}{\partial p^2_{(\alpha)}} \right) + \mathrm{Sp}\left(\frac{\partial^2 L^0}{\partial p^2_{(\beta)}} \frac{\partial^2 S}{\partial x^2_{(\beta)}} \right) - \right.$$

$$\left. - 2\,\mathrm{Sp}\left(\frac{\partial^2 L^0}{\partial x_{(\alpha)} \partial p_{(\beta)}} \frac{\partial^2 S}{\partial p_{(\alpha)} \partial x_{(\beta)}} \right) - 2 \sum_{j=1}^{k} \frac{\partial^2 L^0}{\partial x_{\alpha j} \partial p_{\beta j}} \right]\varphi +$$

$$+ \left(\frac{\partial L}{\partial (i\lambda)^{-1}} \right)_0 \varphi. \tag{5.12}$$

Here

$$L^0 = L\left(-\frac{\partial S(p_{(\alpha)}, x_{(\beta)})}{\partial p_{(\alpha)}}, \; x_{(\beta)}, \; p_{(\alpha)}, \; \frac{\partial S(p_{(\alpha)}, x_{(\beta)})}{\partial x_{(\beta)}}; \; 0 \right). \tag{5.13}$$

The derivative $(\partial L/\partial (i\lambda)^{-1})_0$ is taken for $(i\lambda)^{-1} = 0$ and for the same values of x, p as in L^0.

3. The Transport Equation

We denote

$$\tilde{L}(x, \; p; \; (i\lambda)^{-1}) = L(- x_{(\alpha)}, \; x_{(\beta)}, \; p_{(\alpha)}, \; p_{(\beta)}; (i\lambda)^{-1}). \tag{5.14}$$

The Hamilton–Jacobi equation has the form

$$L^0 = 0. \tag{5.15}$$

We set

$$x' = (p_{(\alpha)}, x_{(\beta)}), \qquad p' = (x_{(\alpha)}, p_{(\beta)}). \tag{5.16}$$

Then the Hamilton system for Equation (5.15) has the form

$$\frac{dx'}{dt} = \frac{\partial \tilde{L}'(x', \; p'; \; 0)}{\partial p'}, \qquad \frac{dp'}{dt} = -\frac{\partial \tilde{L}'(x', \; p'; \; 0)}{\partial x'}, \tag{5.17}$$

$$\tilde{L}'(x', \; p'; \; 0) = \tilde{L}(x, \; p; \; 0).$$

By applying the Liouville formula (3.23) to the system

$$\frac{dx'}{dt} = \frac{\partial \tilde{L}'(x', \; \partial S/\partial x'; \; 0)}{\partial p'},$$

we obtain

$$\frac{d}{dt} \ln J = \mathrm{Sp}\left(\frac{\partial^2 L^0}{\partial p'^2} \frac{\partial^2 S}{\partial x'^2} \right) + \mathrm{Sp}\left(\frac{\partial^2 L^0}{\partial x' \, \partial p'} \right),$$

where

$$J = \det \frac{\partial(p_{(\alpha)}, x_{(\beta)})}{\partial(x_{(\alpha)}, p_{(\beta)})}. \tag{5.18}$$

Further,

$$\left\langle \frac{\partial L^0}{\partial p'}, \frac{\partial \varphi}{\partial x'} \right\rangle = \dot{\varphi},$$

where $\dot{\varphi}$ is the derivative with respect to system (5.17). Consequently, the equation $R\varphi = 0$ takes the form

$$\frac{d}{dt}(\varphi\sqrt{J}) = -(\varphi\sqrt{J})\left(\frac{1}{2} \operatorname{Sp} \frac{\partial^2 \tilde{L}^0}{\partial x' \, \partial p'} + \frac{\partial \tilde{L}^0}{\partial \varepsilon}\bigg|_{\varepsilon=0} \right). \qquad (5.19)$$

We shall formulate the result which is analogous to Theorem 3.10. Let U be a domain in $\mathbf{R}^{n-1}$, $\{x' : x' = x'^0(\gamma)\}$ be a C^∞-manifold of dimension $n-1$ in $\mathbf{R}_x^n$. We consider the Cauchy problem

$$x'|_{t=0} = x'^0(\gamma), \quad p'|_{t=0} = p'^0(\gamma), \quad \gamma \in U, \qquad (5.20)$$

with compatibility condition (3.32), and suppose assumptions (1)–(3) of Theorem 3.10 are fulfilled (the last one for the Cauchy problem (5.19), (5.20)).

Let $\Omega(x_{(\beta)})$ be a small neighbourhood of the point $x_{(\beta)} = x_{(\beta)}^0(\gamma^0)$ and $\Omega \subset \mathbf{R}_x^n$ be a small neighbourhood of the point $(x_{(\alpha)}^0, x_{(\beta)}^0(\gamma^0))$, where $x_{(\alpha)}^0$ is arbitrary. We set

$$S(p_{(\alpha)}(t, \gamma), x_{(\beta)}(t, \gamma)) = S_0(p_{(\alpha)}^0(\gamma), x_{(\beta)}^0(\gamma)) +$$

$$+ \int_0^t \left[\left\langle p_{(\beta)}(t', \gamma), \frac{dx_{(\beta)}(t', \gamma)}{dt'} \right\rangle - \right.$$

$$\left. - \left\langle x_{(\alpha)}(t', \gamma), \frac{dp_{(\alpha)}(t', \gamma)}{dt'} \right\rangle \right] dt'. \qquad (5.21)$$

For small t, this function will be a C^∞-solution of Equation (5.15).

THEOREM 5.3. *Let assumptions* (1)–(3) *of Theorem* 3.10 *be satisfied* (*the last one for the Cauchy problem* (5.19), (5.20)). *Then the function*

$$u(x, \lambda) = F_{\lambda, p_{(\alpha)} \to x_{(\alpha)}}^{-1} \frac{\varphi_0(\gamma)\sqrt{|J(0, \gamma)|}}{\sqrt{|J(t, \gamma)|}} \times$$

$$\times \exp\left[i\lambda S(p_{(\alpha)}, x_{(\beta)}) - \int_0^t \left[\frac{1}{2} \operatorname{Sp}\left(\frac{\partial^2 L^0}{\partial x \partial p} \right) - \frac{\partial L^0}{\partial (i\lambda)^{-1}}\bigg|_{(i\lambda)^{-1}=0} \right] dt' \right]$$

$$(5.22)$$

represents a solution of Equation (3.1) *mod* $O_{-2}^+(\Omega)$.

Function S is the solution of the Hamilton–Jacobi equation (5.15), $\varphi_0(\gamma)$ is an arbitrary function of class $C'(U)$.

All functions $x_{(\alpha)}, p_{(\beta)}, p_{(\alpha)}, x_{(\beta)}$ on the right-hand side of (5.22) are solutions of the Cauchy problem (5.19), (5.20) (i.e. $x_{(\alpha)} = x_{(\alpha)}(t, \gamma)$ etc.); t and γ are such that $x_{(\beta)}(t, \gamma) = x_{(\beta)}$.

The proof of the theorem follows from (5.19) and Theorem 5.2.

The transport equation will be considered again in §§ 11 and 14.

§ 6. The Precanonical Operator (Quantization of the Velocity Field in the Small)

In the previous sections we have constructed the formal asymptotic solutions of Equation (3.1) which have the form of a rapidly oscillating exponential either in x-representation or in p-representation, or in the mixed representation. Any such f.a. solution is associated with Lagrangian manifold Λ^n of maximal dimension in the phase space. Now we take this manifold as the starting point of our considerations. Our task will be to construct a global f.a. solution by patching together all possible f.a. solutions induced by manifold Λ^n. In patching f.a. solutions together, it is necessary, first of all, to take into account that one and the same solution may have various representations (for instance, Λ^n may admit diffeomorphic projections on x-space as well as on p-space). This paragraph will be devoted to put these representations in accordance.

1. Heuristic Considerations

The problem of constructing semi-classical asymptotics for quantum mechanical equations or, in general, for equations of the form

$$L(x, \lambda^{-1}D_x)u(x) = 0$$

leads to a 'classical' object – the Lagrangian manifold in the phase space, as we clarified in the previous paragraphs. Now we shall quantize this object of classical mechanics (the velocity field), i.e. we shall construct an operator K corresponding to manifold Λ^n (the canonical operator).

In this paragraph we discuss the case when Lagrangian manifold Λ^n can be diffeomorphically projected on one of the coordinate Lagrangian planes. The corresponding operator K will be called a *precanonical operator*.

Let manifold Λ^n admit a diffeomorphic projection on x-space. We introduce the operator

$$K : L^2_{\sigma^n}(\Lambda^n) \to L^2(\mathbf{R}^n_x)$$

by means of the formula

$$(K\varphi)(x) = \sqrt{\left|\frac{d\sigma^n}{dx}\right|} \exp[i\lambda S(x)] \, \varphi(x). \tag{6.1}$$

This operator is said to be a *precanonical operator*. Here $d\sigma^n$ is the volume element on Λ^n, $S(x)$ is the generating function of the Lagrangian manifold, i.e. Λ^n is given by the equation $p = \partial S(x)/\partial x$, and the norm in $L^2_{\sigma^n}$ is defined by

$$\|\varphi\|^2 = \int_{\Lambda^n} |\varphi|^2 \, d\sigma^n.$$

Operator K is unitary:

$$\int_{\mathbf{R}^n} |K\varphi|^2 \, dx = \int_{\Lambda^n} |\varphi|^2 \, d\sigma^n.$$

Furthermore, the commutation formula

$$L(\overset{2}{x}, \lambda^{-1}\overset{1}{D}_x)K\varphi = K[L(x, p)\varphi + O(\lambda^{-1})]$$

holds, where $(x, p) \in \Lambda^n$. Thus the λ-p.d. operator is transformed into the operator of multiplication by a function (an operator symbol) with accuracy to $O(\lambda^{-1})$.

If the symbol of operator $\hat{L}$ is zero on manifold Λ^n, then

$$\hat{L}(\overset{2}{x}, \lambda^{-1}\overset{1}{D}_x)K\varphi = O(\lambda^{-1}),$$

i.e. the function $u = K\varphi$ represents a formal asymptotic solution of the equation $\hat{L}u = 0$ with accuracy to $O(\lambda^{-1})$. Thus, by using the precanonical operator, a f.a. solution of the equation $\hat{L}u = 0$ can be constructed.

Let Λ^n admit also a diffeomorphic projection on p-space. Passing to p-representation, the equation of manifold Λ^n takes the form $x = -\partial\tilde{S}(p)/\partial p$, where functions $S(x)$, $\tilde{S}(p)$ are Young-dual. Notice that function $\varphi(r)$ where r is a point of Λ^n, can simultaneously be regarded as a function of x as well as a function of p. For simplicity, let $n = 1$. By applying the stationary phase method (§ 1), we obtain (under the assumption that $S''_{xx}(x) \neq 0$)

$$(F_{\lambda, x \to p}(K\varphi)(x))(p) = \sqrt{\frac{\lambda}{2\pi i}} \int \exp(-i\lambda xp + i\lambda S(x))\, \varphi(x)\, dx =$$

$$= \frac{1}{\sqrt{i}} e^{\pm i(\pi/4)}\, e^{i\lambda \tilde{S}(p)}\, \varphi(x(p)) \frac{1}{\sqrt{|S''(x(p))|}} \sqrt{\left|\frac{d\sigma}{dx}\right|} + O(\lambda^{-1}), \quad (6.1')$$

where $x(p)$ is the solution of the equation $S'_x(x) = p$.

Furthermore, comparing formulae (6.1) and (6.1') we get

$$\left|S''_{xx}(x(p))\right| = \left|\frac{dp}{dx}\right|.$$

Consequently, in p-representation the precanonical operator K has the form

$$(K\varphi)(x) = e^{i\gamma} F^{-1}_{\lambda, p \to x} \left(\sqrt{\left|\frac{d\sigma}{dp}\right|}\, \exp(i\lambda\tilde{S}(p))\, \varphi(x(p)) \right),$$

$$\gamma = \text{const.}, \quad (6.1'')$$

with accuracy to $O(\lambda^{-1})$.

In the mixed representation, the precanonical operator has a similar expression.

2. The Precanonical Operator

Let us summarize the notation to be used in this and the subsequent paragraphs.

(1) Λ^n is an n-dimensional Lagrangian C^∞-manifold in the phase space $\mathbf{R}^{2n} = \{x, p\}$.

(2) r is a point in Λ^n, $(x(r), p(r))$ its coordinates.

(3) $d\sigma^n$ is a differential n-form of class C^∞ on Λ^n (a volume form).

(4) $\Sigma(\Lambda^n)$ is the cycle of singularities (see Definition 6.1).

(5) $C_0^\infty(\Lambda^n)$ is the set of all C^∞-mappings $\varphi : \Lambda^n \to \mathbf{C}$ with compact supports.

(6) $(\alpha), (\beta)$ are disjoint sets the union of which is the set $(1, 2, \dots, n)$. Further $x_{(\alpha)} = (x_{(\alpha_1)}, \dots, x_{(\alpha_k)})$ etc.

(7) $l[r^1, r^2]$ is a smooth curve on Λ^n, $\partial l = r^2 - r^1$.

Next we introduce the preliminary notions of a canonical chart, a non-singular canonical chart and of a canonical atlas.

By the *canonical chart* we understand a simply connected domain $\Omega \subset \Lambda^n$ which admits a diffeomorphic projection on one of the Lagrangian coordinate planes $\{p_{(\alpha)}, x_{(\beta)}\}$. The coordinates $(p_{(\alpha)}, x_{(\beta)})$ will be called *focal coordinates* of the chart.

DEFINITION 6.1 A point $r^0 \in \Lambda^n$ is said to be *non-singular* (with respect to a projection on $\mathbf{R}_x^n$) if it has a neighbourhood admitting a diffeomorphic projection on $\mathbf{R}_x^n$; it is *singular* in the opposite case. The set of all singular points will be denoted by $\Sigma(\Lambda^n)$.

A canonical chart Ω is said to be *non-singular*, if all its points are non-singular.

A *canonical atlas* on Lagrangian manifold Λ^n is a family $\{\Omega^i\}$ where Ω^i are the canonical charts such that: (1) $\{\Omega^i\}$ is a countable covering of Λ^n, (2) each compact subset $F \subset \Lambda^n$ is covered by a finite number of charts.

The existence of canonical atlases follows from Lemma 4.5.

Let Ω be a canonical chart. For $r \in \Omega$ the integral

$$\int_{r^0}^{r} \langle p', dx' \rangle$$

is taken, by definition, along the path

$$l[r^0, r(\Omega)] + l[r(\Omega), r]$$

where r^0 is a fixed point of the chart, and the path lies in Ω. Since the form $\langle p, dx \rangle$ is closed on Λ^n, the integral is a single-valued function of class $C^\infty(\Omega)$.

Now we introduce *operator K* acting from $C_0^\infty(\Omega)$ into $C^\infty(\mathbf{R}_x^n)$.

(1) Suppose that Ω is a non-singular chart. We set (see (6.1'))

$$(K\varphi)(x) = \sqrt{\left| \frac{d\sigma^n(r(x))}{dx} \right|}\ \varphi(r(x)) \exp\left(i\lambda \int_{r^0}^{r(x)} \langle p', dx' \rangle \right). \tag{6.2}$$

Here $d\sigma^n(r(x))/dx$ is the Radon–Nikodým derivative of the measure $d\sigma^n$ with respect to the measure dx at the point $r(x)$. In the local coordinates x we have $d\sigma^n(r(x)) = a(x)\,dx$, so that

$$\left| \frac{d\sigma^n(rx))}{dx} \right| = |a(x)|.$$

(2) Suppose Ω is a singular chart with the focal coordinates $(p_{(\alpha)}, x_{(\beta)})$. Each point $r \in \Omega$ has the form

$$r = r(p_{(\alpha)}, x_{(\beta)}) = (x_{(\alpha)}(p_{(\alpha)}, x_{(\beta)}), x_{(\beta)}, p_{(\alpha)}, p_{(\beta)}(p_{(\alpha)}, x_{(\beta)})),$$

where $x_{(\alpha)}, p_{(\beta)}$ are given functions. We set

$$(K\varphi)(x) = F^{-1}_{\lambda, p_{(\alpha)} \to x_{(\alpha)}} \left[\varphi(r(p_{(\alpha)}, x_{(\beta)})) \sqrt{\left| \frac{d\sigma^n(p_{(\alpha)}, x_{(\beta)})}{dp_{(\alpha)} \, dx_{(\beta)}} \right|} \times \right.$$

$$\left. \times \exp\left[i\lambda \left(\int_{r^0}^{r(p_{(\alpha)}, x_{(\beta)})} \langle p', dx' \rangle - \langle x_{(\alpha)}(p_{(\alpha)}, x_{(\beta)}), p_{(\alpha)} \rangle \right) \right] \right]$$

$$(6.3)$$

Since the integrand depends on the variables $(x_{(\alpha)}, x_{(\beta)}, p_{(\alpha)})$, the integral is a function of $x_{(\alpha)}, x_{(\beta)}$, i.e. of x. Notice that the exponent contains the Legendre transform of function $S(x)$ (up to the sign).

The constructed operator depends, besides $\{\Lambda^n, r^0, d\sigma^n\}$ and $\{\Omega, l[r^0, r]\}$, on the choice of the local coordinates in Ω, since, in general, one can choose different focal coordinates in one and the same chart. For instance it may happen that Ω is uniquely projected on both $\mathbf{R}^n_x$ and $\mathbf{R}^n_p$. We shall show that various representations of operator K for $\lambda \to +\infty$ differ only by a constant multiplier with accuracy to $O(\lambda^{-1})$. This multiplier gives rise to a notion of the index of a curve on a Lagrangian manifold (see § 7).

We shall write $K = K(\Omega; p_{(\alpha)}, x_{(\beta)})$ indicating that we selected $p_{(\alpha)}, x_{(\beta)}$ as the focal coordinates in Ω. Consider the map

$$K(\Omega; p_{(\alpha)}, x_{(\beta)}): C_0^\infty(\Omega) \to C^\infty(\mathbf{R}^n_x \times \mathbf{R}^+_\lambda)/O^+_{-1}(\mathbf{R}^n_x). \qquad (6.4)$$

Class $O^+_{-1}(\mathbf{R}^n_x)$ was introduced in § 1.

LEMMA 6.2 Let Ω be a non-singular chart and let Ω admit a diffeomorphic projection on the plane $(p_{(\alpha)}, x_{(\beta)})$. Then for $\lambda \geqq 1$ we have

$$(K(\Omega, x)\varphi)(x) = \exp\left(-\frac{i\pi}{2} \text{inerdex} \frac{\partial x_{(\alpha)}(p_{(\alpha)}, x_{(\beta)})}{\partial p_{(\alpha)}} \right) \times \qquad (6.5)$$

$$\times (K(\Omega; p_{(\alpha)}, x_{(\beta)})\varphi)(x) \ (\text{mod } O^+_{-1}(\mathbf{R}^n_x)).$$

for any function $\varphi \in C_0^\infty(\Omega)$.

Proof. Function $[K(\Omega; p_{(\alpha)}, x_{(\beta)})\varphi](x)$ is equal to the right-hand side of (6.3). We apply the stationary phase method to it. Let $\tilde{S}(p_{(\alpha)}; x)$ be the function in the exponent. We consider it for fixed x. Then

$$d\tilde{S} = \langle p_{(\alpha)}, dx_{(\alpha)}(p_{(\alpha)}, x_{(\beta)})\rangle + \langle x_{(\alpha)} - x_{(\alpha)}(p_{(\alpha)}, x_{(\beta)}), dp_{(\alpha)}\rangle -$$
$$- \langle p_{(\alpha)}, dx_{(\alpha)}(p_{(\alpha)}, x_{(\beta)})\rangle = \langle x_{(\alpha)} - x_{(\alpha)}(p_{(\alpha)}, x_{(\beta)}), dp_{(\alpha)}\rangle.$$

so that

$$\frac{\partial \tilde{S}}{\partial p_{(\alpha)}} = x_{(\alpha)} - x_{(\alpha)}(p_{(\alpha)}, \ x_{(\beta)}). \tag{6.6}$$

The stationary points are determined from the equation $\partial \tilde{S}/\partial p_{(\alpha)} = 0$, and, due to assumptions on Ω and (6.6), this equation has the unique solution $p_{(\alpha)} = p_{(\alpha)}(x)$. Namely, the equation of Ω can be written in the form $p = p(x) = (p_{(\alpha)}(x), p_{(\beta)}(x))$, which yields the desired solution. At the stationary point we have

$$\tilde{S}_{st} = \int_{r^0}^{r(x)} \langle p', dx' \rangle, \qquad \frac{\partial \tilde{S}}{\partial p_{(\alpha)}} = - \frac{\partial x_{(\alpha)}(p_{(\alpha)}, x_{(\beta)})}{\partial p_{(\alpha)}} \Bigg|_{p_{(\alpha)} = p_{(\alpha)}(x)};$$

consequently,

$$\operatorname{sgn} \frac{\partial^2 \tilde{S}}{\partial p_{(\alpha)}^2} = - \operatorname{sgn} \frac{\partial x_{(\alpha)}(p_{(\alpha)}, x_{(\beta)})}{\partial p_{(\alpha)}},$$

and thus

$$\operatorname{sgn} \frac{\partial^2 \tilde{S}}{\partial p_{(\alpha)}^2} = - \operatorname{sgn} \frac{\partial x_{(\alpha)}(p_{(\alpha)}, x_{(\beta)})}{\partial p_{(\alpha)}} = 2 \operatorname{inerdex} \frac{\partial x_{(\alpha)}(p_{(\alpha)}, x_{(\beta)})}{\partial p_{(\alpha)}} - k,$$

where k is the number of the components of vector (α).

Finally,

$$dx = \pm \det \frac{\partial x_{(\alpha)}(p_{(\alpha)}, x_{(\beta)})}{\partial p_{(\alpha)}} dp_{(\alpha)} \, dx_{(\beta)},$$

so that

$$\sqrt{\left| \frac{d\sigma''(r(x))}{dp_{(\alpha)} \, dx_{(\beta)}} \right|} \, \left| \det \frac{\partial x_{(\alpha)}(p_{(\alpha)}, x_{(\beta)})}{\partial p_{(\alpha)}} \right| = \sqrt{\left| \frac{d\sigma''(r(x))}{dx} \right|}.$$

By using Theorem 1.4, we obtain

$$(K(\Omega ; p_{(\alpha)}, x_{(\beta)})\varphi)(x) = \exp\left(\frac{i\pi}{2}\text{ inerdex }\frac{\partial x_{(\alpha)}}{\partial p_{(\alpha)}}\right)\sqrt{\left|\frac{d\sigma''(r(x))}{dx}\right|} \times$$

$$\times \varphi(r(x))\exp\left(i\lambda \int_{r^0}^{r}\langle p', dx'\rangle\right) + \psi(x, \lambda),$$

(6.5')

where $\varphi \in O^+_{-1}(\mathbf{R}^n_x)$, q.e.d.

REMARK. The estimate of the remainder in (6.5')

$$|\psi(x ; \lambda)| \leqq C\lambda^{-1}$$

on every compact subset $K \subset \mathbf{R}^n$ is uniform with respect to φ belonging to a bounded subset of $C^\infty_0(\Omega)$. This follows from [20].

LEMMA 6.3. *Let chart Ω admit a diffeomorphic projection on the Lagrangian planes $(p_{(\alpha)}, x_{(\beta)})$ and $(p_{(\tilde{\alpha})}, x_{(\tilde{\beta})})$. Then for $\lambda \geqq 1$ we have*

$$(K(\Omega : p_{(\alpha)}, x_{(\beta)})\varphi)(x) = e^{i\pi m/2} K(\Omega ; p_{(\tilde{\alpha})}, x_{(\tilde{\beta})})(\varphi + \lambda^{-1}\psi)(x)$$

(6.7)

for any function $\varphi \in C^\infty_0(\Omega)$. Here m is an integer independent of φ and the point $r \in \Omega$.
    Proof. Let

$$K = K(\Omega ; p_{(\alpha)}, x_{(\beta)}), \quad \tilde{K} = K(\Omega ; p_{(\tilde{\alpha})}, x_{(\tilde{\beta})}).$$

We have

$$(K\varphi)(x) = F^{-1}_{\lambda, p_{(\alpha)} \to x_{(\alpha)}}\, \varphi(r)\sqrt{\left|\frac{d\sigma''(r)}{dp_{(\alpha)}\,dx_{(\beta)}}\right|}\exp(i\lambda S),$$

$$(\tilde{K}\varphi)(x) = F^{-1}_{\lambda, p_{(\tilde{\alpha})} \to x_{(\tilde{\alpha})}}\, \varphi(r)\sqrt{\left|\frac{d\sigma''(r)}{dp_{(\tilde{\alpha})}\,dx_{(\tilde{\beta})}}\right|}\exp(i\lambda \tilde{S}),$$

(6.8)

where S and $\tilde{S}$ are determined by (6.3). We shall show that

$$F_{\lambda, x_{(\tilde{\alpha})} \to p_{(\tilde{\alpha})}}(K\varphi)(x) =$$

$$= \exp\left(\frac{i\pi N}{2}\right)\exp(i\lambda\tilde{S})\varphi(r)\sqrt{\left|\frac{d\sigma''(r)}{dp_{(\tilde{\alpha})}\,dx_{(\tilde{\beta})}}\right|} + \chi(p_{(\tilde{\alpha})}, x_{(\tilde{\beta})}, \lambda),\quad (6.9)$$

where N in an integer independent of φ and of a point $r \in \Omega$, and the remainder χ is such that

$$\chi \in O_{-1}^{+} (\mathbf{R}_{p_{(\tilde{a})}}^{\tilde{k}} \times \mathbf{R}_{x_{(\tilde{\beta})}}^{n-\tilde{k}}). \tag{6.10}$$

This will prove the lemma.

Let ψ be the left-hand side of (6.9). Then

$$\psi = \left(\frac{\lambda}{2\pi}\right)^{(k+\tilde{k})/2} \exp\left[\frac{i\pi}{4}(k - \tilde{k})\right] \times$$

$$\times \int\int \exp(i\lambda\tilde{S})\varphi(r) \sqrt{\left|\frac{d\sigma''(r)}{dp_{(\alpha)}\,dx_{(\beta)}}\right|}\; dp_{(\alpha)}\, dx_{(\tilde{a})}, \tag{6.11}$$

where the integration is performed in the following order: first with respect to $dp_{(\alpha)}$ and then with respect to $dx_{(\tilde{a})}$.

Now we shall apply the stationary phase method: first we shall discuss formal aspects and then we shall substantiate them by rigorous arguments.

(I) *The formal derivation.* Among the numbers $\alpha_j \in (\alpha), \tilde{\alpha}_k \in (\tilde{a})$, equal numbers may appear. We set up the decomposition into disjoint sets

$$\{1, 2, \ldots, n\} = (a) \cup (b) \cup (c) \cup (d)$$

with the property

$$(\alpha) = (a) \cup (b), \qquad (\tilde{a}) = (a) \cup (c).$$

Then

$$(\beta) = (c) \cup (d), \qquad (\tilde{\beta}) = (b) \cup (d).$$

According to the assumption, the equation of manifold Ω can be written in the form

$$x_{(a)} = \varphi_{(a)}(\xi), \qquad x_{(b)} = \varphi_{(b)}(\xi),$$
$$p_{(c)} = \varphi_{(c)}(\xi), \qquad p_{(d)} = \varphi_{(d)}(\xi), \qquad \xi = (p_{(\alpha)}, x_{(\beta)}), \tag{6.12}$$

or in the form

$$x_{(a)} = \tilde{\varphi}_{(a)}(\tilde{\xi}), \qquad x_{(c)} = \tilde{\varphi}_{(c)}(\tilde{\xi}),$$
$$p_{(b)} = \tilde{\varphi}_{(b)}(\tilde{\xi}), \qquad p_{(d)} = \tilde{\varphi}_{(d)}(\tilde{\xi}), \qquad \tilde{\xi} = (p_{(\tilde{a})}, x_{(\tilde{\beta})}). \tag{6.12'}$$

In order to avoid confusion in our notation, stars will be attached to all dummy variables over which we do not integrate, for instance:

$$\xi = (p_{(a)}, p_{(b)}, x_{(c)}, x^*_{(d)}), \qquad \tilde{\xi} = (p_{(b)}, p^*_{(c)}, x^*_{(b)}, x^*_{(d)}).$$

Phase function $\tilde{\tilde{S}}$ has the form

$$\tilde{\tilde{S}} = \int_{r^0}^{r} \langle p', dx' \rangle + \langle \varphi_{(\alpha)}(\xi) - x_{(\alpha)}, p_{(\alpha)} \rangle + \langle x_{(\tilde{\alpha})}, p^*_{(\tilde{\alpha})} \rangle,$$

where $r = r(p_{(\alpha)}, x_{(\beta)})$. Consequently,

$$d\tilde{\tilde{S}} = \langle \varphi_{(\beta)}(\xi), dx_{(\beta)} \rangle + \langle x_{(\alpha)} - \varphi_{(\alpha)}(\xi), dp_{(\alpha)} \rangle + \langle p_{(\alpha)}, dx_{(\alpha)} \rangle -$$
$$- \langle p^*_{(\tilde{\alpha})}, dx_{(\alpha)} \rangle. \tag{6.13}$$

Function $\tilde{\tilde{S}}$ depends on the variables $p_{(a)}, p_{(b)}, x_{(a)}, x_{(c)}$, with respect to which the integration in (6.11) is performed, and on the dummy variables playing the role of parameters. By virtue of relation (6.13), the differential of $\tilde{\tilde{S}}$ expressed as a function of the integration variables (with the fixed dummy variables) is equal to

$$d\tilde{\tilde{S}} = \langle x_{(a)} - \varphi_{(a)}(\xi), dp_{(a)} \rangle + \langle x_{(b)} - \varphi_{(b)}(\xi), dp_{(b)} \rangle +$$
$$+ \langle p_{(a)} - p^*_{(a)}, dp_{(a)} \rangle + \langle \varphi_{(c)}(\xi) - p^*_{(c)}, dx_{(c)} \rangle.$$

This implies that

$$\frac{\partial \tilde{\tilde{S}}}{\partial p_{(a)}} = x_{(a)} - \varphi_{(a)}(\xi), \qquad \frac{\partial \tilde{\tilde{S}}}{\partial p_{(b)}} = x_{(b)} - \varphi_{(b)}(\xi),$$

$$\frac{\partial \tilde{\tilde{S}}}{\partial x_{(a)}} = p_{(a)} - p^*_{(a)}, \qquad \frac{\partial \tilde{\tilde{S}}}{\partial x_{(c)}} = \varphi_{(c)}(\xi) - p^*_{(c)}. \tag{6.14}$$

The stationary points of function $\tilde{\tilde{S}}$ are evaluated from the system

$$\varphi_{(b)}(p_{(a)}, p_{(b)}, x_{(c)}, x^*_{(d)}) = x^*_{(b)},$$
$$\varphi_{(c)}(p_{(a)}, p_{(b)}, x_{(c)}, x^*_{(d)}) = p^*_{(c)}, \tag{6.15}$$
$$p_{(a)} = p^*_{(a)},$$
$$x_{(a)} = \varphi_{(a)}(p_{(a)}, p_{(b)}, x_{(c)}, x^*_{(d)}). \tag{6.15'}$$

We have $p_{(a)} = p^*_{(a)}$. For arbitrary

$$(p^*_{(a)}, p^*_{(c)}, x^*_{(b)}, x^*_{(d)}) \in \pi(p_{(\tilde{\alpha})}, x_{(\tilde{\beta})})\Omega$$

(the projection of Ω on the plane $(p_{(\tilde{a})}, x_{(\tilde{\beta})})$), system (6.15) has the unique solution

$$p_{(b)} = \tilde{\varphi}_{(b)}(p_{(a)}^*, p_{(c)}^*, x_{(b)}^*, x_{(d)}^*),$$
$$x_{(c)} = \tilde{\varphi}_{(c)}(p_{(a)}^*, p_{(c)}^*, x_{(b)}^*, x_{(d)}^*), \tag{6.16}$$

where $\tilde{\varphi}_{(b)}$ and $\tilde{\varphi}_{(c)}$ are the functions from (6.12′) as follows according to the assumption of the lemma. Then the second of the equations in (6.15′) uniquely determines $x_{(a)}$, namely

$$x_{(a)} = \tilde{\varphi}_{(a)}(p_{(a)}^*, p_{(c)}^*, x_{(b)}^*, x_{(d)}^*). \tag{6.16′}$$

Thus the stationary point $Q = Q(p_{(\tilde{a})}^*, x_{(\tilde{\beta})}^*)$ is unique and the value of phase $\tilde{S}$ in this point is equal to

$$\tilde{S}(Q) = \int_{r^0}^{r(p_{(\alpha)}^*, x_{(\beta)}^*)} \langle p', dx' \rangle - \langle \tilde{\varphi}_{(\tilde{a})}(p_{(\tilde{a})}^*, x_{(\tilde{\beta})}^*), p_{(\tilde{a})}^* \rangle = \tilde{S}. \tag{6.17}$$

Consequently, the value $\tilde{S}(Q)$ coincides with function $\tilde{S}$ entering in the definition of operator $\tilde{K}$ (see (6.8)).

Let $A(Q)$ be the matrix composed of the second derivatives of function $\tilde{S}$ at the point Q. Then, by virtue of (6.14),

$$A(Q) = \left\| \begin{array}{cccc} -\partial\varphi_{(a)}/\partial p_{(a)} & -\partial\varphi_{(a)}/\partial p_{(b)} & I & -\partial\varphi_{(a)}/\partial x_{(c)} \\ -\partial\varphi_{(b)}/\partial p_{(a)} & -\partial\varphi_{(b)}/\partial p_{(b)} & 0 & -\partial\varphi_{(b)}/\partial x_{(c)} \\ I & 0 & 0 & 0 \\ \partial\varphi_{(c)}/\partial p_{(a)} & \partial\varphi_{(c)}/\partial p_{(b)} & 0 & \partial\varphi_{(c)}/\partial x_{(c)} \end{array} \right\| \tag{6.18}$$

Here I and 0 are the unit and the null matrices, respectively, the dimension of which is clear from the formula, and all derivatives are taken at the point Q. From here we find that

$$|\det A(Q)| = \left| \det \left\| \begin{array}{cc} \dfrac{\partial\varphi_{(b)}(p_{(\alpha)}, x_{(\beta)})}{\partial p_{(b)}} & \dfrac{\partial\varphi_{(b)}(p_{(\alpha)}, x_{(\beta)})}{\partial x_{(c)}} \\[2ex] \dfrac{\partial\varphi_{(c)}(p_{(\alpha)}, x_{(\beta)})}{\partial p_{(b)}} & \dfrac{\partial\varphi_{(c)}(p_{(\alpha)}, x_{(\beta)})}{\partial x_{(c)}} \end{array} \right\| \right|. \tag{6.19}$$

This determinant is different from zero everywhere in Ω. In fact, one can take $(p_{(\alpha)}, x_{(\beta)})$ and $(p_{(\tilde{\alpha})}, x_{(\tilde{\beta})})$ as the coordinates in Ω, so that the Jacobian

$$\left| \det \frac{\partial(p_{(\alpha)}, x_{(\alpha)})}{\partial(p_{(\tilde{\alpha})}, x_{(\tilde{\alpha})})} \right| = \left| \det \frac{\partial(x_{(b)}, p_{(c)})}{\partial(x_{(c)}, p_{(b)})} \right| \neq 0$$

everywhere in Ω. Consequently,

$$\det A(Q) \neq 0, \qquad \operatorname{sgn} A(Q) = \text{const.} \tag{6.20}$$

for $r \in \Omega$. Moreover, we have

$$\inf_{Q \in \Omega} \lambda_j(A(Q)) > 0$$

on any compact set $F \subset \pi(p_{(\tilde{\alpha})}, x_{(\tilde{\beta})})\Omega$ for all j, as follows from the continuity of eigenvalues of symmetric matrix $A(Q)$ and (6.20).

We shall write down the leading term of the asymptotics of ψ for $\lambda \to \infty$. For this purpose it is necessary to evaluate the Hessian of the phase function at a stationary point (see § 1). Since the order of matrix $A(Q)$ is equal to $k + \tilde{k}$, we have

$$\frac{\pi}{4}(k - \tilde{k} + \operatorname{sgn} A(Q)) = \frac{\pi}{2}(k - \operatorname{inerdex} A(Q)) = \frac{N}{2}, \tag{6.21}$$

where N is an integer independent of Q and φ.

Furthermore, from (6.19) we obtain

$$\sqrt{\left| \frac{d\sigma''(r(Q))}{dp_{(\alpha)} \, dx_{(\beta)}} \right|} \cdot \frac{1}{\sqrt{|\det A(Q)|}} = \sqrt{\left| \frac{d\sigma''(r(Q))}{dp_{(\tilde{\alpha})} \, dx_{(\tilde{\beta})}} \right|}, \tag{6.22}$$

so that by Theorem 1.4 the asymptotics of integral ψ (see (6.11)) has the form

$$\psi = \exp(i\lambda \tilde{S}(Q)) \exp\left(\frac{i\pi N}{2}\right) \sqrt{\left| \frac{d\sigma''(r(Q))}{dp_{(\tilde{\alpha})} \, dx_{(\tilde{\beta})}} \right|} \, \varphi(r(Q)) +$$

$$+ \chi(p_{(\tilde{\alpha})}, x_{(\tilde{\beta})}, \lambda), \tag{6.23}$$

where $\chi = O(\lambda^{-1})$ for $\lambda \to +\infty$.

(II) *The rigorous arguments.* If $|x_{(\tilde{\beta})}| > M$ and M is sufficiently large, then, for $\lambda \leq 1$, we have

$$|D_x^\gamma(K\varphi)(x)| \leq C_{N,\gamma}\lambda^{-N}(1+|x|)^{-N}, \qquad (6.24)$$

where $N \geq 0$ is arbitrary and γ is an arbitrary multi-index. In fact, the stationary points of the integral $(K\varphi)(x)$ are found from the equation (see (6.6))

$$x_{(\alpha)} = \varphi_{(\alpha)}(p_{(\alpha)}, x_{(\beta)}). \qquad (6.25)$$

If $(p_{(\alpha)}, x_{(\beta)}) \in \operatorname{supp} \varphi(r)$, then $x_{(\beta)}$ is finite and the set of all $x_{(\beta)}$ of the form (6.25) is bounded. Estimate (6.24) follows from Lemma 1.5.

We set up a partition of unity: $\eta_1(x) + \eta_2(x) = 1$, $x \in \mathbf{R}^n$, where function $\eta_1(x)$ has compact support and $\eta_1(x) \equiv 1$ for $|x| \leq 2M$. Correspondingly we put (see (6.11)) $\psi = \psi_1 + \psi_2$. Then

$$|\psi_1(p_{(\alpha)}, x_{(\beta)})| \leq C_N\lambda^{-N}$$

($\lambda \geq 1$, N is arbitrary). It is not hard to show by integrating by parts with respect to $dx_{(\tilde{\alpha})}$ (see (1.16)) that

$$|D_{p(\tilde{\alpha})}^\gamma D_{x(\tilde{\beta})}^{\gamma'}\psi_1| \leq C_{N,\gamma,\gamma'}\lambda^{-N}(1+|p_{(\tilde{\alpha})}|)^{-N}(1+|x_{(\tilde{\beta})}|)^{-N}$$

for $\lambda \geq 1$; here $\gamma, \gamma', N \geq 0$ are arbitrary. Consequently, the function

$$F_{\lambda,\ p(\tilde{\alpha})\to x(\tilde{\alpha})}^{-1}\psi_1 \in O_{-k}^+(\mathbf{R}_x^n)$$

for any $k > 0$ (i.e. it has order $O(\lambda^{-\infty})$ for $\lambda \to +\infty$). It therefore remains to investigate integral ψ_1 taken over a finite domain in all variables $dp_{(\alpha)}dx_{(\tilde{\alpha})}$. However, this integral for $|p_{(\tilde{\alpha})}| \geq M$ and M sufficiently large has no stationary points lying on the support of the integrand. By virtue of Lemma 1.5 we have

$$|D_{p(\tilde{\alpha})}^\gamma\psi_1| \leq C_{N,\gamma}(1+|p_{(\tilde{\alpha})}|)^{-N}\lambda^{-N}$$

for $|p_{(\tilde{\alpha})}| \geq 2M$, where $N \geq 0$ and γ are arbitrary. If $|p_{(\tilde{\alpha})}| \leq 2M$, Theorem 1.4 is applicable, which, together with (6.24), yields (6.10).

LEMMA 6.4   Let chart Ω admit diffeomorphic projection on the Lagrangian planes $(p_{(\alpha)}, x_{(\beta)})$ and $(p_{(\tilde{\alpha})}, x_{(\tilde{\beta})})$. Then the quantity

$$\operatorname{inerdex} \frac{\partial x_{(\alpha)}(r)}{\partial p_{(\alpha)}} - \operatorname{inerdex} \frac{\partial x_{(\tilde{\alpha})}(r)}{\partial p_{(\tilde{\alpha})}} \pmod 4$$

is constant on all non-singular points $r \in \Omega$.

Proof. Applying (6.5) to the expressions $(K(\Omega; p_{(\alpha)}, x_{(\beta)})\varphi)(x)$ and

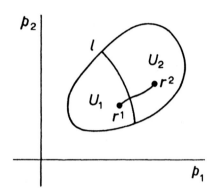

Fig. 8.

$(K(\Omega; p_{(\tilde{a})}, x_{(\tilde{\beta})})\varphi)(x)$ in a non-singular point $r \in \Omega$ and comparing them with (6.7), we obtain that

$$m = \text{inerdex } \frac{\partial x_{(\alpha)}}{\partial p_{(\alpha)}} - \text{inerdex } \frac{\partial x_{(\tilde{a})}}{\partial p_{(\tilde{a})}} + 4k,$$

where k is an integer. By virtue of Lemma 6.3, this quantity is independent of r.

We shall clarify the statement of Lemma 6.4. Let $n = 2$ (for simplicity) and let a singular chart Ω admit a diffeomorphic projection on plane p. Let the projection l of the cycle of singularities $\Sigma(\Lambda^2)$ split the projection U of chart Ω into two domains U_1 and U_2 (Figure 8). Take any two points $r^1, r^2 \in U$ such that $r^j \in U_j$, and consider the integer

$$m = \text{inerdex } \frac{\partial x(r^2)}{\partial p} - \text{inerdex } \frac{\partial x(r^1)}{\partial p}.$$

This number does not depend on the choice of the points r^1, r^2. In fact, the matrix $\partial x(p)/\partial p$ (here $x = x(p)$ is the equation of chart Ω) is non-degenerate for $p \notin l$, and connectedness of the domain U_1 implies that inerdex $\partial x(p)/\partial p \equiv$ const. for $p \in U_1$. Analogously, inerdex $\partial x(p)/\partial p \equiv$ $\equiv$ const. for $p \in U_2$.

Suppose now that the same chart Ω admits a diffeomorphic projection on the plane (p_1, x_2) and that the projection $\tilde{l}$ of the cycle of singularities splits the projection $\tilde{U}$ of chart Ω into two domains $\tilde{U}_1$ and $\tilde{U}_2$ in just the same way as before. Consider the integer

$$\tilde{m} = \text{inerdex } \frac{\partial x_2(r^2)}{\partial p_2} - \text{inerdex } \frac{\partial x_2(r^1)}{\partial p_2},$$

where $r^j \in \tilde{U}_j$. This number does not depend on the choice of the points r^1, r^2. In this case the statement of Lemma 6.4 reads: the number m coincides with $\tilde{m}$ modulo 4,

$$m = \tilde{m} \ (\text{mod } 4).$$

Thus the integer

$$\frac{\partial x_{(\alpha)}(r^2)}{\partial p_{(\alpha)}} - \frac{\partial x_{(\alpha)}(r^1)}{\partial p_{(\alpha)}}$$

is independent of the choice of focal coordinates in a singular chart.

We shall mention still one important consequence of Lemma 6.4. If the focal coordinates $(p_{(\alpha)}, x_{(\beta)})$ can be chosen in singular chart Ω, then the matrix $\partial x_{(\alpha)}/\partial p_{(\alpha)}$ is symmetric. This is due to the fact that manifold Λ^n is Lagrangian; it is, however, not true for an arbitrary manifold. Let, for example, chart Ω admit a diffeomorphic projection on p-space; then the equation of this chart has the form $x = -\partial S(p)/\partial p$. Consequently, $\partial x/\partial p = = -\partial^2 S(p)/\partial p^2$ and the symmetry of matrix S''_{pp} implies the symmetry of matrix $\partial x/\partial p$ in the considered case.

The definition of the Lagrangian manifold is itself by no means connected with dynamics; the same concerns the concept of the precanonical operator. However, in studying asymptotics of solutions of differential equations containing a large parameter, the Lagrangian manifolds appear which are invariant with respect to dynamics; i.e. the Lagrangian manifolds which split into the trajectories of the Hamilton system associated with the equation (see § 3). For such manifolds we can explain more clearly what is the precanonical operator.

Assume that an $(n-1)$-dimensional Lagrangian manifold Λ^{n-1} is given in phase space $\mathbf{R}^{2n}$ and suppose manifold Λ^n is the tube of trajectories of some Hamilton system originating in the initial manifold Λ^{n-1} at $t = 0$. Moreover let the assumptions of Proposition 4.19 be fulfilled; then Λ^n is a Lagrangian manifold. As function $\varphi \in C_0^\infty(\Lambda^n)$ we take a function which identically equals zero outside some smaller tube of trajectories T lying on Λ^n; for clarity we refer to the very sketchy Figure 9.

The points belonging to Λ^n have the form $x = x(t, \alpha)$, $p = p(t, \alpha)$, where

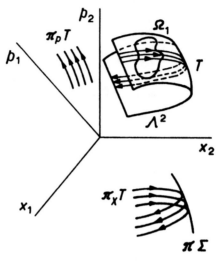

Fig. 9.

t is a parameter along the trajectory (time), and $\alpha = (\alpha_1, \ldots, \alpha_{n-1})$ are the local coordinates on the initial manifold Λ^{n-1} (it is given by equations of the form $x = x^0(\alpha), p = p^0(\alpha)$). We take as volume element $d\sigma$ a volume element invariant under the phase flow; for definiteness we put $d\sigma = dt \; d\alpha$. Furthermore, let manifold Λ^n (it is simultaneously the chart Ω) admit a diffeomorphic projection on p-space. We shall assume that the cycle of singularities $\Sigma(\Lambda^n)$ is a smooth curve and that its projection on $\mathbf{R}_x^n$ – the caustic of the ray family $x = x(t, \alpha)$ – is also a smooth curve (this assumption corresponds to the situation represented in Figure 9 for $n = 2$). We take a chart Ω_1 which admits a diffeomorphic projection on $\mathbf{R}_x^2$ and lies, for difiniteness, on the 'upper sheet' of manifold Λ^2; the distinguished point r^0 will also be taken on the upper sheet.

The precanonical operator K_1 corresponding to chart Ω_1 will be taken in the form (6.2):

$$(K_1 \varphi(r))(x) = \sqrt{\left|\frac{d\sigma}{dx}\right|} \exp(i\lambda S(r))\varphi(r), \qquad r = (x, p).$$

Here $S(r) = \int_{r_0}^r \langle p, dx \rangle$ and, as usual, r is the point of the Lagrangian manifold with the coordinates (x, p). With our choice of volume element dσ we obtain

$$(K_1 \varphi(r))(x) = \sqrt{\left| \frac{J(0, \alpha)}{J(t, \alpha)} \right|} \exp(i\lambda S(r))\varphi(r), \qquad r = (x, p).$$

Here

$$J(t, \alpha) = \det \frac{\partial x(t, \alpha)}{\partial(t, \alpha)}.$$

The phase function S was studied in sufficient detail in the preceding sections; now we shall concentrate on the amplitude. Moreover, we shall establish the connection between the precanonical operator and the formal asymptotic solutions of equation of the form (3.1). The amplitude $\sqrt{|d\sigma/dx|}$ is connected with the following geometrical objects: the tube of trajectories T in the phase space, the ray tube $\pi_x T$ in x-representation ($\pi_x T$ is the projection of tube T on $\mathbf{R}_x^n$) and the ray tube $\pi_p T$ in p-representation (it is the projection of tube T on p-space).

(1) Let $L(x, p)$ be the Hamiltonian of a given dynamical system, $L(x, p) \equiv$ $\equiv 0$ on Λ^n, and let, for simplicity, the corresponding λ-p.d. operator be symmetric (see (3.40)). Then for any function $\varphi \in C_0^\infty(\Omega_1)$ the function $(K\varphi)(x)$ is a f.a. solution of Equation (3.1) with accuracy to $O(\lambda^{-1})$ for $x \in \pi_x \Omega$. This follows from Theorem 3.12.

(2) Function $\varphi(r)$, due to its construction, is localized in the sense that supp φ is contained in the tube of trajectories T. This fact and the definition of the precanonical operator imply that function $(K\varphi)(x)$ is also localized and that supp $K\varphi$ is contained in the projection $\pi_x T$ of the tube of trajectories T.

(3) Let us pass into p-representation, i.e. let us consider the precanonical operator K_2 corresponding to the same chart Ω_1, but in p-representation (see (6.3)). Up to a numerical factor we have

$$(K_2 \varphi(r))(x) = F_{\lambda, \ p \to x}^{-1} \left[\varphi(r) \sqrt{\left| \frac{d\sigma}{dp} \right|} \exp(i\lambda \tilde{S}(r)) \right](x),$$

where

$$\tilde{S}(r) = \int_{r^0}^{r} \langle p', dx' \rangle - \langle x(p), p \rangle.$$

Due to the choice of $d\sigma$ we get

$$(K_2 \varphi(r))(x) = F^{-1}_{\lambda, \, p \to x}\left[\varphi(r)\sqrt{\left|\frac{\tilde{J}(0, \alpha)}{\tilde{J}(t, \alpha)}\right|}\exp(i\lambda\tilde{S}(r)) \right](x),$$

where

$$\tilde{J}(t, \alpha) = \det \frac{\partial p(t, \alpha)}{\partial(t, \alpha)}.$$

As shown in Lemma 6.2, $(K_2 \varphi)(x) = c(K_1 \varphi)(x) + O(\lambda^{-1})$, $c = \mathrm{const}$. Moreover, the following estimate is true outside the projection of $\mathrm{supp}\,\varphi$ on x-space: for any integer N (Lemma 1.5)

$$|(K_2\varphi)(x)| \leqq c_N \lambda^{-N}(1 + |x|)^{-N},$$

where c_N is a constant, i.e. function $(K_2\varphi)(x)$ decreases faster than any power of $|x|$ and any power of λ for $|x| \to \infty$ and for $\lambda \to \infty$, respectively.

We call the x-Fourier transform of a function its *p-representation*. By comparison of the above mentioned formulae we conclude: if a f.a. solution in x-representation is localized in a neighbourhood of the projection $\pi_x T$ of the tube of trajectories T, then, in p-representation, the same f.a. solution is localized in a neighbourhood of the p-projection $\pi_p T$ of tube T (Figure 9).

(4) The quantity $|d\sigma/dx|$ in the language of geometrical optics is nothing else than the intensity of the ray tube [47]. Carrying over this terminology into p-representation we find the intensity of the ray tube in p-representation to be $|d\sigma/dp|$, confirming again the symmetry between x- and p-representations.

Finally, the intensity is infinite in the points lying on the caustic of $\pi_x T$ (since $d\sigma/dx = 0$ in x-representation). The intensity is, however, finite in p-representation; thus, for seeking asymptotic solutions, it is natural to use p-representation.

§ 7. The Index of a Curve on a Lagrangian Manifold

The construction of a global f.a. solution, i.e. the problem of patching this solution from the local ones, leads to the notion of a path index on a Lagrangian manifold. In this paragraph, we introduce the definition of the index, summarize its basic properties and clarify its geometrical meaning.

1. *The Definition of the Index*

Let Λ^n be a Lagrangian manifold, $\{\Omega_i\}$ a canonical atlas on Λ^n. We introduce

ASSUMPTION 7.1.

$$\dim \Sigma(\Lambda^n) \leqq n - 1.$$

This assumption is satisfied by Lagrangian manifolds in 'general position' (Theorem 7.6).

Let canonical charts Ω_i and Ω_j have the focal coordinates $(p_{(\alpha)}, x_{(\beta)})$ and $(p_{(\tilde{\alpha})}, x_{(\tilde{\beta})})$, respectively, and $r \in \Omega_i \cap \Omega_j$ be a non-singular point.

DEFINITION 7.2. Then, the number

$$\gamma(\Omega_i \cap \Omega_j) = \text{inerdex} \, \frac{\partial x_{(\alpha)}(r)}{\partial p_{(\alpha)}} - \text{inerdex} \, \frac{\partial x_{(\tilde{\alpha})}(r)}{\partial p_{(\tilde{\alpha})}} \tag{7.1}$$

is called an *index of the pair of charts* Ω_i, Ω_j. If $\Omega_i \cap \Omega_j = \varnothing$, then $\gamma(\Omega_i \cap \Omega_j) = 0$

PROPOSITION 7.3. *Index γ is a one-dimensional integral cocycle* mod 4 *on* Λ^n.

Proof. By virtue of Lemma 6.4 the number $\gamma(\Omega_i \cap \Omega_j)$ depends neither on the choice of r, nor on the choice of the local coordinates in charts Ω_i, Ω_j modulo 4 so that γ defines a 1-cochain on Λ^n. Due to (7.1), γ is a 1-cocycle.

If a chain of charts $\Omega_{i_0}, \Omega_{i_1}, \Omega_{i_2}, \ldots, \Omega_{i_s}$ is given on Λ^n (with non-void intersections $\Omega_{i_j} \cap \Omega_{i_{j+1}}$), then the number

$$\gamma(\Omega_{i_0}, \Omega_{i_1}, \ldots, \Omega_{i_s}) =$$
$$= \gamma(\Omega_{i_0} \cap \Omega_{i_1}) + \gamma(\Omega_{i_1} \cap \Omega_{i_2}) + \ldots + \gamma(\Omega_{i_{s-1}} \cap \Omega_{i_s}) \tag{7.2}$$

is called an *index of the chain*.

Let $l[r', r'']$ be a smooth curve on $\Lambda^n, \partial l = r'' - r'$ and r', r'' non-singular points. We shall define the *path index* ind l. This notion was first introduced by one of the authors of the present book [59], [60].

DEFINITION 7.4. Let $l \subset \Omega$, where Ω is a canonical chart with the focal coordinates $(p_{(\alpha)}, x_{(\beta)})$. Then

$$\text{ind } l[r', r''] = \text{inerdex } \frac{\partial x_{(\alpha)}(r'')}{\partial p_{(\alpha)}} - \text{inerdex } \frac{\partial x_{(\alpha)}(r')}{\partial p_{(\alpha)}}. \qquad (7.3)$$

If Ω is a non-singular chart, we put ind $l = 0$.

For an arbitrary path, the *index* is defined via additivity:

$$\text{ind } l[r', r''] = \sum_{j=0}^{s} \text{ind } l[r^j, r^{j+1}]. \qquad (7.4)$$

Here $r^0 = r', r^{s+1} = r''$, and the points $r^0, r^1, \ldots$ divide l into the arcs $l[r^k, r^{k+1}]$ satisfying the assumptions of Definition 7.4.

Notice that the index of curve $l[r', r'']$ depends only on its structure in an arbitrarily small neighbourhood of the intersection of this curve with the cycle of singularities $\Sigma(\Lambda^n)$. In fact, if any arc $\bar{l}$ lying in a non-singular chart is taken out of curve l, then the index of the obtained curve will be equal to ind l; this follows from the fact that ind $\bar{l} = 0$ and additivity of the index.

PROPOSITION 7.5. *Index* ind(mod 4) *of a path on a Lagrangian manifold is an integer-valued homotopic invariant.*

We sketch now the proof. If path l lies in one canonical chart, then, by virtue of Lemma 6.4, ind l does not depend on the choice of the focal coordinates in Ω. If $l \subset \Omega_i \cap \Omega_j$, then the definition of ind l does not depend on whether we use chart Ω_i or chart Ω_j (Proposition 7.3). From here one can see that, by using a standard technique [27], ind l of an arbitrary path does not depend on the way we divide path l. Furthermore, if l lies in one chart U, then its index obviously does not change, if the path is slightly deformed; moreover, one of its ends (or even both) can be continuously moved inside U, provided the end-points remain non-singular. As a consequence, ind l is a homotopic invariant. For a rigorous proof see [2], [67].

2. Some Properties of the Index

THEOREM 7.6. [2]. *An arbitrarily small rotation can transform a Lagrangian manifold Λ^n into 'general position' with respect to the projection π_x on $\mathbf{R}_x^n$; the following statements are therefore true:*

(1) $\Sigma(\Lambda^n)$ *consists of an open $(n-1)$-dimensional manifold $\Sigma'(\Lambda^n)$ on which the rank of $d\pi_x$ equals $n-1$ and boundaries $\Sigma(\Lambda^n) \backslash \Sigma'(\Lambda^n)$ of dimensions $< n-2$.*

Thus $\Sigma(\Lambda^n)$ *determines an* $(n-1)$-*dimensional* (*non-oriented*) *cycle in* Λ^n.

(2) *The cycle* $\Sigma(\Lambda^n)$ *lies in* Λ^n '*two-sidedly*'.

The positive side of $\Sigma(\Lambda^n)$ *can be chosen as follows.*

(3) *In a neighbourhood of point* $M \in \Sigma(\Lambda^n)$, *manifold* Λ^n *is given by* n *equations of the form*

$$x_k = x_k(p_{\hat{k}}, x_{\hat{k}}), \qquad p_k = p_k(p_{\hat{k}}, x_{\hat{k}}),$$

where $\hat{k} = 1, 2, \ldots, k-1, k+1, \ldots, n$ *for some* k, $1 \le k \le n$. $\Sigma(\Lambda^n)$ *in a neighbourhood of such a point is given by the equation* $\partial x_k / \partial p_k = 0$.

(4) *In passing through* $\Sigma'(\Lambda^n)$ *the quantity* $\partial x_k / \partial p_k$ *changes sign. We take the side for which* $\partial x_k / \partial p_k > 0$ *as the positive side of* $\Sigma'(\Lambda^n)$.

(5) *This definition of the positive side of* $\Sigma(\Lambda^n)$ *is self-consistent, i.e. independent of the choice of the coordinate system* p_k.

From here on we assume that Λ^n is a Lagrangian manifold in a general position.

For the proof of Theorem 7.6 see [2].

Let the end-points of curve $l[r', r'']$ be non-singular, i.e. $r', r'' \notin \Sigma(\Lambda^n)$ and $l[r', r'']$ be transversal to the cycle of singularities.

DEFINITION 7.7 [2]. The intersection index of a curve $\gamma(M_1, M_2)$ with the cycle of singularities $\Sigma'(\Lambda^n)$ is called an *index of curve* $\gamma(M_1, M_2)$ *on a Lagrangian manifold*, $\mathrm{ind}\gamma(M_1, M_2)$; it is equal to the number v_+ of the transition points from the negative to the positive side minus the number v_- of transition points from the positive to the negative side:

$$\mathrm{ind}\,\gamma(M_1, M_2) = v_+ - v_-. \tag{7.5}$$

PROPOSITION 7.8. *Definitions 7.4 and 7.7 of the index are equivalent.*

Thus consider a curve $l[r', r'']$ such that:

(1) l lies inside a singular chart Ω;

(2) the points r', r'' lie on different sides of the cycle of singularities;

(3) l is transversal to $\Sigma'(\Lambda^n)$ and intersects $\Sigma'(\Lambda^n)$ in exactly one point.

Then, according to Definition 7.7, $\mathrm{ind}\,l = \pm 1$; for definiteness let $\mathrm{ind}\,l = +1$. We choose the focal coordinates in Ω like in Theorem 7.6, (3), $k = 1$. Then

$$\mathrm{sgn}\,\frac{\partial x_1(r'')}{\partial p_1} = +1, \qquad \mathrm{sgn}\,\frac{\partial x_1(r')}{\partial p_1} = -1,$$

$$\text{inerdex } \frac{\partial x_1 (r'')}{\partial p_1} - \text{inerdex } \frac{\partial x_1 (r')}{\partial p_1} = +1,$$

i.e. the values of index ind l evaluated with the aid of the Definitions 7.4 and 7.7 coincide.

The case ind $l = -1$ is treated analogously. Hence Definitions 7.4 and 7.7 are equivalent.

The well-known topological properties of the intersection index lead to

THEOREM 7.9 *The index* ind *of a path on a Lagrangian manifold in a general position is an integer-valued homotopic invariant.*

It was shown in [2] that the index of a closed curve on a Lagrangian manifold is the Grassmann characteristic class. In this paragraph, we restrict our considerations to the simplest definitions and the simplest formulae for calculation of the index; various algebraic and topological problems connected with the concept of index ind are discussed in the works [2], [9], [26], [67], [68], [70], [71], [79].

It is a simple task to calculate the index of a path lying on a Lagrangian manifold in a general position. For example, let Λ^1 be the circle $x^2 + p^2 = 1$ (Figure 10). We compute the indexes of the arcs $l_1 = l[r^1, r^2]$, $l_2 = l[r^2, r^3]$, $l_3 = l[r^3, r^4]$, $l_4 = l[r^4, r^1]$ and ind Λ^1 (the circle is oriented anti-clockwise) as follows. In the points r^1, r^2, r^3, r^4 we have sgn $(\partial x/\partial p) = = +1, -1, +1, -1$, respectively. Thus ind $l_1 = +1$. Furthermore, since path l_2 lies in a non-singular chart, ind $l_2 = 0$; it is completely inessential

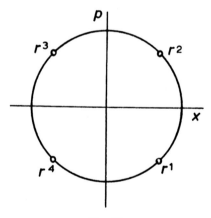

Fig. 10.

what are the signs of the function $\partial x/\partial p$ at the end-points of path l_2.
Similarly, $\mathrm{ind}\, l_3 = +1$; $\mathrm{ind}\, l_4 = 0$, $\mathrm{ind}\, \Lambda^1 = +2$.

The index of a path on a Lagrangian manifold is determined quite
independently of the dynamics; the index is a topological invariant and
has no connection with the trajectories of the Hamilton system. However,
the Lagrangian manifolds which arise in determining asymptotic solutions
of the differential equations are always connected with some Hamilton
system and split into the trajectories of the system. The Lagrangian
manifolds invariant under classical dynamics belong therefore to the
most interesting topics of mathematical physics; we shall consider
various examples of them.

3. Examples

EXAMPLE 7.10. Lagrangian manifold Λ^1 in a general position in
the plane $\mathbf{R}^2$ is a smooth curve, the tangents of which are parallel to the
p-axis only in a finite number of points.

EXAMPLE 7.11. Consider the eikonal equation

$$(\nabla S(x))^2 = 1 \qquad (7.6)$$

in the plane $x = (x_1, x_2)$, and the Cauchy problem

$$S|_\Gamma = 0, \qquad \frac{\partial S}{\partial n}\bigg|_\Gamma = 1, \qquad (7.7)$$

where Γ is the parabola $x_2^2 = 2ax_1$, $(a > 0)$, and $\partial/\partial n$ is the derivative in
the direction of the outward normal to Γ. The Hamilton system has the
form

$$\frac{dx}{dt} = 2p, \qquad \frac{dp}{dt} = 0, \qquad \frac{dS}{dt} = 2. \qquad (7.8)$$

By solving the Cauchy problem we get

$$p_1 = -\frac{a}{(a^2 + \alpha^2)^{1/2}}, \qquad p_2 = \frac{\alpha}{(a^2 + \alpha^2)^{1/2}}, \qquad (7.9)$$

$$x_1 = -\frac{2at}{(a^2 + \alpha^2)^{1/2}} + \frac{\alpha^2}{2a}, \qquad x_2 = \frac{2at}{(a^2 + \alpha^2)^{1/2}} + \alpha, \qquad S = 2t.$$

These equations give a parametric representation of the Lagrangian
manifold $\Lambda^2 (-\infty < t, \alpha < +\infty)$. This manifold admits a diffeomorphic

projection on the plane (x_1, p_2), since it is a C^∞-manifold and

$$\det \frac{\partial(x_1, p_2)}{\partial(\alpha, t)} = 2a^3(a^2 + \alpha^2)^{-2} \neq 0.$$

The cycle of singularities Σ can be found from the equation

$$\det \frac{\partial(x_1, x_2)}{\partial(\alpha, t)} = 0,$$

which leads to the following relation between α and t :

$$2a^2 t + (a^2 + \alpha^2)^{3/2} = 0.$$

From here we obtain the equation of Σ :

$$x_1 = a + \frac{3\alpha^2}{2a}, \qquad x_2 = \frac{\alpha^3}{a^2},$$

$$p_1 = -\frac{a}{(a^2 + \alpha^2)^{1/2}}, \qquad p_2 = \frac{\alpha}{(a^2 + \alpha^2)^{1/2}},$$

$$-\infty < \alpha < +\infty.$$

Since

$$\frac{dp_2}{d\alpha} = a^2(a^2 + \alpha^2)^{3/2} \neq 0,$$

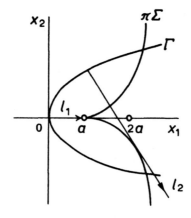

Fig. 11.

Σ is a curve of class C^∞ and manifold Λ^2 is in a general position. The projection $\pi\Sigma$ of cycle Σ on the plane (x_1, x_2) is a semi-cubic parabola given by

$$\pi\Sigma: \quad x_2^2 = \frac{8}{27a}(x_1 - a)^3, \tag{7.10}$$

and represents the evolute of the original parabola. The last fact follows from general considerations (see Example 3.17).

From (7.9) we obtain that Λ^2 is given by the equations

$$x_2 = \frac{p_2}{(1 - p_2^2)^{1/2}}\left(a - x_1 + \frac{ap_2^2}{2(1 - p_2^2)}\right), \tag{7.11}$$

$$p_1 = -(1 - p_2^2)^{1/2},$$

where $x_1 \in \mathbf{R}, -1 < p_2 < 1$. We find from here that

$$\frac{\partial x_2}{\partial p_2} = (1 - p_2^2)^{-5/2}\left(a - x_1 + x_1 p_2^2 + \frac{ap_2^2}{2}\right) \tag{7.12}$$

on Λ^2. Now we compute the index of two concrete curves on Λ^2 (Figure 11).

(1) The first curve $l[r^1, r^2]$ is given by the Equations (7.9) where $\alpha = 0, -2a \leq t \leq 0$, i.e. $p_1 = -1, p_2 = 0, x_1 = -2t, x_2 = 0$, and the point r^1 corresponds to $t = 0$. The projection l_1 on x-plane is the segment $[0, 2a]$. From (7.12) we find that

$$\left.\frac{\partial x_2}{\partial p_2}\right|_{r^1} = a > 0, \quad \left.\frac{\partial x_2}{\partial p_2}\right|_{r^2} = -a < 0.$$

As a consequence, $\mathrm{ind}\, l = -1$.

(2) The second curve $l[r^1, r^2]$ is given by the Equations (7.9), where $\alpha = \alpha_0 > 0$ and $-t_0 \leq t \leq 0$. The projection l_1 on x-plane is the segment πl with the initial point $x_1 = \alpha^2/2a, x_2 = \alpha$, lying on the upper half of the original parabola, and with the end-point

$$x_1 = \frac{\alpha^2}{2a} + \frac{2at_0}{(a^2 + \alpha^2)^{1/2}}, \quad x_2 = \frac{2\alpha t_0}{(a^2 + \alpha^2)^{1/2}} + \alpha.$$

We take $t_0 > 0$ sufficiently large; then segment πl will touch the curve $\pi\Sigma$ and the points πr^1 and πr^2 will lie on different sides of the contact point. We find from (7.12) that $[p_2 = \alpha/((a^2 + \alpha^2)^{1/2})]$

$$\frac{\partial x_2}{\partial p_2}\bigg|_{r^1} = a(1 - p_2^2)^{-5/2} > 0,$$

$$\frac{\partial x_2}{\partial p_2}\bigg|_{r^2} = (1 - p_2^2)^{-5/2}\left(a - \frac{2a^3 t_0}{a^2 + \alpha^2}\right) < 0$$

for $t_0 \gg 1$. Consequently, ind $l = -1$.

Thus the index of a ray touching the caustic $\pi\Sigma$ is equal to ± 1 (in accordance with the orientation of the ray).

Yet to be noticed that the singularity of the caustic 'beak' in this example corresponds to the point in which the wave front curvature is maximal. Namely, the top of the 'beak' is situated in the centre of the osculating circle touching the vertex of the parabola. If we take a smooth curve Γ instead of the parabola, then the 'beaks' of the caustic will correspond again to the points of maximal curvature of the curve.

From the point of view of geometrical optics, parabola Γ represents a wave front of the constructed ray family (since eikonal $S(x)$ is constant on Γ). The curve $\pi\Sigma$ is the caustic of this ray family and has a singularity ('beak') in the point $x_1 = a, x_2 = 0$ (Figure 11).

4. *The Lagrangian Singularities (Caustics)*

The structure of the cycle of singularities $\Sigma(\Lambda^n)$ of various Lagrangian manifolds in a general position was investigated in the work [3] where, in particular, the complete classification of singularities was obtained for $n < 6$. We shall formulate these results for Lagrangian manifolds of dimensions $n = 1, 2, 3$.

Let, as usual, $(\alpha) \cup (\beta) = \{1, \dots, n\}$ be a decomposition of the set $(1, \dots, n)$ into two disjoint subsets. Denote the generating function of Lagrangian manifold Λ^n by the symbol $S(x_{(\beta)}, p_{(\alpha)})$; the equation of Λ^n can be written as

$$p_{(\beta)} = \frac{\partial S}{\partial x_{(\beta)}}, \qquad x_{(\alpha)} = -\frac{\partial S}{\partial p_{(\alpha)}}. \tag{7.13}$$

THEOREM 7.12. [3]. *Any Lagrangian manifold Λ^n can be transformed by a slight distorsion (in the class of Lagrangian maps) into a manifold which, in a neighbourhood of each of its points, can be reduced by a*

Lagrangian equivalence to one of the following normal forms:
for $n = 1$:

$$A_1 : S = p_1^2, \qquad A_2 : S = \pm p_1^3\,;$$

for $n = 2$, *in addition to* A_1, A_2:

$$A_3 : S = \pm p_1^4 + x_2 p_1^2\,;$$

for $n = 3$, *in addition to* A_1, A_2, A_3:

$$A_4 : S = \pm p_1^5 + x_3 p_1^3 + x_2 p_1^2,$$
$$D_4 : S = \pm p_1^2 p_2 \pm p_2^3 + x_3 p_2^2.$$

(Here it is assumed that the origin of coordinates is translated to the considered point.)

For $n = 1$, case A_1, the Lagrangian manifold in a neighbourhood of the point $x_1 = 0$, $p_1 = 0$ is given by the equation $x_1 = -2p_1$ and admits a diffeomorphic projection on x_1-axis. Analogously, the point of type A_1 does not belong to the cycle of singularities for all n.

Let $n = 1$, the origin of coordinates be a point of type A_2, and Ω be its small neighbourhood. Then the equation of the manifold has the form

$$x_1 = \mp 3p_1^2$$

(Figure 3). If the equation is written as $p_1 = \partial S(x_1)/\partial x_1$, then $S(x_1) \sim$
$\sim \mathrm{const.}\,(\mp x_1)^{3/2}$ for $x_1 \to 0$.

For $n = 2$ — a point of type A_3, the equation of the Lagrangian manifold has the form (in the positive sign case)

$$x_1 = -4p_1^3 - 2p_1 x_2, \qquad p_2 = p_1^2$$

in a neighbourhood of the origin. The projection $\pi\Sigma(\Lambda^2)$ of the cycle of singularities on x-plane is given by

$$27x_1^2 = 8x_2^3, \tag{7.14}$$

i.e. is of the form of a 'beak'-like curve. Locally, for small x, the curve looks like the projection of the cycle of singularities in Example 7.11. The curve $\pi\Sigma(\Lambda^2)$ in the negative sign case has exactly the same form.

5. *Connection with the Morse Index*

Consider the Hamilton system

$$\frac{dx}{dt} = \frac{\partial H(x, p)}{\partial p}, \qquad \frac{dp}{dt} = -\frac{\partial H(x, p)}{\partial x} \tag{7.15}$$

with a real Hamiltonian $H \in C^\infty(\mathbf{R}^{2n})$, and let $\{x(t, y), p(t, y)\}$ be the solution of the Lagrangian Cauchy problem

$$x|_{t=0} = x^0(y), \qquad p|_{t=0} = p^0(y), \qquad y \in U. \tag{7.16}$$

Here U is a domain in $\mathbf{R}_y^n$ and $\Lambda^n = \{(x, p): x = x^0(y), p = p^0(y), y \in U\}$ is a Lagrangian C^∞-manifold of dimension n. We shall also assume that the stronger Legendre condition holds:

$$\frac{\partial^2 H(x, p)}{\partial p^2} > 0, \qquad (x, p) \in \mathbf{R}^{2n}. \tag{7.17}$$

Let $x(y; 0, t)$ be the trajectory $x = x(y, \tau), 0 \leq \tau \leq t$,

$$J(\tau, y) = \det \frac{\partial x(\tau, y)}{\partial y}. \tag{7.18}$$

The points on this trajectory in which the Jacobian vanishes are called *focal points*; by definition, the multiplicity of a focal point is equal to the multiplicity of the zero of the Jacobian. A focal point is said to be *simple*, if the Jacobian has simple zero in that point. It was proved by Morse [69] that the multiplicity of the zero of Jacobian $J(t, y^0)$ in a focal point is equal to the co-rank of the matrix $\partial x(t^0, y^0)/\partial y$.

Morse [69] introduced the notion of an index μ of a trajectory $l: x = x(y; 0, t)$ with non-focal end-points. The Morse index μ is equal to the number of the focal points (including their multiplicity) on trajectory l. There exists an extensive literature connected with the Morse index ([66], [80] and others).

Now we show that the Morse index μ of trajectory l is equal to the index ind of a trajectory lying on a certain $(n + 1)$-dimensional Lagrangian manifold in the $(2n + 2)$-dimensional phase space, the projection on $\mathbf{R}_x^n$ of which coincides with the trajectory.

Consider the $(2n + 2)$-dimensional phase space $\mathbf{R}^{2n+2} = \{(x_0, x); (p_0, p)\}$ and the set

$$M^{n+1} = \{(x_0, p_0, x, p): x_0 = \tau, p_0 = -H(x, p), x = x(\tau, y),$$
$$p = p(\tau, y); 0 \leq \tau \leq t, y \in \Omega\}.$$

If the assumptions of Proposition 4.19 are satisfied (i.e., in particular, $H(x^0(y), p^0(y)) = M \equiv \text{const.}$), then M^{n+1} is a Lagrangian C^∞-manifold of dimension $n + 1$ in $\mathbf{R}^{2n+2}$. Consider the curve γ_t on M^{n+1}:

$$\gamma_t = \{(x_0, p_0, x, p): x_0 = \tau, p_0 = -H(x, p), x = x(\tau, y^0),$$
$$p = p(\tau, y^0), 0 \leq \tau \leq t\}.$$

THEOREM 7.13. *Let the assumptions formulated above be satisfied. Then*

$$\operatorname{ind} \gamma_t = \mu(r(y^0, 0, t)). \tag{7.19}$$

Proof. If $J(t^0, y^0) \neq 0$, then the point on M^{n+1} in which $x_0 = t^0, x = x(t^0, y^0)$, is non-singular with respect to the projection on $\mathbf{R}^{n+1}_{x_0, x}$. There is at most a finite number of focal points on the trajectory $x(y; 0, t)$ [66]. Due to the additivity of the indexes $\operatorname{ind} \mu$, it is sufficient to consider only the case when the trajectory $r(y^0, 0, t)$ contains one focal point $x = x(t^0, y^0)$. We shall restrict ourselves to the case with a simple focal point; the general case is treated in [59], Chap. 8, § 2. We shall show that $\operatorname{ind} \gamma_t = +1$, thus proving the theorem.

Since the rank of matrix $\partial x(t^0, y^0)/\partial y$ is equal to $n-1$, it is possible to assume without losing generality, that

$$\det \frac{\partial x'(t^0, y^0)}{\partial y'} \neq 0,$$

where

$$x' = (x_2, \ldots, x_n), \qquad y' = (y_2, \ldots, y_n).$$

The local coordinates on M^{n+1} are (t, y). In a neighbourhood of the point $Q^* \in M^{n+1}$ corresponding to (t^0, y^0), the variables (x', t) can be taken as a part of the local coordinates. Since M^{n+1} is Lagrangian, then, according to Lemma 4.5, it is possible to take (t, p_1, x') as the local coordinates in the neighbourhood of point Q^*. Thus the equation of the cycle of singularities $\Sigma(M^{n+1})$ in the neighbourhood of point Q^* has the form $\partial x_1/\partial p_1 = 0$, where $x_1 = x_1(x^0, p_1, x')$. Point Q^* belongs to the cycle of singularities, because its equation can be written in the form $\det(\partial(x^0, x)/\partial(t, y)) = 0$, i.e. $J(t, y) = 0$.

Instead of y, we introduce the parameters $\alpha = (\alpha_1, \ldots, \alpha_n)$, $\{y = y_0\} \leftrightarrow \{\alpha = 0\}$, such that $\det(\partial\alpha/\partial y)|_{\alpha=0} \neq 0$. Then $J(t^0, 0)$ will have a simple zero. Namely, we put

$$p_1(t^0, y) = \alpha_1, \qquad x'(t^0, y) = \alpha' = (\alpha_2, \alpha_3, \ldots, \alpha_n)$$

for small $|y - y^0|$. Then

$$x_1(t^0, y) = \varphi_1(t^0, \alpha), \quad p'(t^0, y) = \varphi'(t^0, \alpha), \quad \varphi' = (\varphi_2, \ldots, \varphi_n).$$

We shall denote by $\{x(t, \alpha), p(t, \alpha)\}$ the solution of the Hamilton system with the Cauchy data

$$p_1|_{t=t^0} = \alpha, \quad x'|_{t=t^0} = \alpha', \quad x_1|_{t=t^0} = \varphi_1(t^0, \alpha),$$
$$p'|_{t=t^0} = \varphi'(t^0, \alpha), \quad \alpha \in V,$$

where V is a small neighbourhood of the point $\alpha = 0$. It suffices to consider the arcs of the curves γ_t and $x(t, \alpha)$ for $|t - t^0| \leq \delta$, where $\delta > 0$ is arbitrarily small; denote the arc of curve γ_t by $\gamma_{t,\delta}$. From (7.15) we find

$$\frac{\partial}{\partial t}\left(\frac{\partial x_1(t, \alpha)}{\partial \alpha_1}\right) = \sum_{j=1}^{n} \frac{\partial^2 H(x, p)}{\partial p_1 \partial p_j} \frac{\partial p_j}{\partial \alpha_1} + \sum_{j=1}^{n} \frac{\partial^2 H(x, p)}{\partial p_1 \partial x_j} \frac{\partial x_j}{\partial \alpha_1}, \quad (7.20)$$

where $x = x(t, \alpha)$, $p = p(t, \alpha)$. The equation of the cycle of singularities in the neighbourhood of point Q^* has the form $\det (\partial(x_0, x)/\partial(t, \alpha)) = 0$, i.e. $\det (\partial x/\partial \alpha) = 0$. Since $\partial x_j(t^0, 0)/\partial \alpha_k = \delta_{jk}$ $(j = 2, ..., n; k = 1, ..., n)$, and $Q^* \in \Sigma(M^{n+1})$, we have $\partial x_1(t^0, 0)/\partial \alpha_1 = 0$. We perform a linear canonical transformation in the phase space:

$$p_0^* = p_0, \quad p_1^* = p_1, \qquad\qquad p_j^* = p_j - a_j p_1, \, j = 2, ..., n,$$

$$x_0^* = x_0, \quad x_1^* = x_1 + \sum_{j=0}^{n} a_j x_j, \quad x_j^* = x_j, \qquad j = 2, ..., n,$$

where $a_j = \partial p_j(t^0, 0)/\partial \alpha_1$. Then

$$H(x, p) \to H^*(x^*, p^*), \quad M^{n+1} \to M^{*n+1},$$

and manifold M^{*n+1} is Lagrangian. Condition (7.17) is valid for H^*, and the Hamilton system takes the form

$$\frac{dx^*}{dt} = \frac{\partial H^*(x^*, p^*)}{\partial p^*}, \quad \frac{dp^*}{dt} = -\frac{\partial H^*(x^*, p^*)}{\partial x^*}.$$

Due to the choice of a_j we have

$$\frac{\partial x_1^*}{\partial \alpha_1} = 0, \quad \frac{\partial x_j^*}{\partial \alpha_k} = \delta_{jk}, \quad j = 2, ..., n; k = 1, 2, ..., n, \quad (7.22)$$

$$\frac{\partial p_1^*}{\partial \alpha_1} = 1, \quad \frac{\partial p_j^*}{\partial \alpha_1} = 0, \quad j = 2, ..., n$$

for $t = t^0$, $\alpha = 0$. Therefore from (7.20) we find

$$\frac{\partial}{\partial t}\left(\frac{\partial x_1^*}{\partial \alpha_1}\right)\bigg|_{t=t^0, \alpha=0} = \frac{\partial^2 H^*(x, p)}{\partial p_1^2}, \quad x = x(t^0, 0), p = p(t^0, 0),$$

and, by virtue of condition (7.17), we have

$$A = \frac{\partial}{\partial t}\left(\frac{\partial x_1^*}{\partial \alpha_1}\right)\bigg|_{t=t^0, \alpha=0} > 0. \quad (7.23)$$

The equation of the cycle of singularities $\Sigma(M^{*n+1})$ has, because of (7.21), the form $\partial x_1^*/\partial p_1^* = 0$. Furthermore, we have

$$\frac{\partial x_1^*(t, 0)}{\partial \alpha_1} = (t - t^0)A + O((t - t^0)^2) \tag{7.24}$$

along $\gamma_{t,\delta}^*$. The local coordinates on M^{*n+1} in the neighbourhood of the point Q^{**} corresponding to $Q^* \in M^{n+1}$ are (p^*, x'^*), so that $x_1^* = \varphi_1^*(p_1^*, x'^*)$. We have

$$\frac{\partial \varphi_1^*(p^*, x'^*)}{\partial \alpha_1} = \frac{\partial x_1(t, \alpha)}{\partial \alpha_1} = \frac{\partial \varphi_1^*}{\partial p_1^*} \frac{\partial p_1^*}{\partial \alpha_1} + \sum_{j=2}^{n} \frac{\partial \varphi_1^*}{\partial x_j^*} \frac{\partial x_j^*}{\partial \alpha_1}.$$

Owing to (7.22), we get

$$\frac{\partial p_1^*(t, 0)}{\partial \alpha_1} = 1 + O(t - t^0), \qquad \frac{\partial x_j^*(t, 0)}{\partial \alpha_1} = O(t - t^0),$$

and from (7.24) we find

$$\frac{\partial \varphi_1^*}{\partial p_1^*} = A(t - t^0)(1 + O(t - t^0))$$

along γ_t^* for small $|t - t^0|$. Consequently, $\partial x_1^*/\partial p_1^* > 0$ for $t > t^0$ on $\gamma_{t,\delta}^*$, and $\partial x_1^*/\partial p_1^* < 0$ for $t < t^0$ on $\gamma_{t,\delta}^*$, so that ind $\gamma_{t,\delta}^* = +1$, q.e.d.

Let Λ^n be the Lagrangian manifold $x = x^0(y), p = p^0(y), y \in U$ (see (7.16)), and $g^t : \mathbf{R}^{2n} \to \mathbf{R}^{2n}$ be the phase flow induced by the Hamilton system (7.15).

COROLLARY 7.14. *For any curve* $l \subset \Lambda^n$

$$\text{ind } g^t l - \text{ind } l = \mu(g^\tau r^+) - \mu(g^\tau r^-), \qquad 0 \leq \tau \leq t, \tag{7.25}$$

where r^- *and* r^+ *are the initial and the final point of curve* l, *respectively, and* $g^\tau r^-$ *and* $g^\tau r^+$ *their corresponding trajectories.*

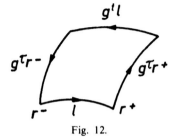

Fig. 12.

Proof. The tetragon $l, g^{\tau}r^{+}, g^{l}l, g^{\tau}r^{-}$ (Figure 12) on M^{n+1} is homologous to zero, so its index is equal to zero and (7.25) follows from Theorem 7.13.

EXAMPLE 7.15. Consider the Hamilton system corresponding to the quantum mechanical oscillator (Example 3.6); let $m = 1$, $\omega = 1$. The initial data are taken in the form

$$x|_{t=0} = x^0 = (\tilde{y}, 0), \qquad p|_{t=0} = p^0, \qquad \tilde{y} \in \mathbf{R}^{n-1},$$

where $\tilde{y} = (y_1, \ldots, y_{n-1}), p^0$ is a constant vector, and $p_n^0 \neq 0$. Then the phase trajectories are given by the equations (see (3.77))

$$x_k = p_k^0 \sin t - y_k \cos t, \qquad p_k = p_k^0 \cos t + y_k \sin t,$$
$$k = 1, \ldots, n-1,$$
$$x_n = p_n^0 \sin t, \qquad p_n = p_n^0 \cos t.$$

The Jacobian $J(t, \tilde{y}) = \det [\partial x(t, \tilde{y})/\partial(t, \tilde{y})]$ is in this case equal to

$$J(t, \tilde{y}) = (-1)^{n-1} p_n^0 (\cos t)^n.$$

The focal points of the ray $l(\tilde{y}): x = x(t, \tilde{y})$ are the points in which the Jacobian J vanishes, i.e. $t_k = (\pi/2) + k\pi, k = 0, \pm 1, \pm 2, \ldots$. Notice that all rays collect at one point $x = (-1)^k p^0$ for $t = t_k$ independently of the initial data x^0; in other words, a complete ray focusation takes place. The zero of the Jacobian J at $t = t_k$ has multiplicity n. Now if $l(\tilde{y})$ is the ray $x = x(t, \tilde{y}), 0 \leq t \leq \pi$, then its Morse index is equal to n, $\mu(l(\tilde{y})) = n$. In our example the Hamiltonian is $H = p^2 + x^2$, hence the stronger Legendre condition (7.17) is satisfied, and, according to Theorem 7.13 we get ind $l(\tilde{y}) = n$.

§ 8. The Canonical Operator
(Global Quantization of the Velocity Field)

To a Lagrangian manifold Λ^n – an object of classical mechanics, there corresponds a quantum object – the canonical operator K_{Λ^n}:

$$\Lambda^n \to K_{\Lambda^n}.$$

This operator is made of the precanonical operators.

If symbol $L(x, p) \equiv 0$ on Λ^n, and if this manifold is invariant with respect to the dynamical system $dx/dt = \partial L/\partial p, dp/dt = -\partial L/\partial x$, then the commutation formula

$$L(\overset{2}{x}, \lambda^{-1}\overset{1}{D}_x) K_{\Lambda^n}\varphi(x) =$$
$$= \frac{1}{i\lambda} K_{\Lambda^n}\left(\frac{d}{dt} - \frac{1}{2}\sum_{j=1}^{n}\frac{\partial^2 L(x,p)}{\partial x_j\,\partial p_j}\right)\varphi + O(\lambda^{-2})$$

holds, where d/dt is the derivative along trajectory. Especially, if $\hat{L}$ is a symmetric operator, we have

$$\hat{L}K_{\Lambda^n}\varphi = \frac{1}{i\lambda}K_{\Lambda^n}\frac{d\varphi}{dt} + O(\lambda^{-2}).$$

Thus operator $\hat{L}$ is transformed (with the help of the canonical operator) into an operator of differentiation along the trajectory.

1. Definition of the Canonical Operator

The canonical operator

$$K_{\Lambda^n}^{r^0} : C_0^\infty(\Lambda^n) \to C^\infty(\mathbf{R}_x \times \mathbf{R}_\lambda^+) \tag{8.1}$$

is defined by specifying the following objects:

(1) A Lagrangian manifold Λ^n with an n-dimensional volume element $d\sigma^n$ and with a distinguished point r^0.

We assume that $\dim \Sigma(\Lambda^n) \le n - 1$, which is always true for a Lagrangian manifold in a general position.

(2) A canonical atlas $\{\Omega_j\}$ and a partition of unity: $1 = \sum e_j(r)$. Here $e_j \in C_0^\infty(\Lambda^n), 0 \le e_j \le 1$, $\operatorname{supp} e_j \subset \Omega_k, k = k(j)$, and, as usual, $r = (x, p)$ denotes a point of manifold Λ^n.

Let Ω_i be a chart with the local coordinates $\{p_{(\alpha)}, x_{(\beta)}\}$, and φ a function from $C_0^\infty(\Omega_i)$. We introduce an operator equal to the precanonical one up to a numerical factor (cf. (6.3)):

$$(K_{\Lambda^n}^{r^0}(\Omega_i)\varphi)(x) = \exp\left\{-\frac{i\pi}{2}\gamma(C(r))\right\} \times$$

$$\times F_{\lambda, p_{(\alpha)} \to x_{(\alpha)}}^{-1}\left|\frac{d\sigma^n(r(p_{(\alpha)}, x_{(\beta)}))}{dp_{(\alpha)}\,dx_{(\beta)}}\right|^{1/2}\varphi(r(p_{(\alpha)}, x_{(\beta)})) \times$$

$$\times \exp\left\{i\lambda\left(\int_{r^0}^{r(p_{(\alpha)}, x_{(\beta)})}\langle p', dx'\rangle - \langle x_{(\alpha)}(p_{(\alpha)}, x_{(\beta)}), p_{(\alpha)}\rangle\right)\right\}. \tag{8.2}$$

Here $C(r)$ is a chain of charts $\{\Omega_{j_k}\}, 0 \leq k \leq s$, such that $\Omega_{j_0} \ni r^0, j_s = i$ and $\gamma(C(r))$ is its index.

The canonical operator (8.1) can be introduced by the formula

$$(K_{\Lambda^n}^{r^0} \varphi)(x) = \sum_j (K_{\Lambda^n}^{r^0}(\Omega_j)(e_j \varphi))(x). \tag{8.3}$$

Thus operator K_{Λ^n} depends on the choice of the canonical atlas, on the choice of the local coordinates in charts and on the partition of unity.

It is evident that formula (8.3) determines a multi-valued function of x, since both the integral $\int_{r^0}^r \langle p, dx \rangle$, and the index $\gamma(C(r))$ may depend on our choice of a path joining the points r^0 and r. In order to define a single-valued function by formula (8.3), it is necessary that the integral $\int_{r^0}^r \langle p, dx \rangle$ and the index $\gamma(C(r))$ are independent of the chosen path connecting the points r^0, r.

2. The Invariance of the Canonical Operator

THEOREM 8.1. *The canonical operator does not depend (with accuracy to $O(\lambda^{-1})$):*

(a) *on the choice of the canonical atlas and of the local coordinates in the charts;*

(b) *on the partition of unity,*

if and only if

(1) $\int_l \langle p, dx \rangle = 0$ *along any closed path l;*

(2) ind $l = 0$ *for any closed path l.*

Thus the characteristic class introduced in § 7 must be trivial.

Proof. The necessity of these conditions is obvious. We shall prove their sufficiency.

We fix chart Ω_j and focal coordinates $\{p_{(\alpha)}, x_{(\beta)}\}$ therein. First we show that operator (8.2) depends only on chart Ω_j. By hypothesis, $\int_{r^0}^r \langle p, dx \rangle$ does not depend on the integration path, and the quantity $\gamma(C(r))$ for $r \in \Omega_j$ does not depend on our choice of the chain of charts. We set $\gamma(C(r)) = \gamma, r \in \Omega_j$. According to Lemma 6.4, operator (8.2) does not depend on the choice of the focal coordinates in chart Ω_j. As a consequence, formula (8.3) defines the map (8.1). The operator thus constructed depends, generally speaking, on the choice of the canonical atlas and on the partition of unity; we indicate it by writing $K(\{\Omega_j\}, \{e_j\})$.

Now we show that K is independent of the partition of unity for the fixed atlas $\{\Omega_j\}$. Let $\{f_k\}$ be another partition of unity. For each k there exists $j(k)$ such that $\operatorname{supp} f_k \subset \Omega_{j(k)}$. If $\operatorname{supp} f_k$ is contained in several charts $\Omega_{j_1(k)}, \Omega_{j_2(k)}, \ldots$, then $j(k)$ can be chosen to be any number from

$j_1(k), j_2(k), \ldots$. We put $K_0 = K(\{\Omega_j\}, \{e_j\})$ (i.e. operator $K\{\Omega_j\}, \{e_j\})$ is determined by the canonical atlas $\{\Omega_j\}$ and by the partition of unity $\{e_j\}$); for shortness, the arguments of functions will be omitted. We have

$$K_0 \varphi = \sum_j K(\Omega_j)(e_j \varphi),$$

$$K_1 \varphi = \sum_k K(\Omega_{j(k)})(f_k \varphi),$$

where $K(\Omega_j)$ are precanonical operators (see (8.2)). Since $f_k = \sum_j e_j f_k$, we obtain

$$K_1 \varphi = \sum_j \left(\sum_k K(\Omega_{j(k)})(e_j f_k \varphi) \right). \tag{8.4}$$

We shall show that

$$\sum_k K(\Omega_{j(k)})(e_j f_k \varphi) = K(\Omega_j)(e_j \varphi). \tag{8.4'}$$

As $e_j = \sum_k e_j f_k$, we have

$$K(\Omega_j)(e_j \varphi) = \sum_k (K(\Omega_j)(e_j f_k \varphi). \tag{8.4''}$$

If $\psi \in C_0^\infty(\Lambda^n)$ and $\operatorname{supp} \psi \subset \Omega_1 \cap \Omega_2$, then, according to Lemma 6.4, $K(\Omega_1)\psi = K(\Omega_2)\psi$ (let us recall that all equalities are understood mod $O_{-1}^+(\mathbf{R}_x^n)$). Consequently, $K(\Omega_j)(e_j f_k \varphi) = K(\Omega_{j(k)})(e_j f_k \varphi)$, and (8.4'') implies (8.4'). Inserting (8.4') in (8.4) we obtain $K_0 = K_1$.

Finally we prove that K is independent of the choice of a canonical atlas. Let two canonical atlases $\{\Omega_j\}$, $\{\Theta_k\}$ be given and let the corresponding partitions of unity be $\{e_j\}$, $\{f_k\}$. We set $K_0 = K(\{\Omega_j\}, \{e_j\})$, $K_1 = (\{\Theta_k\}, \{f_k\})$. We introduce the atlas of the charts $U_{jk} = \Omega_j \cap \Theta_k$, the corresponding partition of unity g_{jk} and the operator $K_2 = K(\{U_{jk}\}, \{g_{jk}\})$. According to the facts proved above, $K_0 = K(\{\Omega_j\}, \{g_{jk}\})$, $K_1 = K(\{\Theta_k\}, \{f_k\})$. If set U_{jk} is non-void and $\operatorname{supp} \psi \subset U_{jk}$, then, according to Lemma 6.3,

$$K(\Omega_j)\psi = K(U_{jk})\psi.$$

It follows from here that $K_0 = K_2$. Analogously $K_1 = K_2$.

3. The Commutation Formulae

The canonical operator allows to construct a global f.a. solution for equations of the form (3.1). The central rôle in the canonical operator method is played by the commutation formula for a λ-pseudodifferential operator L and a canonical operator K, that roughly looks like

$$LK = K\tilde{L}.$$

The structure of operator $\tilde{L}$ after commutation is essentially simpler than the structure of operator L. In what follows, it is assumed that Lagrangian manifold Λ^n satisfies the conditions (1), (2) of Theorem 8.1.

THEOREM 8.2. *The following commutation formula holds:*

$$L(\overset{2}{x}, \lambda^{-1}\overset{1}{D}_x; (i\lambda)^{-1})K_{\Lambda^n}\varphi = K_{\Lambda^n}(L(x,p;0)\varphi) + \lambda^{-1}\psi. \tag{8.5}$$

Here L is a λ-p.d. operator with the symbol of class $T^m_+(\mathbf{R}^n_x)$, function $\varphi \in C^\infty_0(\Lambda^n)$ and function $\psi \in O^+_0(\Lambda^n)$.

Proof. Let $\Omega \subset \Lambda^n$ be a canonical chart, $K_{\Lambda^n}(\Omega)$ the precanonical operator and $\varphi \in C^\infty_0(\Omega)$.

Let Ω be a non-singular chart. Then $K_{\Lambda^n}(\Omega)$ has the form (6.1), i.e. $K_{\Lambda^n}\varphi = \exp(i\lambda S(x))|d\sigma^n/dx|^{1/2}\varphi$. According to Theorem 2.6 we have

$$\hat{L}K_{\Lambda^n}(\Omega)\varphi =$$

$$= \exp(i\lambda S(x))\left|\frac{d\sigma^n}{dx}\right|^{1/2} L\left(x, \frac{\partial S}{\partial x}; 0\right)\varphi + \lambda^{-1}\psi =$$

$$= K_{\Lambda^n}(\Omega)(L(x,p;0)\varphi) + \lambda^{-1}\psi, \tag{8.6}$$

since $p = \partial S(x)/\partial x$ on Λ^n and the remainder $\psi \in O^+_0(\mathbf{R}^n)$.

Let Ω be a singular chart with the focal coordinates $(p_{(\alpha)}, x_{(\beta)})$. Then $K_{\Lambda^n}(\Omega)$ has the form (6.3), and

$$\hat{L}K_{\Lambda^n}(\Omega)\varphi =$$

$$= F^{-1}_{\lambda, p_{(\alpha)} \to x_{(\alpha)}} L(-\lambda^{-1}\overset{2}{D}_{p_{(\alpha)}}, \overset{2}{x}_{(\beta)}, \overset{1}{p}_{(\alpha)}, \overset{1}{D}_{x_{(\beta)}}; (i\lambda)^{-1}) \times$$

$$\times \left|\frac{d\sigma^n(r)}{dp_{(\alpha)}\,dx_{(\beta)}}\right|^{1/2} \exp(i\lambda S(p_{(\alpha)}, x_{(\beta)}))\varphi(r),$$

where $r = r(p_{(\alpha)}, x_{(\beta)})$. Function S has the form

$$S(p_{(\alpha)}, x_{(\beta)}) = \int\limits_{r^0}^r \langle p, dx \rangle - \langle p_{(\alpha)}, x_{(\alpha)}(p_{(\alpha)}, x_{(\beta)}) \rangle,$$

so that

$$\frac{\partial S}{\partial p_{(\alpha)}} = -x_{(\alpha)}(p_{(\alpha)}, x_{(\beta)}), \qquad \frac{\partial S}{\partial x_{(\alpha)}} = p_{(\alpha)}.$$

By applying Theorem 5.2 we obtain

$$\hat{L}K_{\Lambda^n}(\Omega)\varphi =$$
$$= F^{-1}_{\lambda,\,p_{(\alpha)}\to x_{(\alpha)}}\left(\left|\frac{\mathrm{d}\sigma^n(r)}{\mathrm{d}p_{(\alpha)}\,\mathrm{d}x_{(\beta)}}\right|^{1/2}\exp\left(i\lambda S(p_{(\alpha)},x_{(\beta)})\right)\varphi(r)\times\right.$$
$$\left.\times L(x_{(\alpha)}(p_{(\alpha)},x_{(\beta)}),x_{(\beta)},p_{(\alpha)},p_{(\beta)}(p_{(\alpha)},x_{(\beta)});0)+\lambda^{-1}\psi\right),$$

where the remainder $\psi\in O_0^+(\Lambda^n)$. Consequently, formula (8.6) holds.

Let $\varphi\in C_0^\infty(\Lambda^n)$. Then (8.5) follows from (8.6) and (8.3). The commutation formula in Theorem 8.2 was established under the assumption that the differential operators act first, and the operators of multiplication by the independent variables act second. Exactly the same formula is valid even if the mentioned operators act in the reverse order.

COROLLARY 8.3.   *Under the assumptions of Theorem* 8.2 *the commutation formula*

$$L(\overset{1}{x},\lambda^{-1}\overset{2}{D}_x;(i\lambda)^{-1})(K_{\Lambda^n}\varphi)(x)=$$
$$= K_{\Lambda^n}(L(x,p;0)\varphi(x))+\lambda^{-1}\psi \tag{8.7}$$

holds, where $\psi\in O_0^+(\Lambda^n)$.
    Thus

$$\hat{L}K\varphi = KR\varphi + O(\lambda^{-1}),$$

where R is the operator of multiplication by function $L(x,p;0)$.

Now we derive the commutation formula for the case when the Lagrangian manifold iş invariant with respect to a displacement along trajectories of the Hamilton system of operator L.

THEOREM 8.4.   *Let operator L and Lagrangian manifold* Λ^n *fulfil the assumptions of Theorem* 8.2 *and moreover the conditions*:
    (1) *Function* $L(x,p;0)$ *is real-valued, and the equation*

$$L(x,p;0)=0 \tag{8.8}$$

determines a $(2n-1)$-*dimensional* C^∞-*manifold* $M^{2n-1}(L)$ *in the phase space*.
    (2) $\Lambda^n\subset M^{2n-1}(L)$.
    (3) *Manifold* Λ^n *and volume element* $\mathrm{d}\sigma^n$ *are invariant with respect to the dynamical system*

$$\frac{\mathrm{d}x}{\mathrm{d}t}=\frac{\partial L(x,p;0)}{\partial p},\qquad \frac{\mathrm{d}p}{\mathrm{d}t}=-\frac{\partial L(x,p;0)}{\partial x}. \tag{8.9}$$

(4) *There exists a solution of system (8.9) for all $t \in \mathbf{R}$, it is unique and infinitely differentiable for arbitrary initial data* $(x, p) \in M^{2n-1}(L)$.
Then the commutation formula

$$L(\overset{2}{x}, \lambda^{-1}\overset{1}{D}_x; (i\lambda)^{-1})(K_{\Lambda^n}\,\varphi(r))(x) =$$

$$= \frac{1}{i\lambda} K_{\Lambda^n}\left(\frac{\mathrm{d}}{\mathrm{d}t} - \tfrac{1}{2}\,\mathrm{Sp}\left(\frac{\partial^2 L(x, p; 0)}{\partial x\,\partial p}\right) + \right.$$

$$\left. + \frac{\partial L(x, p; \varepsilon)}{\partial \varepsilon}\bigg|_{\varepsilon = 0}\right)\varphi + \lambda^{-2}\chi(x) \qquad (8.10)$$

is true, where $\chi \in O_0^+(\Lambda^n)$.
Here $\mathrm{d}/\mathrm{d}t$ is the derivative with respect to the Hamilton system (8.9), i.e.

$$\frac{\mathrm{d}\varphi(r)}{\mathrm{d}t} = -\left\langle\frac{\partial L(x, p; 0)}{\partial x_{(\alpha)}}, \frac{\partial \varphi(r)}{\partial p_{(\alpha)}}\right\rangle + \left\langle\frac{\partial L(x, p; 0)}{\partial p_{(\beta)}}, \frac{\partial \varphi(r)}{\partial x_{(\beta)}}\right\rangle, \quad (8.11)$$

provided a neighbourhood of the point $r \in \Lambda^n$ admits a diffeomorphic projection on the plane $(p_{(\alpha)}, x_{(\beta)})$.
Proof. (1) Let r^0 be a non-singular point. We shall construct an $(n - 1)$-dimensional Lagrangian C^∞-manifold Λ^{n-1} such that $r^0 \in \Lambda^{n-1}$ and Λ^{n-1} is transversal to the trajectories of the system (8.9). We shall assume that $\Lambda^{n-1} = \{(x, p): x = x^0(y), p = p^0(y), y \in U\}$, where U is a domain in $\mathbf{R}_y^{n-1}$. Let $x = x^0(y, t), p = p^0(y, t)$ be the solution of the Cauchy problem $x|_{t=0} = x^0(y), p|_{t=0} = p^0(y)$ for the system (8.9). We take $\delta > 0$ and U so small that the set

$$\Omega = \{(x, p); x = x(t, y), p = p(t, y), y \in U, |t| < \delta\}$$

is a non-singular canonical chart on Λ^n; let $\pi_x \Omega$ be the projection of Ω on $\mathbf{R}_x^n$. Then

$$\mathrm{d}\sigma^n(r(x)) = a(x)\,\mathrm{d}x = a(x(t, y))J(t, y)\,\mathrm{d}t\,\mathrm{d}y,$$

where we denoted

$$J(t, y) = \det\frac{\partial x(t, y)}{\partial(t, y)}.$$

Since the volume element $\mathrm{d}\sigma^n$ is invariant under displacements along bicharacteristics, the function $a(x(t, y))\,J(t, y)$ depends on y only. Thus the volume element has the form $d\sigma'' = b(y) \, dt \, dy, b(y) > 0$. Consequently,

$$\left|\frac{d\sigma''(x)}{dx}\right|^{1/2} = \frac{(b(y))^{1/2}}{(J(t,y))^{1/2}}.$$ (8.12)

Let $K_{\Lambda^n}(\Omega)$ be the precanonical operator which corresponds to the chart Ω (see § 6) and $\varphi \in C_0^\infty(\Omega)$. Then $K\varphi = |d\sigma''/dx|^{1/2} e^{i\lambda S} \varphi$. By applying Theorem 2.6 and by taking into account that $L(x, p; 0) \equiv 0$ on Λ^n, we obtain

$$L(\overset{2}{x}, \lambda^{-1} \overset{1}{D}_x; (i\lambda)^{-1})(K_{\Lambda^n}\varphi(r))(x) =$$

$$= \frac{1}{i\lambda} \exp(i\lambda S(x))(R_1 \psi)(x) + \lambda^{-2}\chi(x, \lambda),$$ (8.13)

where $\chi \in O_0^+(\pi\Omega)$. Here the notation

$$S(x) = \int_{r^0}^{r} \langle p, dx \rangle, \qquad \psi(x) = \left|\frac{d\sigma''}{dx}\right|^{1/2} \varphi(x)$$

was used. It follows from (3.39) that

$$(R_1 \psi)(x) =$$

$$= \frac{1}{\sqrt{J}} \frac{d}{dt}(\sqrt{J}\,\psi(x)) - \tfrac{1}{2} \mathrm{Sp}\left(\frac{\partial^2 L(x, p; 0)}{\partial x \, \partial p}\right)\psi(x) +$$

$$+ \frac{\partial L(x, p; \varepsilon)}{\partial \varepsilon} \psi(x), \qquad x = x(t, y).$$ (8.14)

Since $\psi(x) = \sqrt{b(y)}(\varphi(x)/\sqrt{J})$ by virtue of 8.12, we finally get

$$(R_1 \psi)(x) = \frac{1}{\sqrt{J}}\left(\frac{d\varphi}{dt} - \tfrac{1}{2}\mathrm{Sp}\left(\frac{\partial^2 L(x, p; 0)}{\partial x \, \partial p}\right)\varphi + \frac{\partial L(x, p; \varepsilon)}{\partial \varepsilon}\bigg|_{\varepsilon=0}\varphi\right),$$

and the theorem is proved for operator $K_{\Lambda^n}(\Omega)$ corresponding to the non-singular chart Ω.

(2) Let $r^0 \in \Lambda^n$ be a singular point, some neighbourhood of which admits a diffeomorphic projection on the coordinate Lagrangian plane $(p_{(\alpha)}, x_{(\beta)})$. As in (1), we construct a canonical chart $\Omega \ni r^0$ in which the focal coordinates $(p_{(\alpha)}, x_{(\beta)})$ can be introduced; let $K_{\Lambda^n}(\Omega)$ be the precanonical operator corresponding to this chart (see § 6). We set

$$J(t, y) = \det \frac{\partial(p_{(\alpha)}(t, y), x_{(\beta)}(t, y))}{\partial(t, y)}.$$

It follows from the invariance of volume element $d\sigma''$ with respect to displacements along the bicharacteristics that

$$\left| \frac{d\sigma''(r)}{dp_{(\alpha)} \, dx_{(\beta)}} \right|^{1/2} = \frac{\sqrt{b(y)}}{\sqrt{J(t,y)}},$$

where $b(y) > 0$. We deduce from (8.13) and (5.10) that

$$L(\overset{2}{x}, \lambda^{-1} \overset{1}{D}_x ; (i\lambda)^{-1})(K_{\Lambda^n} \varphi(r))(x) =$$
$$= F^{-1}_{\lambda, \, p_{(\alpha)} \to x_{(\alpha)}} L(-\lambda^{-1} \overset{2}{D}_{p_{(\alpha)}}, \overset{2}{x}_{(\beta)}, \overset{1}{p}_{(\beta)}, \lambda^{-1} \overset{1}{D}_{x_{(\alpha)}} ; (i\lambda)^{-1}) \times$$
$$\times (\psi \exp(i\lambda S)),$$

where the notation

$$\psi = \left| \frac{d\sigma''}{dp_{(\alpha)} \, dx_{(\beta)}} \right|^{1/2} \varphi,$$

$$S(p_{(\alpha)}, x_{(\beta)}) = \int_{r^0}^{r} \langle p, dx \rangle - \langle p_{(\alpha)}, x_{(\alpha)}(p_{(\alpha)}, x_{(\beta)}) \rangle \qquad (8.15)$$

(see (6.2)) was used. By applying Theorem 5.2 we obtain

$$L(K_{\Lambda^n} \varphi)(x) =$$
$$= F^{-1}_{\lambda, \, p_{(\alpha)} \to x_{(\alpha)}} \left(\exp(i\lambda S) \left(R_0 \psi + \frac{1}{i\lambda} R_1 \psi \right) + \lambda^{-2} \chi \right), \qquad (8.16)$$

where $\chi \in O_0^+(\Omega)$. From (5.11) we have

$$R_0 \psi = L\left(-\frac{\partial S}{\partial p_{(\alpha)}}, x_{(\beta)}, p_{(\alpha)}, \frac{\partial S}{\partial x_{(\beta)}} ; 0 \right) \psi.$$

It follows from the form of function S that

$$\frac{\partial S(p_{(\alpha)}, x_{(\beta)})}{\partial p_{(\alpha)}} = -x_{(\alpha)}(p_{(\alpha)}, x_{(\beta)}),$$

$$\frac{\partial S(p_{(\alpha)}, x_{(\beta)})}{\partial x_{(\beta)}} = p_{(\beta)}(p_{(\alpha)}, x_{(\beta)}),$$

so that on Λ^n

$$R_0 \psi = L(x, p; 0) \psi = 0$$

due to assumption (2). By virtue of (5.12) and (5.13), the operator R_1 has the form

$$
R_1 \psi = \left\langle \frac{\partial L_1^0(x, p'; 0)}{\partial p'}, \frac{\partial \psi(x')}{\partial x'} \right\rangle +
$$

$$
+ \left[\frac{1}{2} \operatorname{Sp} \left(\frac{\partial^2 L_1^0(x', p'; 0)}{\partial p'^2} \frac{\partial^2 S(x')}{\partial x'^2} \right) - \right.
$$

$$
\left. - \operatorname{Sp} \left(\frac{\partial^2 L_1^0(x', p'; 0)}{\partial x' \, \partial p'} \right) + \frac{\partial L_1^0(x', p'; \varepsilon)}{\partial \varepsilon} \right|_{\varepsilon = 0} \right] \psi. \tag{8.17}
$$

Here

$$
x' = (p_{(\alpha)}, x_{(\beta)}), \qquad p' = (x_{(\alpha)}, p_{(\beta)}),
$$
$$
L_1^0(x', p'; 0) = L(- x_{(\alpha)}, x_{(\beta)}, p; 0),
$$

and the index 0 indicates that the values of $\partial L_1/\partial p'$ etc. are taken at the point

$$
(x'^0, p'^0) = (p_{(\alpha)}, x_{(\beta)}, - \partial S/\partial p_{(\alpha)}, \partial S/\partial x_{(\beta)})
$$

(i.e. at the point

$$
(x, p) = (x_{(\alpha)}(p_{(\alpha)}, x_{(\beta)}), x_{(\beta)}, p_{(\alpha)}, x_{(\beta)}(p_{(\alpha)}, x_{(\beta)})) \text{ on } \Lambda^n).
$$

It follows from the Hamilton system (8.9) that

$$
\frac{dp_{(\alpha)}}{dt} = \frac{\partial L_1^0(x', p'; 0)}{\partial x_{(\alpha)}}, \qquad \frac{dx_{(\beta)}}{dt} = \frac{\partial L_1^0(x', p'; 0)}{\partial p_{(\beta)}}. \tag{8.18}
$$

Therefore

$$
\left\langle \frac{\partial L_1^0(x', p'; 0)}{\partial p'}, \frac{\partial \psi}{\partial x'} \right\rangle = - \left\langle \frac{\partial L_1^0(x', p'; 0)}{\partial x_{(\alpha)}}, \frac{\partial \psi}{\partial p_{(\alpha)}} \right\rangle +
$$

$$
+ \left\langle \frac{\partial L_1^0(x', p'; 0)}{\partial p_{(\beta)}}, \frac{\partial \psi}{\partial x_{(\beta)}} \right\rangle = \frac{d\psi}{dt},
$$

where d/dt is the derivative with respect to the system (8.9). By applying the Liouville formula (3.28) to system (8.18) we obtain that the expression in the square brackets in (8.17) equals

$$
\frac{d}{dt} (\ln J) - \frac{1}{2} \operatorname{Sp} \left(\frac{\partial^2 L(x, p; 0)}{\partial x \, \partial p} \right) + \frac{\partial L(x, p; \varepsilon)}{\partial \varepsilon} \bigg|_{\varepsilon = 0}.
$$

Finally, we get

$$(R_1 \psi)(p_{(\alpha)}, x_{(\beta)}) = \frac{1}{\sqrt{J}} \frac{d}{dt} (\sqrt{J}\, \psi) -$$

$$- \tfrac{1}{2} \operatorname{Sp} \left(\frac{\partial^2 L(x, p; 0)}{\partial x\, \partial p} \right) \psi + \left. \frac{\partial L(x, p; \varepsilon)}{\partial \varepsilon} \right|_{\varepsilon = 0} \psi.$$

By taking into account formulae (8.14) and (8.16) we obtain (8.10).

(3) Both the representation (8.5) and the validity of (8.10) for precanonical operators yield the theorem, if the following fact is taken into account. Let $r^* \in \Lambda^n$ and $\Omega_{j_1}, \ldots, \Omega_{j_s}$ be canonical charts such that $\Omega_{j_l} \ni r^*$. Then

$$\sum_{l=1}^{s} e_{j_l}(r) \equiv 1, \qquad \sum_{l=1}^{s} \frac{d}{dt}(e_{j_l}(r)) \equiv 0$$

in a neighbourhood of point r^*, so

$$\sum_{l=1}^{s} \left[K_{\Lambda^n}(\Omega_{j_l}) \left(\frac{de_{j_l}(r)}{dt} \varphi(r) \right) \right](x) =$$

$$= \left[K_{\Lambda^n}(\Omega_{j_l}) \left(\sum_{l=1}^{s} \frac{de_{j_l}(r)}{dt} \varphi(r) \right) \right](x) + O(\lambda^{-1}) = O(\lambda^{-1})$$

(explicit indication of the local coordinates in Ω_{j_l} is omitted).

COROLLARY 8.5.   *Under the assumptions of Theorem* 8.4 *the formula*

$$\tfrac{1}{2} \big[L(\overset{2}{x}, \lambda^{-1}\overset{1}{D}_x ; (i\lambda)^{-1}) + L(\overset{1}{x}, \lambda^{-1}\overset{2}{D}_x ; (i\lambda)^{-1}) \big] (K_{\Lambda^n} \varphi)(x) =$$

$$= \frac{1}{i\lambda} \left(K_{\Lambda^n} \left(\frac{d}{dt} + \left. \frac{\partial L(x, p; \varepsilon)}{\partial \varepsilon} \right|_{\varepsilon = 0} \right) \varphi(r) + \lambda^{-1} \chi \right)(x)$$

is true.

We shall consider an important special case of Theorem 8.4. Let

$$L = \lambda^{-1} D_t + H(t, \overset{2}{x}, \lambda^{-1}\overset{1}{D}_x ; (i\lambda)^{-1}) \tag{8.19}$$

where operator H satisfies the assumptions H3.1–H3.3 of § 3, Section 5. We put $x' = (x^0, x)$, $p' = (E, p)$, where $x^0 = t$, E (energy) is the variable conjugate to t; we consider the Hamilton system corresponding to L in phase space $\mathbf{R}^{2n+2}$ with the coordinates (x', p'):

$$\frac{dx}{dt} = \frac{\partial H(t, x, p; 0)}{\partial p}, \qquad \frac{dp}{dt} = -\frac{\partial H(t, x, p; 0)}{\partial x},$$

$$\frac{dx^0}{dt} = 1, \qquad \frac{dE}{dt} = -\frac{\partial H(t, x, p; 0)}{\partial t}. \tag{8.20}$$

Let $g^t: \mathbf{R}^{2n+2} \to \mathbf{R}^{2n+2}$ be the displacement operator along trajectories of system (8.20). We construct an $(n + 1)$-dimensional Lagrangian manifold Λ^{n+1} in $\mathbf{R}^{2n+2}$ which is invariant with respect to the dynamical system (8.20).

Let Λ_0^n be an n-dimensional Lagrangian C^∞-manifold in $\mathbf{R}^{2n+2}$ with an n-dimensional volume element $d\sigma_0(x, p)$.

We set

$$\Lambda^n = \{(x', p'): (x, p) \in \Lambda^n, \quad t = 0, \quad E = -H(0, x, p; 0)\}. \tag{8.21}$$

Then Λ^n is an n-dimensional Lagrangian C^∞-manifold in $\mathbf{R}^{2n+2}$. Let $r^0(x, p)$ be a point in Λ^n, where (x', p') have the form (8.21). Furthermore, by virtue of Proposition 4.19, the set

$$\Lambda^{n+1} = \bigcup_{0 < t < +\infty} g^t \Lambda^n \tag{8.22}$$

is a Lagrangian C^∞-manifold in $\mathbf{R}^{2n+2}$ of dimension $n + 1$. We take

$$d\sigma(g^t(r(x, p))) = dt \, d\sigma_0(r(x, p))$$

as the $(n + 1)$-dimensional volume element on Λ^{n+1}.

Thus we constructed the $(n + 1)$-dimensional Lagrangian manifold Λ^{n+1} invariant with respect to the dynamical system (8.20) and the $(n + 1)$-dimensional volume $d\sigma$ invariant under this dynamical system as well. If Λ_0^n is simply connected, then Λ^{n+1} is simply connected, too, and $\oint_\gamma \langle p', dx' \rangle = 0$ for any closed path γ lying on manifold Λ^{n+1}. According to Theorem 8.4 we have

$$(\lambda^{-1} D_t + H(t, x, p; (i\lambda)^{-1})) K_{\Lambda^n} \varphi =$$

$$= \frac{1}{i\lambda} K_{\Lambda^n} \left(\frac{d}{dt} - \mathrm{Sp} \left(\frac{\partial^2 H(t, x, p; 0)}{\partial x \, \partial p} \right) + \frac{\partial H(t, x, p; \varepsilon)}{\partial \varepsilon} \bigg|_{\varepsilon = 0} \right) \varphi. \tag{8.23}$$

The value of the function on the right-hand side of this equality is taken at the point $g'(r(x, p))$, and d/dt is the derivative with respect to system (8.20), i.e.

$$\frac{d\varphi}{dt} = \left\langle \frac{\partial \varphi}{\partial x}, \frac{\partial H}{\partial p} \right\rangle - \left\langle \frac{\partial \varphi}{\partial p}, \frac{\partial H}{\partial x} \right\rangle + \frac{\partial \varphi}{\partial t} - \frac{\partial \varphi}{\partial E} \frac{\partial E}{\partial t}. \tag{8.24}$$

In deriving formula (8.23), two facts were used:

$$\frac{\partial^2 L(x, p, t; 0)}{\partial p^0 \, \partial t} \equiv 0 \quad \text{and} \quad p^0 + H(t, x, p; 0) \equiv 0 \quad \text{on} \quad \Lambda^n.$$

4. The Canonical Operator on a Riemannian Manifold

This operator is constructed by the same method as before. Thus we restrict ourselves only to a brief description of the construction and to an explanation of the results. Let M be an n-dimensional Riemannian C^∞-manifold with the metric tensor $g = (g_{ij})$. Let TM and T^*M be its tangent and cotangent bundles, respectively, and $\pi: T^*M \to M$ be a projection. If $q = (x^1, \ldots, x^n)$ denotes a local coordinate system in a domain (open connected subset) $U \subset M$, then

$$(x, p) = (x^1, \ldots, x^n, p_1, \ldots, p_n),$$

where $p_i = \partial/\partial x^i$ form the system of local coordinates in the domain $V = \pi^{-1}(U)$ on T^*M. Here, $T^{**}M$ and TM are identified in the standard way.

There is a 2-form ω^2 invariantly defined on manifold T^*M, which in the local coordinates has the form $\omega^2 = \sum_{i=1}^{n} dp_i \wedge dx^i$. Two vectors $\xi, \eta \in T_\alpha T^*M$ $(\alpha \in T^*M)$ are said to be *skew-orthogonal*, if $\omega^2(\xi, \eta) = 0$. A submanifold $\Lambda \subset T^*M$ is called *Lagrangian*, if any two vectors tangent to Λ are skew-orthogonal. Below we shall consider only n-dimensional Lagrangian manifolds $\Lambda^n \subset T^*M$.

In a neighbourhood of any point $r^0 \in \Lambda^n$ it is possible to take one of the sets $(p_{(\alpha)}, x_{(\beta)})$ as the local coordinates. The notion of a canonical atlas remains the same as before.

Let $d\sigma^n(r)$ be a smooth n-dimensional volume element on Λ^n. A *pre-canonical operator* in a canonical chart $\Omega \subset \Lambda^n$ with the focal coordinates $(p_{(\alpha)}, x_{(\beta)})$ is defined by the formula

$$(K\varphi)(x) = F^{-1}_{\lambda, p_{(\alpha)} \to x_{(\alpha)}} \times$$
$$\times \left[\varphi(r) \left| \frac{d\sigma^n(r)}{dp_{(\alpha)} \, dx_{(\beta)}} \right|^{1/2} (\det \tilde{g}(r))^{-1/4} \exp(i\lambda \tilde{S}(r)) \right]. \tag{8.25}$$

Here $\varphi \in C_0^\infty(\Omega)$, $r = r(p_{(\alpha)}, x_{(\beta)})$, function $\tilde{S}$ is precisely the same as the function in the exponent of (6.3), and $\tilde{g}(r) = (g \circ \pi)(r)$. Formulae (8.25) and (6.3) are evidently identical, if $M = \mathbf{R}_x^n$.

The *canonical operator* on Λ^n is constructed in terms of the precanonical operators exactly in the same way as before; the proof of its invariance runs as previously. The commutation formula (8.10) takes, in this case, the form

$$L(\overset{2}{x}, \lambda^{-1}\overset{1}{D}_x ; (i\lambda)^{-1})(K_{\Lambda^n}\varphi(r))(x) =$$

$$= \frac{1}{i\lambda}K_{\Lambda^n}\left(\frac{d}{dt} - \frac{1}{2}\sum_{j=1}^n \frac{\partial^2 L(x,p;0)}{\partial x^j \, \partial p_j} - \frac{1}{4}\frac{d}{dt}(\ln \det \tilde{g}(r)) + \right.$$

$$\left. + \frac{\partial L(x,p;\varepsilon)}{\partial \varepsilon}\Bigg|_{\varepsilon=0}\right) \varphi(r)(x) + \lambda^{-2}\chi(x,\lambda). \qquad (8.26)$$

5. On the Structure of the Canonical Operator

We write down more explicit formulae for the values $(K_{\Lambda^n}\varphi)(x)$, $\varphi \in C_0^\infty(\Lambda_0^n)$.

(1) The point $x^0 \in \mathbf{R}^n$ does not lie on a caustic (recall that the caustic is the projection $\pi_x \Sigma(\Lambda^n)$ of the cycle of singularities on $\mathbf{R}_x^n$). Then there arise two possibilities:

(A)

$$x^0 \notin \pi\Lambda^n.$$

Then

$$(K_{\Lambda^n}\varphi)(x) = O(\lambda^{-1}), \qquad (\lambda \to +\infty)$$

uniformly with respect to x from some neighbourhood Ω of the point x^0. It will be shown in §9 that, by reexpressing the canonical operator, the relation

$$(K_{\Lambda^n}\varphi)(x) = O(\lambda^{-N}), \qquad (\lambda \to +\infty) \qquad (8.27)$$

can be derived for any integer $N > 0$.

From the point of view of geometrical optics, the point x^0 in the case (1), (A) lies in the shadow region.

(B)

$$x^0 \in \pi\Lambda^n.$$

We assume that only a finite number of points $r^1, \ldots, r^s \in \Lambda^n$ is projected into the point x^0, $\pi r^j = x^0$ (Figure 13). Then one may choose a neighbour-

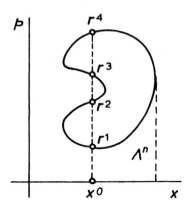

Fig. 13.

hood Ω of point x^0 in such a manner that neighbourhoods Ω_j of points r^j admit diffeomorphic projections on Ω: $\pi\Omega_j = \Omega$, $1 \leq j \leq s$, so that Ω_j are non-singular canonical charts. The operators $K(\Omega_j)$ can therefore be taken in the form (6.2), i.e. they are operators of multiplication by a function. Consequently, we have

$$(K_{\Lambda^n} \varphi)(x) = \sum_{j=1}^{s} \exp\left(-\frac{i\pi}{2} \operatorname{ind} l[r^0, r^j] \right) \left| \frac{d\sigma^n(r^j(x))}{dx} \right|^{1/2} \times$$

$$\times \exp\left(i\lambda \int_{r^0}^{r^j(x)} \langle p, dx \rangle \right) \varphi(r^j)(x) + O(\lambda^{-1}) \quad (8.28)$$

for $x \in \Omega$.

Thus the canonical operator in the case (1), (B) is equal to the sum of rapidly oscillating exponentials.

(2) The point x^0 lies on a caustic (i.e. $x^0 \in \pi\Sigma(\Lambda^n)$).

Suppose, for simplicity, that the preimage $\pi^{-1}(U)$, where U is a small neighbourhood of point x^0, is contained in one canonical chart Ω. Furthermore, we assume that the origin of coordinates is translated into the point $r(x^0)$, and that $r^0 \in \Omega$. According to formula (6.3), $K\varphi$ is a λ-Fourier transform of a rapidly oscillating exponential so that it is hard to give more precise information on the structure of the canonical operator in a general case. However, for some singularities in a general position this is possible.

(A). *The singularity of type* A_2 (see §7). In this case we can take

$(p_1, x_2, \ldots, x_n)$ as the focal coordinates. The manifold Λ^n in a neighbourhood of the origin of coordinates is given by the equations

$$x_1 = p_1^2, \qquad p' = \frac{\partial S_0(x')}{\partial x'},$$

where $x' = (x_2, \ldots, x_n)$, $p' = (p_2, \ldots, p_n)$. We set $S_0(0) = 0$. Further

$$\int_{r^0}^{r(x)} \langle p, dx \rangle - \langle x_1(p_1), p_1 \rangle = S_0(x') - \tfrac{2}{3} p_1^3.$$

Consequently,

$$(K\varphi)(x) = \exp\left[i\lambda S_0(x') \right] F^{-1}_{\lambda, p_1 \to x_1} \times$$
$$\times \left[(\varphi \circ r)(p_1, x') \left| \frac{d\sigma''(p_1, x')}{dp_1 \, dx'} \right|^{1/2} \exp\left(-\tfrac{2}{3} i p_1^3 \right) \right]. \qquad (8.29)$$

The last integral is equal to

$$\left(-\frac{\lambda}{2\pi i} \right)^{1/2} \int_{-\infty}^{\infty} \varphi \left| \frac{d\sigma''}{dp_1 \, dx'} \right|^{1/2} \exp\left[i\lambda(p_1 x_1 - \tfrac{2}{3} p_1^3) \right] dp_1. \qquad (8.30)$$

The phase function $p_1 x_1 - \tfrac{2}{3} p_1^3$ in the integral is exactly the same as in the Airy–Fock integral

$$v(x) = \frac{1}{2\sqrt{\pi}} \int_{-\infty}^{\infty} \exp\left[i\left(\frac{t^3}{3} + xt \right) \right] dt.$$

The asymptotics of integral (8.30) for $\lambda \to +\infty$ is uniform with respect to x_1, and, for small x_1, is expressible in terms of the Airy–Fock functions [17].

Namely, in a neighbourhood of the origin the asymptotic formula

$$(K\varphi)(x) = e^{i\pi/4} 2^{-5/6} \lambda^{1/6}(f_+ + f_-)v(-2^{-1/3}\lambda^{2/3} x_1) +$$
$$+ e^{-i\pi/4} 2^{-7/6} \lambda^{-1/6} \frac{f_+ - f_-}{\sqrt{x_1}} v'(-2^{-1/3}\lambda^{2/3} x_1) +$$
$$+ O(\lambda^{-1}) \qquad (8.31)$$

is valid. Here $f_\pm$ are the values of the function $(\varphi |d\sigma''/dp_1 \, dx'|^{1/2})(r)$ in the points $r^\pm = (\pm\sqrt{x_1}, x_2, \ldots, x_n)$. In particular, for $x_1 = 0$, the leading term of the asymptotics is equal to const. $\varphi(0)\lambda^{1/6}$, i.e. the function $(K\varphi)(x)$ is increasing near the caustic for $\lambda \to +\infty$.

The singularity of type A_2 considered above is the simplest and belongs to the singularities in a general position (see §7). As an example we consider the eikonal equation $(\nabla S(x))^2 = 1$ corresponding to the Helmholtz equation $(\Delta + k^2)u(x) = 0$ for $n = 2$.

Let $S(x)$ be a solution of the eikonal equation; the line $S(x) = \text{const.}$ is the wave front of some ray family. The singularities of the caustic correspond to the points in which the wave front curvature attains its local maximum. If x^0 is one of these points, and moreover, it is a non-degenerate maximum (i.e., x^0 is not an inflexion point of the wave front), then there is a 'beak' of the caustic corresponding to it (Figure 3). It lies in the centre of an osculating circle touching the wave front at the point x^0. For the wave front of a parabolic form this fact was established in § 7, Example 7.11; the proof in a general case can be done similarly.

Suppose that the focal coordinates $(p_1, x_2, \ldots, x_n)$ in a neighbourhood of the point $r^0 = (x^0, p^0) \in \Omega$ can be chosen in such a way that a precanonical operator can be defined as the λ-Fourier transform in one variable p_1 only. We shall show that even in this case, the function $K\varphi$ is not expressible in terms of the Airy–Fock or any other special function. Let $n = 1$ for simplicity, and the Lagrangian manifold Λ^1 be given by the equation

$$\Lambda^1: \quad x = \frac{p^k}{k}, \qquad p \in \mathbf{R}, \tag{8.32}$$

where $k \geq 3$ is an integer. We notice that such a manifold is associated with the ordinary differential equation

$$\frac{1}{k}\left(\frac{1}{i\lambda}\frac{d}{dx}\right)^k y = xy. \tag{8.33}$$

Actually, in this case the Hamilton–Jacobi equation has the form

$$x = p^k, \qquad p = \frac{dS(x)}{dx}.$$

The Lagrangian manifold–curve Λ^1– admits a diffeomorphic projection on $\mathbf{R}_p^n$. We have

$$\frac{d\sigma}{dp} = \frac{ds}{dp} = \sqrt{1 + p^{2k-2}}$$

(ds is the arc length of curve Λ^1). Furthermore

$$S(r) = \int_{r^0}^{r} \langle p, dx \rangle - \langle x(p), p \rangle = -\frac{k}{k+1} p^{k+1}.$$

(Here $r = (x, p)$, $r^0 = (0, 0)$).

The canonical operator $K = K_{\Lambda^1}$ can be taken in the form (6.3), i.e.

$$(K\varphi(r))(x) = F^{-1}_{\lambda, p \to x}\left[\left|\frac{d\sigma}{dp}\right|^{1/2} \exp(i\lambda S(r))\varphi(r(p))\right] =$$

$$= \left(-\frac{\lambda}{2\pi i}\right)^{1/2} \int (1 + p^{k-2})^{1/4} (\varphi \circ r)(p) \times$$

$$\times \exp\left[i\lambda\left(-\frac{k}{k+1}p^{k+1} + xp\right)\right] dp. \tag{8.34}$$

The asymptotics of the integral of form (8.34) for $\lambda \to +\infty$ was evaluated in [15]. We shall reduce our description to a qualitative picture.

The stationary points of the phase function are found from the equation

$$kp^k = x, \tag{8.35}$$

Let $x \neq 0$ be fixed. For k odd, Equation (8.35) has one real solution, and the asymptotics of integral (8.34) is given by the contribution of the stationary point $p = (x/k)^{1/k}$. For k even, there are two real stationary points $p = \pm(x/k)^{1/k}$ for $x > 0$, and the asymptotics of integral (8.34) is given by the sum of contributions of these points. For $x < 0$, all stationary points are non-real, hence $(K\varphi)(x) = O(\lambda^{-\infty})(\lambda \to +\infty)$.

(B). *The singularity of type A_3* (see § 7). We can choose $(p_1, x_2, \ldots, x_n)$ as the focal coordinates; the equation of manifold Λ^n has the form

$$x_1 = -4p_1^3 - 2p_1 x_2, \qquad p_2 = p_1^2, \qquad p_j = \frac{\partial S_0(x_3, \ldots, x_n)}{\partial x_j},$$
$$j = 3, \ldots, n.$$

We shall restrict ourselves to the case $n = 2$. Then

$$(K\varphi)(x) = F^{-1}_{\lambda, p_1 \to x_1}\left[(\varphi \circ r)(p_1, x_2)\left|\frac{d\sigma}{dp_1}\right|^{1/2} \exp\left[i\lambda(p_1^4 + 2p_1^2 x_2)\right]\right]. \tag{8.36}$$

As a consequence, the asymptotics of $(K\varphi)(x)$ for $\lambda \to +\infty$ can be written in terms of a standard integral of the form

$$\sqrt{\lambda} \int_{-\infty}^{\infty} \exp\left[i\lambda(p^4 + 2p^2 x_2 + px_1)\right] dp, \qquad (8.37)$$

which, however, is not expressible in terms of the special functions. Notice that

$$(K\varphi)(0) \sim \text{const. } \varphi(0)\lambda^{1/4}, \qquad (\lambda \to +\infty).$$

Integral (8.37) itself turns out to be a new special function naturally arising in three-dimensional problems of optics and of quantum mechanics.

§ 9. Global Quantization of the Velocity Field. Higher Approximations

1. *Higher Approximations in One Chart*

In the present paragraph we shall construct the canonical operator not with accuracy to $O(\lambda^{-1})$ as in § 8, but with accuracy to $O(\lambda^{-m})$ ($m > 0$ is arbitrary) following [44], [61], and thus improve the commutation formula (8.10). We suppose that the assumptions of Theorem 8.4 are satisfied.

First of all let us give a more precise definition of the canonical operator. Let Ω be a canonical chart on a Lagrangian manifold Λ^n, and a point $r^0 \in \Omega$. Suppose we can choose the coordinates $I = (p_{(\alpha)}, x_{(\beta)})$ and $\tilde{I} = (p_{(\tilde{\alpha})}, x_{(\tilde{\beta})})$ as local coordinates in Ω. Then the canonical operator on functions $\varphi \in C_0^\infty(\Omega)$ can be defined either by the formula

$$K(\Omega, I)\varphi = e^{i\delta} F_{\lambda, \, p_{(\alpha)} \to x_{(\alpha)}}^{-1} \left[\varphi \left| \frac{d\sigma}{dp_{(\alpha)} \, dx_{(\beta)}} \right|^{1/2} \times \right.$$

$$\left. \times \exp\left[i\lambda \left(\int_{r^0}^{r} \langle p', dx' \rangle - \langle x_{(\alpha)}, p_{(\alpha)} \rangle \right) \right] \right], \qquad (9.1)$$

where the arguments of all functions are the same as in (6.3), or by the formula $K(\Omega, \tilde{I})\varphi$; number δ is determined analogously to (8.2). As shown in § 6,

$$K(\Omega, \tilde{I})\varphi = K(\Omega, I)[\varphi + O(\lambda^{-1})], \quad (\lambda \to +\infty).$$

LEMMA 9.1. *Suppose that I and $\tilde{I}$ can be chosen as local coordinates in a canonical chart Ω. Then there exists a differential operator*

$$V(I, \tilde{I}) = 1 + \sum_{j=1}^{M-1} (i\lambda)^{-j} V^j(I, \tilde{I}) \tag{9.2}$$

for any $M \geq 1$, such that

$$K(\Omega, \tilde{I})\varphi = K(\Omega, I)V(I, \tilde{I})\varphi \ (\text{mod } O^+_{-M}(\Omega)) \tag{9.3}$$

for arbitrary function $\varphi \in C_0^\infty(\Omega)$.

Here V^j are linear differential operators of order $2j$ in the space $C_0^\infty(\Omega)$ with the coefficients of class $C^\infty(\Omega)$, and, moreover, V^j are independent of M.

Proof is more or less contained in the proof of Lemma 6.3. Namely, in evaluating the asymptotics of integral ψ (see (6.11)) we took into account only the leading term. As shown in Lemma 6.3, ψ is an integral over a finite domain in the space $(p_{(\alpha)}, x_{(\beta)})$ with an integrand having compact support, with accuracy to a summand of class O^+_{-k} for any $k \geq 0$. Writing down the next terms of the asymptotic expansion, we obtain according to Theorem 1.1:

$$\psi = \exp\left[i\lambda\tilde{S}(Q) + \frac{i\pi N}{2} \right] \left| \frac{d\sigma}{dp_{(\tilde{\alpha})} \, dx_{(\tilde{\beta})}} \right|^{1/2} \times$$

$$\times \left[\varphi(r(Q)) + \sum_{j=1}^{M-1} (i\lambda)^{-j} W^j \varphi \right] + \chi_M,$$

(cf. (6.23)), where $F^{-1}_{\lambda, \, p_{(\tilde{\alpha})} \to x_{(\tilde{\alpha})}} \chi_M \in O^+_{-M}(\mathbf{R}^n_x)$ (we kept the notation of Lemma 6.4). By virtue of Theorem 1.1, operators W^j possess the same properties as V^j, and differ from the required operators V^j by a numerical factor only.

LEMMA 9.2. *Let Ω be a canonical chart with the coordinates I, and let assumptions of Theorem 8.4 be fulfilled. Then there exists a differential operator*

$$R_L(\Omega, I) = (i\lambda)^{-1} \left[R_L^0 + \sum_{j=1}^{M-1} (i\lambda)^{-j} R_L^j(\Omega, I) \right] \tag{9.4}$$

for any integer $M \geq 1$, *such that for arbitrary function* $\varphi \in C_0^\infty(\Omega)$ *the commutation formula*

$$L(\overset{2}{x}, \lambda^{-1}\overset{1}{D}_x ; (i\lambda)^{-1})K(\Omega, I)\varphi =$$
$$= K(\Omega, I)R_L(\Omega, I)\varphi \ (\text{mod } O^+_{-M}(\Omega)). \tag{9.5}$$

is true.

Here R_L^j *are linear differential operators of order* $j + 1$ *with the coefficients of class* $C^\infty(\Omega)$, *and* R_L^0 *is a transport operator.*

For the proof it is sufficient, in formula (8.16), to write down not two but the M terms of the asymptotic expansion derived in Theorem 5.2, and to keep in mind that operators R_j in expansion (5.9) are linear homogeneous differential operators.

It is worthwhile to note that the explicit evaluation of the operators $V(I, \tilde{I})$ and $R_L(\Omega, I)$ leads to very cumbersome formulae.

2. Global Higher Approximations

Let $\Lambda = \Lambda^n$ be an n-dimensional Lagrangian C^∞-manifold in the phase space. Suppose the assumptions of Theorem 8.4 are satisfied. Fix a point $r^0 \in \Lambda$, specify a canonical atlas $\{\Omega_j\}$, local coordinates I_j in each of the charts Ω_j and fix a partition of unity $\{e_j\}$. We construct the canonical operator according to the formula (8.3):

$$K_\Lambda \varphi = \sum_j K(\Omega_j, I_j)(e_j \varphi),$$

where $\varphi \in C_0^\infty(\Lambda)$. Let the operator

$$L = -i\lambda^{-1}\frac{\partial}{\partial t} + H(t, x, \lambda^{-1}D_x ; (i\lambda)^{-1})$$

satisfy the assumptions of Theorem 8.4.

THEOREM 9.3. *There exist differential operators* $V^l = V^l(H)$ *of order* j *on a manifold* Λ *such that*

$$\left(-i\lambda^{-1}\frac{\partial}{\partial t} + H\right)K_\Lambda \varphi = K_\Lambda \left(\sum_{l=1}^{M-1} \lambda^{-l}V^l\right)\varphi \ (\text{mod } O^+_{-M}(\Lambda))$$

for functions $\varphi \in C_0^\infty(\Lambda)$ *and for any integer* $M \geq 1$.

Proof. By virtue of Lemma 9.2, we have

$$LK_\Lambda \varphi = \sum_j K(\Omega_j, I_j)R_L(\Omega_j, I_j)(e_j \varphi) \ (\text{mod } O^+_{-M}(\Lambda)). \tag{9.6}$$

We transform the right-hand side of this equation to the form (9.5).

Following [45], we introduce the operator $B^j = \sum\limits_k V(I_j, I_k)e_k$ in the space $C_0^\infty(\Omega_j)$. Due to (9.2), we obtain

$$B^j = 1 + \sum_{l=1}^{M-1} B_l^j \lambda^{-l}, \qquad B_l^j = \sum_k V^l(I_j, I_k)e_k.$$

The operator B_l^j is a differential operator of order l with the coefficients of class $C^\infty(\Omega_j)$. We look for a formal right-inverse operator $(B^j)^{-1}$ in the form

$$(B^j)^{-1} = 1 + \sum_{l=1}^{M-1} \tilde{B}_l^j \lambda^{-l},$$

where $\tilde{B}_l^j$ are differential operators of order l with the coefficients of class $C^\infty(\Omega_j)$. For this purpose we consider the formal equation

$$B^j\left(1 + \sum_{l=1}^{\infty} \tilde{B}_l^j \lambda^{-l}\right) = 1$$

and set the coefficients at powers λ^{-m}, $m = 1, \ldots, N-1$, equal to zero. We obtain a system of recursion relations from which operators $\tilde{B}_l^j, l = 1, 2, \ldots, N-1$, can be derived step by step. Operator $(B^j)^{-1}$ constructed in this way will satisfy the relation

$$B^j(B^j)^{-1} = 1 + \chi^{-M} \sum_{l=0}^{M-2} \lambda^{-l} \tilde{\tilde{B}}_l^j,$$

where $\tilde{\tilde{B}}_l^j$ are differential operators of order l with the coefficients of class $C^\infty(\Omega_j)$.

Multiplying the functions $R_L(e_j \varphi)$ in formula (9.5) by $B^j(B^j)^{-1}$ from the left, and using relation (9.3), we get

$$\sum_j K(\Omega_j, I_j) R_L(\Omega_j, I_j)(e_j \varphi) =$$

$$= \sum_j K(\Omega_j, I_j)\left[\sum_k V(I_j, I_k) e_k \times \right.$$

$$\left. \times \left(\sum_m V(I_j, I_m) \right)^{-1} R_L(\Omega_j, I_j) e_j \varphi \; (\mathrm{mod}\; O_{-M}^+(\Lambda)) \right] =$$

$$= \sum_k K(\Omega_k, I_k) e_k \left\{ \sum_j \left(\sum_m V(I_j, I_m) e_m \right)^{-1} \times \right.$$

$$\left. \times R_L(\Omega_j, I_j) e_j \right\} \varphi \; (\mathrm{mod}\; O_{-M}^+(\Lambda)).$$

The expression in the curly brackets can, by its construction, be represented in the form

$$1 + \sum_{l=1}^{M-1} \lambda^{-l} Q_j^l,$$

where Q_j^l are differential operators; this proves the theorem.

Note that commutation formula (9.5), in contradistinction to formula (8.10), was proved for a fixed canonical atlas and for a fixed partition of unity.

REMARK.   If supp φ does not contain singular points, the commutation formulae become more transparent; for instance, formula (8.5) takes the form

$$\hat{L}K\varphi = KL\varphi + O(\lambda^{-1}).$$

Later on, we shall use for brevity just this shorthand notation for the commutation formulae.

PART II

SEMI-CLASSICAL APPROXIMATION FOR
NON-RELATIVISTIC AND RELATIVISTIC
QUANTUM MECHANICAL EQUATIONS

§ 10. The Cauchy Problem with Rapidly Oscillating Initial Data for Scalar Hamiltonians

Consider the Cauchy problem

$$[\lambda^{-1} D_t + H(t, x, \lambda^{-1} D_x)] u = 0,$$

$$u|_{t=0} = u_0(x) \exp [i\lambda S_0(x)],$$

where $S_0(x)$ is a real-valued function. The Cauchy data induce an n-dimensional Lagrangian manifold Λ_0^n in the phase space (x, p). The asymptotic solutions of this Cauchy problem 'in the small', i.e. for a short time t, were constructed in § 3. In the present paragraph, the asymptotic solutions of the Cauchy problem will be constructed 'in the large', i.e. for any finite time $T (0 \leqq t \leqq T)$.

The global asymptotic solution of the Cauchy problem is constructed via the following scheme:

$$
\begin{array}{ccc}
\Lambda_0^n & \xrightarrow{\;g^t\;} & g^t \Lambda_0^n \\
\downarrow & & \downarrow \\
K_{\Lambda_0^n} & \xrightarrow{\;L_t\;} & K_{g^t \Lambda_0^n} = u.
\end{array}
$$

Here g^t is the displacement along trajectories of the dynamical system with Hamiltonian H. Thus the asymptotic solution $u(t, x)$ for $\lambda \to + \infty$ is expressed in terms of the canonical operator $K_{g^t \Lambda_0^n}$ associated with the Lagrangian manifold $g^t \Lambda_0^n$ obtained by the displacement of the initial manifold Λ_0^n through time t.

1. Formulation of the Problem

Consider the Cauchy problem

$$L u(t, x) = 0, \tag{10.1}$$

$$u|_{t=0} = u_0(x) \exp [i\lambda S_0(x)] \tag{10.2}$$

with rapidly oscillating initial data. Here L is a scalar λ-p.d. operator of the form

$$L = \lambda^{-1} D_t + H(t, \overset{2}{x}, \lambda^{-1} \overset{1}{D}_x ; (i\lambda)^{-1}), \qquad (10.3)$$

$t \geq 0, x \in \mathbf{R}^n$; H is a λ-p.d. operator. Variable t in operator H plays the rôle of a parameter. It is assumed that the initial data $u_0(x) \in C_0^\infty(\mathbf{R}^n)$, $S_0(x) \in C^\infty(\mathbf{R}^n)$, and that function $S_0(x)$ is real-valued.

We assume validity of the condition

H10.1 The symbol $H(t, x, p; (i\lambda)^{-1})$ is real-valued for $-\infty < t < +\infty$, $\lambda \geq 1, (x, p) \in \mathbf{R}^{2n}$ and belongs to class $T_+^m(\mathbf{R}_x^n)$ for each fixed t, $0 \leq t \leq T$, where m is independent of t. The constants $C_{\alpha\beta}$ in estimate (2.6) can be chosen to be independent of t.

The choice (10.2) of the initial data is not arbitrary, and is imposed by the nature of the problem. The asymptotics of a solution cannot be described in terms of classical mechanics, i.e. in terms of the Hamilton–Jacobi equation, for all initial data.

For this purpose it is required that the initial data should satisfy the *correspondence principle*, i.e. that all physically meaningful quantities of quantum mechanics should tend to the corresponding classical counterparts for $h = \lambda^{-1} \to 0$. In particular, the momentum $p = -\lambda^{-1} D_x u$ ought to have the classical limit for $\lambda \to +\infty$. This condition is satisfied for the initial data of the form (10.2):

$$\lim_{\lambda \to +\infty} \lambda^{-1} D_x u|_{t=0} = u_0(x) \frac{\partial S_0(x)}{\partial x}.$$

More general initial data will be discussed below.

It is evident that not all initial data satisfy the correspondence principle. For example, if we consider the Cauchy problem

$$\psi|_{t=0} = \varphi(x) \exp\left(\frac{ix}{h^2}\right), \qquad \varphi \in C_0^\infty(\mathbf{R})$$

for the Schrödinger equation

$$ih \frac{\partial \psi}{\partial t} = -\frac{h^2}{2m} \frac{\partial^2 \psi}{\partial x^2},$$

where $\varphi \not\equiv 0$, then the mean value of momentum, i.e.

$$-\int_{-\infty}^{+\infty} \overline{\psi(t, x)} ih \frac{\partial \psi(t, x)}{\partial x} dx,$$

tends to infinity for $h \to 0$.

Initial data (10.2) have the following remarkable property: they induce the n-dimensional Lagrangian manifold

$$\Lambda_0^n = \left\{ (x, p): p = \frac{\partial S_0(x)}{\partial x}, \quad x \in \mathbf{R}^n \right\}$$

in phase space $\mathbf{R}_{x,p}^{2n}$. We choose the form $d\sigma^n(x) = dx$ as a volume element on Λ_0^n and fix a point $r^0 = (x^0, p^0) \in \Lambda_0^n$. Then the Cauchy data (10.2) take the form

$$u|_{t=0} = \exp\left[i\lambda S(x^0) \right] K_{\Lambda_0^n} u_0(x), \tag{10.2'}$$

where K is the canonical operator corresponding to a non-singular chart (see (6.2)).

As a natural generalization of the Cauchy problem (10.1), (10.2) we may consider a problem with the initial data

$$u|_{t=0} = K_{\Lambda_0^n} u_0(x), \tag{10.4}$$

where $u_0 \in C_0^\infty(\Lambda_0^n)$ and K is the canonical operator. Such Cauchy data are called *canonical*. The Cauchy data of the type (10.4) form a larger class than those in (10.2), since the canonical operator cannot be reduced to the operator of multiplication by a function.

The Hamilton system corresponding to operator L has the form (see § 3)

$$\frac{dx}{dt} = \frac{\partial H(t, x, p; 0)}{\partial p}, \quad \frac{dp}{dt} = -\frac{\partial H(t, x, p; 0)}{\partial x}. \tag{10.5}$$

Furthermore

$$\frac{dS}{dt} = \left\langle p, \frac{\partial H(t, x, p; 0)}{\partial p} \right\rangle - H(t, x, p; 0), \tag{10.6}$$

$$\frac{dE}{dt} = -\frac{\partial H(t, x, p; 0)}{\partial t}, \quad \frac{dt}{dt} = 1, \tag{10.7}$$

where E is the variable conjugate to t. System (10.5) is the truncated Hamilton system; namely, instead of the $(2n + 2)$-dimensional phase space (t, x, E, p) we take the $2n$-dimensional phase space (x, p). Cauchy data (10.2) induce the Lagrangian Cauchy problem

$$x|_{t=0} = y, \quad p|_{t=0} = \frac{\partial S_0(y)}{\partial y}, \quad y \in \mathbf{R}_y^n, \tag{10.8}$$

for the Hamilton system (10.5). The set

$$\Lambda_0^n = \left\{ (x, p) : x = y, \ p = \frac{\partial S_0(y)}{\partial y}, y \in \mathbf{R}^n \right\}$$

is an n-dimensional Lagrangian C^∞-manifold in phase space $\mathbf{R}^{2n}$, which admits a diffeomorphic projection on x-space.

The construction procedure of the semi-classical asymptotics for the canonical Cauchy problem is illustrated by the following diagram:

$$
\begin{array}{ccc}
\Lambda_0^n & \xrightarrow{\ g^t\ } & g^t \Lambda_0^n \\
\downarrow & & \downarrow \\
K_{\Lambda_0^n} & \xrightarrow{\ L_t\ } & K_{g^t \Lambda_0^n} u_0 = u(t).
\end{array}
$$

All factors inessential for understanding, as $\exp [i\lambda S_0(x)]$, are left out.

(1) A quantum object (Hamiltonian L) induces a classical object (the Hamilton system and the displacement g^t).

(2) The canonical Cauchy data for Equation (10.1) consist of a pair (Λ_0^n, u_0), where Λ_0^n is an n-dimensional Lagrangian manifold in the phase space and u_0 is a function on Λ_0^n.

(3) The time-evolution of manifold Λ_0^n follows the laws of classical mechanics. Namely, the displacement along trajectories of the Hamilton system is $g^t : \Lambda_0^n \to g^t \Lambda_0^n$, where $g^t \Lambda_0^n$ is also an n-dimensional Lagrangian manifold.

(4) Now we come back from classical to quantum objects. Canonical operator $K_{\Lambda_0^n}$ is constructed in accordance with the initial data (Λ_0^n, u_0). The time-evolution of the function $K_{\Lambda_0^n} u_0$ (i.e. of the Cauchy data) is governed by operator L_t, and that of its asymptotics by operator g^t, i.e. $u_0(x) \to u(t, x) \approx K_{g^t \Lambda_0^n} u_0$. The exact formulae are given below (see (10.17)).

Another approach is based on the whole $(2n + 2)$-dimensional phase space. The Cauchy data (10.2) generate an n-dimensional Lagrangian manifold $\tilde{\Lambda}_0^n$ in the $(2n + 2)$-dimensional phase space with the coordinates (t, x, E, p). Namely,

$$\tilde{\Lambda}_0^n = \{ (\tilde{x}, \tilde{p}) : (x, p) \in \Lambda_0^n, t = 0, E = -H(0, x, p; 0) \},$$

where $\tilde{x} = (t, x), \tilde{p} = (E, p)$. Let $G^t : \mathbf{R}^{2n+2} \to \mathbf{R}^{2n+2}$ be the displacement operator along trajectories of the complete Hamilton system (10.5), (10.7). Then the set

$$\Lambda^{n+1} = \bigcup_{-\infty < t < +\infty} G^t \tilde{\Lambda}_0^n$$

is a Lagrangian $(n + 1)$-dimensional manifold in $\mathbf{R}^{2n+2}$ provided the assumptions of Proposition 4.19 are fulfilled. Manifold Λ^{n+1} is invariant under the displacements along trajectories of the complete Hamilton system by construction. The asymptotic solutions of the Cauchy problem (10.1), (10.2) can be obtained with the help of the canonical operator $K_{\Lambda^{n+1}}$.

2. The Asymptotic Solution of the Cauchy Problem

Besides H10.1 we introduce another assumption

H10.2. A solution of the Cauchy problem (10.8) for the Hamilton system (10.5) exists and is infinitely differentiable in the strip $\Pi_T : y \in \mathbf{R}^n$, $0 \leq t < T$.

Now we construct the asymptotic solution of the Cauchy problem (10.1), (10.2). Let us note that we actually construct a f.a. solution, i.e. a function which exactly fulfils the Cauchy data and approximately satisfies the equation $Lv = O(\lambda^{-2})$. A rigorous derivation of the asymptotic formulae requires an estimate of the norm of inverse operator; this estimate will be presented in Section 3 for the simplest case. Let us consider the Cauchy problem for the transport equation

$$\frac{d\varphi}{dt} - M\varphi = 0, \qquad \varphi|_{t=0} = u_0(y), \tag{10.9}$$

with the notation

$$M = \frac{1}{2} \sum_{j=1}^{n} \frac{\partial^2 H^0}{\partial x_j \partial p_j} - \frac{\partial H^0}{\partial \varepsilon}, \qquad H^0 = H(t, x, p; 0). \tag{10.10}$$

In these formulae $x = x(t, y)$, $p = p(t, y)$ (i.e. the point (x, p) lies on the phase trajectory passing through the point $(y, \partial S_0(y)/\partial y)$ at the initial time $t = 0$). The solution of (10.9) is

$$\varphi = \exp\left(\int_0^t M \, dt'\right) u_0(y). \tag{10.11}$$

Here and later on

$$\frac{\partial H^0}{\partial \varepsilon} = \frac{\partial H(t, x, p; \varepsilon)}{\partial \varepsilon}\bigg|_{\varepsilon=0}.$$

Since $u_0(y)$ is a function with compact support, function φ is, for $0 \leq t \leq T$, different from zero only on the compact subset $\mathscr{F}_T = \bigcup_{0 \leq t \leq T} g^t r$ of the phase space, where $r = (y, \partial S_0(y)/\partial y)$, $y \in \operatorname{supp} u_0$.

Consider now an $(n+1)$-dimensional Lagrangian manifold in the phase space (t, x, E, p), and introduce the notations:

$$r(y) = \left(y, \frac{\partial S_0(y)}{\partial y} \right), \qquad r_t(y) = g^t r(y),$$

$$\tilde{r}(y) = (0, -H(0, r^0(y); 0), r^0(y)),$$

$$r^0 = r(y^0), \qquad \tilde{r}^0 = \tilde{r}(y^0),$$

where y^0 is a fixed point. We choose

$$d\sigma^{n+1}(G^t \tilde{r}(y)) = dt\, dy$$

as a volume element on Λ^{n+1}, invariant under displacement G^t.

Consider the function

$$v(t, x) = \exp\left[i\lambda S_0(x^0) \right] K_{\Lambda^{n+1}} \varphi, \tag{10.12}$$

where φ is determined by (10.11) for $0 \leq t \leq T$. As mentioned above, function φ is non-zero only on some compact set $\mathscr{F}_T$, the projection of which on $\mathbf{R}^{2n}_{x,p}$ coincides with $\mathscr{F}_T$. We choose a partition of unity with the help of which the canonical operator on Λ^{n+1} is constructed, such that only a finite number of functions e_j are non-zero on $\mathscr{F}_T$; then function v will be identically equal to zero outside some compact set $\tilde{\mathscr{F}}_T$ in $\mathbf{R}^n_x \times \mathbf{R}_t$. Further, $\operatorname{supp} u_0$ can be included into a non-singular chart, and the canonical operator in it can be defined by the simplest formula (6.2).

THEOREM 10.1.   *Function v given by (10.12) satisfies the initial condition (10.2). Furthermore,*

$$Lv = \lambda^{-2} \psi(t, x, \lambda) \tag{10.13}$$

and the estimate

$$|\psi(t, x, \lambda)| \leq C_N (1 + |x|)^{-N} \tag{10.14}$$

holds for $\lambda \geq 1$, $0 \leq t \leq T$ and for any integer $N \geq 0$.

Proof. We have

$$v(0, x, \lambda) = \exp\left[i\lambda S_0(x^0) \right] K_{\Lambda^{n+1}} \varphi \big|_{t=0} =$$

$$= \exp\left[i\lambda \left(S_0(x^0) + \int_{\tilde{r}^0}^{\tilde{r}(y)} (\langle p, dx \rangle - H\, dt) \right) \right] \varphi \big|_{t=0} =$$

$$= \varphi \big|_{t=0} = u_0(y).$$

Here we exploited the fact that the integral

$$\int (\langle p, dx \rangle - H dt) \tag{10.15}$$

taken along a path lying on Λ^{n+1} is path-independent; so we can choose from the paths connecting the points $\tilde{r}^0, \tilde{r}(y)$ that on which $t = 0$. Then

$$\int_{\tilde{r}^0}^{\tilde{r}(y)} \langle p, dx \rangle = \int_{\tilde{r}^0}^{\tilde{r}(y)} dS_0(y) = S_0(y) - S_0(y^0)$$

(here and later on $y^0 = x^0$).

Furthermore, it follows from the commutation formula (8.10) and from the fact that function φ satisfies the transport equation (10.9) that

$$L K_{\Lambda^{n+1}} \varphi = O(\lambda^{-2})$$

uniformly with respect to $x \in \mathbf{R}^n, 0 \leq t \leq T$. The finer estimate (10.14) follows from the fact that function φ has compact support in the variables x, p and from Theorems 2.6, 5.2, and Lemma 6.3.

We express function v in terms of the canonical operators $K_{\Lambda^n_t}$ (Figure 2); this yields more lucid asymptotic formulae for v. The generating function $\tilde{S}$ associated with manifold Λ^{n+1} has the form (10.15). The Lagrangian manifold Λ^n_t is parametrized by the variables y. Let $\tilde{r}_t(y) = G^t \tilde{r}(y)$. Since integral (10.15) is path-independent on Λ^{n+1}, we have

$$\int_{\tilde{r}^0}^{\tilde{r}_t(y)} (\langle p, dx \rangle - H dt) =$$

$$= \int_{\tilde{r}^0}^{\tilde{r}_t(y^0)} (\langle p, dx \rangle - H dt) + \int_{\tilde{r}_t(y^0)}^{\tilde{r}_t(y)} \langle p, dx \rangle, \tag{10.16}$$

where the first integral on the right-hand side is taken along the phase trajectory in $\mathbf{R}^{2n+2}$ and the second along the curve lying on Λ^n_t. The former is a function only of t, the latter is the action on manifold Λ^n_t.

Furthermore, the definition of canonical operators $K_{\Lambda^n_t}$ and $K_{\Lambda^{n+1}}$ involves the index of a curve. Replacing the curve on Λ^{n+1} connecting the points $\tilde{r}^0$ and $\tilde{r}_t(y)$, by the sum of two curves, as previously, then, due to additivity of the index and simple connectedness of the considered manifold, we obtain for the index of the curve the expression

$$\text{ind } \tilde{\gamma}[\tilde{r}^0, \tilde{r}_t(y^0)] + \text{ind } \gamma[\tilde{r}_t(y^0), \tilde{r}_t(y)].$$

The second index is included in the operator $K_{\Lambda_t^n}$ in the natural way, and the first gives a numerical factor. Thus, finally, we have obtained the following 'working' asymptotic formula:

$$v(t, x, \lambda) = e^{i\delta} K_{\Lambda^{n+1}} \varphi, \tag{10.17}$$

$$\delta = \lambda \left[S_0(y^0) + \int_{\tilde{r}^0}^{\tilde{r}_t(y^0)} (\langle p, dx \rangle - H \, dt) - \frac{\pi}{2} \text{ind } \tilde{\gamma}[\tilde{r}^0, \tilde{r}_t(y^0)] \right].$$

REMARK 10.2.   It is clear that Theorem 10.1 is also true if the assumptions H10.1 and H10.2 are satisfied for $0 \leq t \leq T$.

REMARK 10.3.   Theorem 10.1 is also valid for the Lagrangian Cauchy problem (10.1), (10.2′) with function $u_0 \in C_0^\infty(\mathbf{R}^n)$.

COROLLARY 10.4.   *Let the assumptions of Theorem 10.1 be satisfied and let $\hat{H}$ be an operator with the symmetric symbol:*

$$\hat{H} = \tfrac{1}{2}[H(t, \overset{2}{x}, \lambda^{-1} \overset{1}{D}_x) + H(t, \overset{1}{x}, \lambda^{-1} \overset{2}{D}_x)].$$

Then

$$v(t, x, \lambda) = e^{i\delta} K_{\Lambda_t^n} u_0. \tag{10.18}$$

Now we shall give the explicit formula for f.a. solutions in the non-focal points. We introduce the assumption

H10.3. Suppose the point (t, x) is non-focal and has the property: there exists a finite number of phase trajectories, the projections of which (the rays) meet in x at time t.

This means that the problem

$$X\big|_{\tau=0} = y, \qquad P\big|_{\tau=0} = \frac{\partial S_0(y)}{\partial y}, \qquad X\big|_{\tau=t} = x \tag{10.19}$$

for the Hamilton system (10.5) is solvable for a finite number of the values $y = y^j$, $1 \leq j \leq N$. We denote the corresponding phase trajectories by

$$l_j(t, x) = \{X = X(\tau, y^j), P = P(\tau, y^j), 0 \leq \tau \leq t\},$$

so that $X(t, y^j) = x$. We set

$$J(t, y) = \det \frac{\partial X(t, y)}{\partial y};\qquad (10.20)$$

then $J(t, y^j) \neq 0$ by assumption. We introduce the notation

$$S_j(t, x) = S_0(y^j) + \int_0^t (\langle p, dx \rangle - H \, dt'),\qquad (10.21)$$

where the integral is taken along the trajectory $l_j(t, x)$,

$$m_j(t, x) = \operatorname{ind} l_j(t, x).\qquad (10.22)$$

THEOREM 10.5.   *Under the assumptions* H10.1–H10.3, *the f.a. solution of the Cauchy problem* (10.1), (10.2) *has the form*

$$v(t, x, \lambda) = \sum_{j=1}^n \frac{\varphi_j(t, x)}{\sqrt{|J(t, y^j)|}} \exp\left[i\lambda S_j(t, x) - \frac{i\pi}{2} m_j(t, x) \right].\qquad (10.23)$$

Here $\varphi_j(t, x)$ is determined by formula (10.11) where the integral is taken along the phase trajectory $l_j(t, x)$.

The proof follows from (10.17) and the definition of the canonical operator.

3. *On the Estimation of the Inverse Operator*

The formal asymptotic solution of the Cauchy problem (10.1), (10.2) was constructed in Section 2. Namely, the function S was derived (see (10.17)) which satisfied the Cauchy data, and, when inserted in Equation (10.1), gave a small discrepancy (i.e. smaller than $v = O(\lambda^{-1})$ for $\lambda \to +\infty$). However, an open question remained: is the 'approximate solution' v close to the exact solution u?

This question is very important indeed, and extremely hard to answer. There are examples in which the differential equation has a formal asymptotic solution, but there is no exact solution with such an asymptotics [88].

We shall mention one result which allows to derive the asymptotics rigorously, i.e. which allows to prove the closeness of f.a. and exact solutions. Let $\mathscr{H}$ be a Hilbert space with the norm $\|\cdot\|$, $A : \mathscr{H} \to \mathscr{H}$ a self-adjoint operator with the domain of definition $D(A)$, and $f(t), 0 \leq t \leq T$, a function with the values in $\mathscr{H}$ such that the norm $\|f(t)\|$ is continuous for $0 \leq t \leq T$. Consider the Cauchy problem (see [25])

$$i \frac{du}{dt} = Au + f(t), \qquad 0 \leq t \leq T,$$

$$u(0) = u_0 \in D(A). \tag{10.24}$$

PROPOSITION 10.6. *The solution of the Cauchy problem* (10.24) *exists, is unique and fulfils the estimate*

$$\|u(t)\| \leq \|u_0\| + \int_0^t \|f(\tau)\| \, d\tau \tag{10.25}$$

for $0 \leq t \leq T$.

Proof follows from the explicit form of the solution

$$u(t) = \exp(-itA)u_0 + \int_0^t \exp[i(\tau - t)A] f(\tau) \, d\tau$$

and unitarity of the operator $\exp(i\alpha A)$ for $\alpha \in \mathbf{R}$.

4. *The Cauchy problem with rapidly oscillating initial data and the discontinuities of fundamental solutions of hyperbolic systems*

Consider a strictly hyperbolic system of N partial differential equations

$$\mathcal{L}u(t, x) = f(t, x). \tag{10.26}$$

Here $u = (u_1, \ldots, u_N)$ (column vector), $x \in \mathbf{R}^n, f$ is a column vector,

$$\mathcal{L} = L(\overset{2}{x}, \overset{1}{D}_t, \overset{1}{D}_x), \tag{10.27}$$

and L is an $(N \times N)$-matrix. It is assumed that the following assumption holds:

L10.1. The elements of the matrix function $L(x, p_0, p)$ are polynomials of degree $\leq m$ in the variables $(p_0, p) \in \mathbf{R}_{p_0} \times \mathbf{R}_p^n$ with the coefficients of class $C^\infty(\mathbf{R}_x^n)$. The coefficient at p_0^m is the unit $(N \times N)$-matrix I.

Thus the symbol L is equal to

$$L(x, p_0, p) = p_0^m I + \sum_{k=0}^{m-1} p_0^k L_k(x, p),$$

where matrices L_k are polynomials in p with the coefficients of class $C_0^\infty(\mathbf{R}_x^n)$.

Let us formulate the condition of strict hyperbolicity for the system (10.26)(see L10.2). Let $L^0(x, p_0, p)$ be the highest homogeneous part of the matrix $L(x, p_0, p)$, i.e.

$$L(x, p_0, p) = L^0(x, p_0, p) + L^1(x, p_0, p),$$

where the matrix elements of L^0 and L^1 are homogeneous polynomials of degrees m and $\leq m - 1$, respectively, in the variables (p_0, p).

L10.2.   The equation

$$\Delta(x, p_0, p) \equiv \det L^0(x, p_0, p) = 0 \qquad (10.28)$$

has exactly mN roots $p_{0j} = p_{0j}(x, p)$ with respect to p_0 for all $x \in \mathbf{R}^n$, $p \in \mathbf{R}^n \setminus \{0\}$. All these roots are real and distinct for all $x \in \mathbf{R}^n, p \in \mathbf{R}^n \setminus \{0\}$.

In particular, it follows from this condition that all roots $p_{0j}(x, p)$, $1 \leq j \leq mN$, are infinitely differentiable for $x \in \mathbf{R}^n, p \in \mathbf{R}^n \setminus \{0\}$. The roots $p_{0j}(x, p)$ are homogeneous functions of p of degree 1.

As an example of a strictly hyperbolic equation ($N = 1$) we may take the wave equation

$$\frac{\partial^2 u}{\partial t^2} = c^2(x) \Delta u,$$

where the coefficient $c(x) \in C^\infty(\mathbf{R}^n)$ and $c(x) \geq \delta > 0$ for $x \in \mathbf{R}^n$. If a first order differential operator with smooth coefficients,

$$a(x) \frac{\partial}{\partial t} + \sum_{j=1}^{n} b_j(x) \frac{\partial}{\partial x_j} + d(x)$$

is added to the wave operator, the resulting equation will again be strictly hyperbolic.

In addition, it is necessary to require that the strict hyperbolicity condition is satisfied uniformly with respect to the variables x. Let us also recall that in the present book differential operators are considered with the symbols of class T. The simplest (but very rough) expression of this fact is the assumption.

L10.3.   There exists $a > 0$ such that the matrix $L(x, p_0, p)$ does not depend on x for $|x| \geq a$ and is homogeneous with respect to the variables (p_0, p).

It is sufficient to demand that, instead of assumption L10.3, e.g. the coefficients (at powers of p_0, p in matrix L) tend sufficiently fast to constants for $|x| \to \infty$, and that a sufficiently large number of their derivatives tends to zero for $|x| \to \infty$. Here the limit symbol $L(\infty, p_0, p)$ is understood to be strictly hyperbolic.

We shall formulate the Cauchy problem for equation (10.26):

$$D_t^k u\big|_{t=0} = u_k(x), \qquad 0 \leq k \leq m - 1. \tag{10.29}$$

The theory of the Cauchy problem for hyperbolic equations was developed in works [72], [50], [29]. We shall use only its one relatively coarse result.

Let the Cauchy data $u_k(x) \in C^\infty(\mathbf{R}^n)$ and the right-hand side of (10.26), $f(t, x) \in C^\infty$, for $0 \leq t \leq T, x \in \mathbf{R}^n$. Then (under the assumptions L10.1–L10.3) the solution of the Cauchy problem (10.26), (10.29) exists, is unique and infinitely differentiable for $0 \leq t \leq T, x \in \mathbf{R}^n$.

The system (10.26) does not involve a large parameter λ; however, such a parameter will appear somewhat later.

Now we introduce the notion of a *Green matrix* $G(t, x, x^0)$ *of the Cauchy problem* (or simply a Green matrix) *for operator* $\mathscr{L}$. The *Green matrix* is an $(N \times N)$-matrix $G(t, x, x^0)$ which satisfies the homogeneous matrix equation

$$\mathscr{L}G = 0 \tag{10.30}$$

for $x \in \mathbf{R}^n, t > 0$, and the Cauchy data

$$D_t^k G\big|_{t=0} = 0, \qquad 0 \leq k \leq m - 2;$$
$$D_t^{m-1} G\big|_{t=0} = \delta(x - x^0)I. \tag{10.31}$$

Here $x^0 \in \mathbf{R}^n$, I is the unit $(N \times N)$-matrix and δ is the Dirac delta-function. If $\mathscr{L}$ is a scalar operator ($N = 1$), then G is a scalar function, called a *fundamental solution* of the Cauchy problem [82].

The Green matrix G is a distribution. It is possible to regard G as a functional over the space $D = C_0^\infty(\mathbf{R}_x^n)$ depending on t as on a parameter (the matrix G is extended to the half-space $t < 0$ by putting its elements equal to zero there). Our next task is to investigate the discontinuities of the Green matrix.

Let $G^0(t, x, x^0)$ be an $(N \times N)$-matrix satisfying the equation

$$\mathscr{L}G^0 = f(t, x) \qquad (0 \leq t \leq T, x \in \mathbf{R}^n) \tag{10.32}$$

and the Cauchy data

$$D_t^k G^0\big|_{t=0} = u_k(x), \qquad 0 \leq k \leq m - 2;$$
$$D_t^{m-1} G^0\big|_{t=0} = \delta(x - x^0)I + u_{m-1}(x). \tag{10.33}$$

If all functions $u_k(x), f(t, x) \in C^\infty$ (for $x \in \mathbf{R}^n, 0 \leq t \leq T$), then the difference

$$G(t, x, x^0) - G^0(t, x, x^0) \in C^\infty$$

for $0 \leq t \leq T, x \in \mathbf{R}^n$. Indeed, the difference $\tilde{G} = G - G^0$ satisfies the Equation (10.32) and the Cauchy data (10.29). Hence, as already mentioned above, the theory of hyperbolic equations yields $\tilde{G} \in C^\infty$ in the strip $0 \leq t \leq T, x \in \mathbf{R}^n$.

Thus the structure of singularities of Green matrix G is exactly the same as that of matrix G^0 and, therefore, investigation of singularities of a Green matrix is reduced to the construction of matrix G^0. We shall present only a formal construction of matrix G^0. We shall show how the fundamental solution G of the Cauchy problem and the solution of the Cauchy problem with rapidly oscillating initial data are mutually related.

Let $u(t, x, p, x^0)$ be the solution of the Cauchy problem (for $0 \leq t \leq T$, $x \in \mathbf{R}^n$)

$$\mathscr{L}u = f(t, x, p),$$

$$D_t^k u|_{t=0} = 0, \quad 0 \leq k \leq m - 2,$$

$$D_t^{m-1} u|_{t=0} = \varphi(x) \exp\left[i \langle p, x - x^0 \rangle\right]. \tag{10.34}$$

Here $p \in \mathbf{R}^n, x^0 \in \mathbf{R}^n$, point x^0 is fixed, matrix function $\varphi(x) \in C_0^\infty(\mathbf{R}^n)$ and $\varphi(x) \equiv I$ in some neighbourhood of point x^0; the matrix function f will be defined below.

We introduce the distribution $G^0(t, x, x^0)$ by the formula

$$G^0(t, x, x^0) = (2\pi)^{-n} \int_{\mathbf{R}^n} \eta(p) u(t, x, p, x^0) \, dp. \tag{10.35}$$

Function $\eta(p)$ is a scalar function, $\eta(p) \in C_0^\infty(\mathbf{R}^n)$, is equal to zero for $|p| \leq a$, is equal to one for $|p| \geq a$, and is a function of $|p|$ only. Then the difference $\tilde{G} = G - G^0$ satisfies the equation

$$\mathscr{L}\tilde{G} = g(t, x) \tag{10.36}$$

and the Cauchy data

$$D_t^k \tilde{G}|_{t=0} = 0, \quad 0 \leq k \leq m - 2;$$

$$D_t^{m-1} \tilde{G}|_{t=0} = \varphi_{m-1}(x). \tag{10.37}$$

Matrix function $\varphi_{m-1}(x)$ is equal to

$$\varphi_{m-1}(x) = \delta(x - x^0)I - (2\pi)^{-n}\varphi(x)\int \eta(p)\exp\left[i\langle p, x - x^0 \rangle\right]dp =$$

$$= (2\pi)^{-n}\int (1 - \eta(p))\exp\left[i\langle p, x - x^0 \rangle\right]dp$$

(the integrals are taken over $\mathbf{R}_p^n$), since

$$\delta(x - x^0) = (2\pi)^{-n}\int \exp\left[i\langle p, x - x^0 \rangle\right]dp$$

and the matrix function $\varphi(x) \equiv I$ for x close to x^0. The function $(1 - \eta(p))\in C_0^\infty(\mathbf{R}^n)$ so that $\varphi_{m-1}(x)\in C^\infty(\mathbf{R}_x^n)$.

LEMMA 10.7.   *Let the estimates*

$$\left|D_{(t,x)}^\alpha f(t, x, p)\right| \leq C_\alpha |p|^{-n-\varepsilon} \tag{10.38}$$

be true for $|p| \geq a, 0 \leq t \leq T$, $x\in\mathbf{R}^n$, *with* $\varepsilon > 0$ *and for an arbitrary multi-index* α. *Then the matrix function*

$$G(t, x, x^0) - G^0(t, x, x^0)\in C^\infty$$

for $0 \leq t \leq T, x\in\mathbf{R}^n$.

   Proof. It follows from condition (10.38) that the right-hand side of Equation (10.36)

$$g(t, x) = (2\pi)^{-n}\int f(t, x, p)\,dp$$

is infinitely differentiable for $0 \leq t \leq T, x\in\mathbf{R}^n$. The Cauchy data (10.37) belong to $C_0^\infty(\mathbf{R}^n)$, and the statement of the lemma follows from strict hyperbolicity of operator $\mathscr{L}$.

   Formula (10.35) describes the structure of the singularities of Green matrix G, since the difference $G - G^0 = \tilde{G}\in C^\infty$. To complete the construction of matrix G^0, it remains to study the asymptotics of the solution of the Cauchy problem with rapidly oscillating initial data (10.34).

   The rôle of a large parameter is played by $|p|$. Now we introduce this large parameter into operator $\mathscr{L}$. Dividing operator $\mathscr{L}$ by $|p|^m$ we obtain

$$|p|^{-m}\mathscr{L} = \sum_{k=0}^{m}(i|p|)^{-k}L^k(\overset{2}{x}, |p|^{-1}\overset{1}{D}_t, |p|^{-1}\overset{1}{D}_x), \tag{10.39}$$

where $L^k(x, p_0, p)$ are homogeneous matrix polynomials of degree $m - k$ of the variables (p_0, p). The symbol of the $|p|$-differential operator

$$\hat{L} = |p|^{-m} \mathscr{L} \tag{10.40}$$

equals

$$L(x, p_0, p) = \sum_{k=0}^{m} (i|p|)^{-k} L^k(x, p_0, p) \tag{10.41}$$

and belongs to class T_+^0 (see the assumptions L10.1–L10.3).

Notice that we can assume that $|p| \geq a$ in the Cauchy data (10.34) (see (10.35)). The number $a > 0$ can be taken arbitrarily large, but fixed.

We construct a f.a. solution of the Cauchy problem (10.34) for the equation $\hat{L}u = 0$ (i.e. for the equation $\mathscr{L}u = 0$) with accuracy to $O(|p|^{-M})$, where $M > n$. Namely, we find a matrix function $u(t, x, p, x^0)$ which exactly satisfies the Cauchy data (10.34) and fulfils the equation

$$\hat{L}u = |p|^{-M} f_M(t, x, p),$$

where $f_M \in O_{0,T}$. The classes $O_{s,T}$ were introduced in Section 2 of this paragraph. Thus all assumptions of Lemma 10.7 are fulfilled, and matrix function G^0 given by formula (10.35) has the same singularities as Green matrix G.

It remains to derive explicit formulae for the singularities of matrix function G^0 (i.e. for singularities of Green matrix G) by using the explicit asymptotic formulae for solutions $u(t, x, p, x^0)$. We shall give only qualitative arguments; for the explicit formulae see [11], [59], [61], [78]. The bibliography of the subject can be found in [84].

In § 11, the asymptotic formulae for solutions of the Cauchy problem with rapidly oscillating initial data are given for $m = 1$, i.e. for the first order systems. For small $t \geq 0$, the solution u is a sum

$$u = \sum_{\alpha=1}^{N} u^\alpha$$

of the terms

$$u^\alpha(t, x, p, x^0) = \sum_{j=0}^{k} (i|p|)^{-j} \psi_j^\alpha(t, x, \omega) \exp\left[iS_\alpha(t, x, p)\right]. \tag{10.42}$$

Here $\omega = p/|p|$, functions S_α are homogeneous functions of p of degree 1.

Function S_α is the solution of the Cauchy problem

$$\frac{\partial S_\alpha}{\partial t} + h_\alpha\left(x, \frac{\partial S_\alpha}{\partial x}\right) = 0,$$

$$S_\alpha\big|_{t=0} = \langle x - x^0, p \rangle,$$

where $h_\alpha(x, q)$ are the roots of the equation

$$\det L^0(x, -h, q) = 0.$$

Take one of the summands (10.42) corresponding to the value $j = 0$ and insert it in the integral (10.35). Then we obtain the integral

$$\int_{\mathbf{R}^n} \exp\left[iS_\alpha(t, x, p)\right]\psi_0^\alpha\left(t, x, \frac{p}{|p|}\right) \eta(p)\, dp \tag{10.43}$$

with the rapidly oscillating phase function S_α. Singularities of this integral are contained in the projection of the set $\{(t, x, p): \partial S_\alpha/\partial p = 0\}$ on $\mathbf{R}_{t,x}^{n+1}$. Analogous expressions are obtained if the remaining summands (10.42) are substituted into integral (10.35).

We shall show where the singularities of Green matrix G are located. There are N Hamilton systems associated with system (10.1) (recall that we consider the case $m = 1$):

$$\frac{dx^\alpha}{d\tau} = -\frac{\partial h_\alpha}{\partial p^\alpha}, \qquad \frac{dp_\alpha}{d\tau} = \frac{\partial h_\alpha}{\partial x^\alpha},$$

$$\frac{dt^\alpha}{d\tau} = 1, \qquad \frac{dp_0^\alpha}{d\tau} = 0, \tag{10.44}$$

where $1 \le \alpha \le N$. The Cauchy data induced by the Cauchy problem (10.34) have the form

$$x^\alpha\big|_{\tau=0} = y, \qquad p^\alpha\big|_{\tau=0} = p,$$

$$t\big|_{\tau=0} = 0, \qquad p_0^\alpha\big|_{\tau=0} = -h_\alpha(y, p), \qquad y \in \operatorname{supp}\varphi.$$

Let us fix the direction ω of vector p. Let $\Pi_T^\alpha(p)$ be a domain in the phase space filled by the system trajectories with the label α, for $0 \le \tau \le T$ (a tube of trajectories), $\Pi_T(p) = \bigcup_{\alpha=1}^N \Pi_T^\alpha(p)$ and $\tilde{\Pi}_T(p)$ is the projection of the union of tubes of trajectories $\Pi_T(p)$ on $\mathbf{R}_{t,x}^{n+1}$. Thus $\tilde{\Pi}_T(p)$ is the union of the ray tubes ($1 \le \alpha \le N$) which have origin in supp φ for $t = 0$.

As shown in § 11, solution u of the Cauchy problem (10.34) decreases faster than any power of $|p|$ for $|p| \to \infty$ outside an arbitrary fixed neighbourhood of the tube $\tilde{\Pi}_T(\omega)$ (uniformly with respect to $t \in [0, T], \omega$, and x lying in a compact set). Consequently, the matrix $G^0 \in C^\infty$ outside $\bigcup_{\omega \in S^{n-1}} \tilde{\Pi}_T(\omega)$. Remember that matrix function $\varphi(x) \in C_0^\infty(\mathbf{R}^n)$ (see (10.34)) is an arbitrary matrix function such that $\varphi(x) \equiv I$ in a neighbourhood of point x^0, so that $\operatorname{supp} \varphi$ may lie in an arbitrary small neighbourhood of point x^0.

The singularities of the Green matrix are therefore contained in the set $\mathcal{K}_T$ for $0 \le t \le T, x \in \mathbf{R}^n$; set $\mathcal{K}_T$ can be constructed in the following way. For each of the systems (10.44) formulate the Cauchy problem

$$x^\alpha|_{\tau=0} = x^0, \qquad p^\alpha|_{\tau=0} = p,$$
$$t|_{\tau=0} = 0, \qquad p_0^\alpha|_{\tau=0} = -h_\alpha(x^0, p),$$

where $|p| \ge 1$. Then the set $\mathcal{K}_T$ is the union of projections on $\mathbf{R}_{t,x}^{n+1}$ of all trajectories ($1 \le \alpha \le N$) during time $0 \le \tau \le T$, where p runs through $\mathbf{R}^n$.

A similar result holds for hyperbolic systems of any order m. As an example consider the wave equation

$$\frac{\partial^2 u}{\partial t^2} = c^2 \Delta u,$$

where $c > 0$ is a constant. The Hamilton system has the form

$$\frac{dx}{d\tau} = 2cp, \qquad \frac{dp}{d\tau} = 0, \qquad \frac{dt}{d\tau} = -2p_0, \qquad \frac{dp_0}{d\tau} = 0.$$

For the Cauchy problem

$$x|_{\tau=0} = x^0, \qquad p|_{\tau=0} = p, \qquad t|_{\tau=0} = 0, \qquad p_0|_{\tau=0} = p_0,$$

where $p_0^2 = c^2 p^2$, the rays are given by

$$x = x^0 + 2\tau cp, \qquad t = -2\tau p_0.$$

Eliminating τ, we obtain the known result: the set of singularities of the fundamental solution of the wave equation is contained on the light cone

$$(x - x^0)^2 = c^2 t^2.$$

5. *Discontinuities of the Green matrix, short-wave asymptotics of the resolvent for stationary problems, asymptotic solutions of non-stationary problems for $t \to +\infty$*

We shall discuss the following three topics.

(I) The asymptotics with respect to smoothness of the Green matrix for a non-stationary problem.

(II) The short-wave asymptotics (for $k \to +\infty$) of the resolvent for a stationary problem.

(III) The asymptotics of solutions of a non-stationary problem for $t \to +\infty$.

Problem (I) was already considered in Section 4. It turns out that problems (I)–(III) can be investigated in the logical sequence

$$(I) \to (II) \to (III).$$

Consider the Cauchy problem (10.26), (10.29) and the corresponding stationary problem

$$(\mathcal{L}_k u)(x) = f(x), \qquad x \in \mathbf{R}^n. \tag{10.45}$$

The operator

$$\mathcal{L}_k = L(\overset{2}{x}, k, \overset{1}{D}_x) \tag{10.46}$$

is derived from operator $\mathcal{L}$ by the substitution $D_t \to k$.

For that we shall give some heuristic arguments. Let $G(t, x, x^0)$ be the Green matrix of the Cauchy problem extended by zeros into the half-space $t < 0$. Applying formally the Fourier transformation in t, we obtain that the matrix function

$$\tilde{G}(x, k, x^0) = \int_0^\infty e^{-ikt} G(t, x, x^0) \, dt$$

satisfies the equation

$$\mathcal{L}_k \tilde{G} = \delta(x - x^0)I.$$

Let the points x, x^0 be fixed; then $\tilde{G}$ is an integral of a rapidly oscillating function for $k \gg 1$. The asymptotics of this integral can be evaluated by the stationary phase method. It is well known that the main contribution to the asymptotics of integral $\tilde{G}$ is due to the singularities of the Green matrix $G(t, x, x^0)$ (considered as a function of t).

Thus the connection between the discontinuities of Green matrix G of the Cauchy problem and the short-wave asymptotics of Green matrix $\tilde{G}$ of the stationary problem is formally established.

A rigorous derivation of such direct interrelation encounters a serious obstacle – the question of convergence of the integral representing matrix $\tilde{G}$, i.e. the question about the asymptotics of the Green matrix G for

$t \to +\infty$. This difficulty is so essential that the simple proposed scheme is practically never realized.

Now we shall review the results related to the problems (I)–(III), derived in the work [84], where also an extensive list of references can be found.

Let $\Delta(x, p_0, p) = \det L^0(x, p_0, p)$. The Hamilton system associated with Equation (10.26) is given by

$$\frac{dt}{d\tau} = \frac{\partial \Delta}{\partial p_0}, \qquad \frac{dx}{d\tau} = \frac{\partial \Delta}{\partial p},$$

$$\frac{dp_0}{d\tau} = 0, \qquad \frac{dp}{d\tau} = -\frac{\partial \Delta}{\partial x} \qquad (10.47)$$

in the $(2n + 2)$-dimensional phase space with the coordinates (t, x, p_0, p).

In addition to the assumptions L10.1–L10.3 the following supplementary assumptions are introduced.

L10.4. For any $x \in \mathbf{R}^n$,

$$\Delta(x, 0, p) \neq 0$$

for $p \neq 0$.

This assumption is equivalent to ellipticity of operator $\mathscr{L}_k$.

L10.5. The trajectories of Hamilton system (10.47) sent out from the ball $|x| \leq a$, and on which $\Delta = 0$, go to infinity.

More precisely, for any $R < \infty$ there exists $T(R) < \infty$ such that the trajectories of system (10.45) with the Cauchy data

$$t(0) = 0, \qquad x(0) = x^0, \qquad p_0(0) = p_0^0, \qquad p(0) = p^0,$$

where $|x^0| \leq a$, $\Delta(x^0, p_0^0, p^0) = 0$, lie in the domain $|x| \leq R$ for $t \geq T(R)$. Number a was specified in assumption L10.3.

As follows from these assumptions, the discontinuities of the Green matrix $G(t, x, x^0)$, where $|x^0| \leq a$, go to infinity for $t \to +\infty$.

Let $H^s(\Omega)$ denote the Sobolev space of functions on domain $\Omega \subset \mathbf{R}^n$, the derivatives of which are square-integrable up to the order s. We assume below that $s \geq 0$ is an integer. We shall denote by the same symbol $H^s(\Omega)$ the spaces of vector functions and of matrix functions, all elements of which belong to space $H^s(\Omega)$.

We introduce the resolvent R_k of the stationary problem and study its analytic properties (with respect to parameter k). Consider an operator R_k

(the resolvent):

$$R_k : H^s(\mathbf{R}^n) \to H^{s+m}(\mathbf{R}^n)$$

in the half-plane $\mathrm{Im}\, k > 0$. This operator acts according to the formula

$$R_k f = u,$$

where $f \in H^s(\mathbf{R}^n)$, and u is the solution of the equation $\mathcal{L}_k u = f$, belonging to $H^{s+m}(\mathbf{R}^n)$. Resolvent R_k is a finitely meromorphic function of k in the half-plane $\mathrm{Im}\, k > 0$ [83].

All points of the real axis k belong to the continuous spectrum of the resolvent, hence operator R_k cannot be analytically continued from the upper ($\mathrm{Im}\, k > 0$) to the lower half-plane. It is possible to associate an operator $\hat{R}_k$ with the resolvent R_k such that $\hat{R}_k$ admits an analytic continuation into the lower half-plane.

Let $H^s_a \subset H^s(\mathbf{R}^n)$ be the space of functions identically equal to zero for $|x| \geq a$. We introduce the operator

$$\hat{R}_k : H^s_a \to H^{s+m}(|x| \leq b),$$

where $b \geq a$ is a fixed constant. By definition, we have

$$\hat{R}_k f(x) \equiv R_k f(x), \qquad (|x| \leq b),$$

if function $f(x) \in H^s_a$. Thus operator $\hat{R}_k$ is obtained from the resolvent R_k by restricting its domain of definition from $H^s(\mathbf{R}^n)$ to H^s_a and by extending its range of values from $H^{s+m}(\mathbf{R}^n)$ to $H^{s+m}(|x| \leq b)$.

The operator $\hat{R}_k$ admits a continuation into the whole complex plane as a *finitely meromorphic* function if n is odd (n is the dimension of the space), and into the Riemann surface of the function $\ln k$, if n is even [83].

We derive the asymptotics of operator $\hat{R}_k$ for $|k| \to \infty$ in the domain

$$U_{\alpha,\beta} = \{k : |\mathrm{Im}\, k| < \alpha \ln |\mathrm{Re}\, k| - \beta\},$$

where α, β are constants. More precisely, we find the asymptotics of the function $\hat{R}_k f$, where function f belongs to space H^s_a (i.e. it has compact support). The asymptotics of the kernel $\tilde{G}(x, k, x^0)$ of the resolvent cannot be derived by this approach.

We construct the matrix $G_M(t, x, x^0)$ satisfying the equation

$$\mathcal{L} G_M = f_M(t, x, x^0), \qquad t > 0,$$

and the Cauchy data

$$D_t^j G_M\big|_{t=0} = 0, \qquad 0 \le j \le m-2;$$
$$D_t^{m-1} G_M\big|_{t=0} = \delta(x - x^0)I.$$

Here $f_M \in C^M (M \ge 2)$ with respect to the variables (t, x, x^0) and for any $R > 0$

$$G_M(t, x, x^0) \equiv 0, \qquad (|x| \le R, t \ge T(R) + 1).$$

The number $T(R)$ was defined in the assumption L10.5.

It is evident that the matrix function

$$G_M(t, x, x^0) = G^0(t, x, x^0) h(t, x)$$

can be taken as G_M, where $h \in C^\infty$ for $t \ge 0$, $x \in \mathbf{R}^n$, $h \equiv 1$ for $t < T(R) + \frac{1}{2}$, and $h \equiv 0$ for $t > T(R) + 1$. The construction of matrix G^0 was described in Section 4; it is obtained with the help of the canonical operator.

We denote by $\hat{R}_{M,k}$ the Fourier transform of the integral operator with kernel G_M, i.e.

$$\hat{R}_{M,k} \varphi = \int\limits_0^\infty \int\limits_{|x^0| \le a} G_M(t, x, x^0) \varphi(x^0) e^{ikt}\, dx^0\, dt, \qquad (10.48)$$

where the integral is understood in the sense of distributions. It turns out that operator $\hat{R}_{M,k} : H_a^s \to H^{s+m}(|x| \le b)$ is close to operator $\hat{R}_k$ for $|k| \to \infty$, $k \in U_{\alpha,\beta}$.

THEOREM 10.8. *Let n be odd. If M is sufficiently large, the operator $\hat{R}_{M,k}$ is bounded for all k and is an entire function of k. For $|k| \gg 1$, the estimates*

$$\|\hat{R}_{M,k} f\|_{s+m-j,(b)} \le C |k|^{1-j} \exp(\gamma |\mathrm{Im}\, k|)\, \|f\|_{s,a},$$
$$0 \le j \le m+1, \qquad (10.49)$$

hold, where γ is a constant. For $k \in U_{\alpha,\beta}$ the estimates

$$\|(\hat{R}_k - \hat{R}_{M,k}) f\|_{s+m-j,(b)} \le C |k|^{-j} \exp(\gamma |\mathrm{Im}\, k|)\, \|f\|_{s,a} \quad (10.50)$$

are true.

Here $\|f\|_{s,a}$, $\|f\|_{s,(b)}$ are the norms in spaces H_a^s, $H^s(|x| \le b)$, respectively, and estimate (10.50) is valid for some $\alpha, \beta, \gamma > 0$. Similar results can be obtained for n even.

This theorem yields the asymptotics of the function $(\hat{R}_k f)(x)$ (i.e. of the solution of Equation (10.45) with a compactly supported right-hand side $f \in H_a^s$), when $|k| \to \infty$ in a domain of the form $U_{\alpha,\beta}$ and x lies in the compact set $|x| \leq b$. In particular, the estimates (10.49) and (10.50) together with formula (10.48) yield the asymptotics of the resolvent $(R_k f)(x)$ on the compact set $|x| \leq b$ for $|k| \to \infty$.

Moreover, Theorem 10.8 enables the evaluation of the asymptotic solution (for $t \to +\infty$) of the Cauchy problem for the homogeneous Equation (10.26) with compactly supported initial data. Let us consider the Cauchy problem

$$\mathscr{L}u = 0; \qquad D_t^k u\big|_{t=0} = 0, \qquad 0 \leq k \leq m-2;$$
$$D_t^{m-1} u\big|_{t=0} = \varphi(x), \tag{10.51}$$

where $\varphi(x) \in H_a^s$.

Let n be odd; then operator $\hat{R}_k$ has no more than a finite number of poles in each strip $c_1 < \text{Im } k < c_2$ of a finite width.

THEOREM 10.9. *Let n be odd and let there be no poles of operator $\hat{R}_k$ on the straight line $\text{Im } k = q$. Then the following asymptotic expansion of the solution of the Cauchy problem (10.51) is valid:*

$$u(t,x) = -i \sum_{\text{Im } k_j < q} \text{res}_{k=k_j} (\hat{R}_k \varphi(x) e^{-ikt}) + u_q(t,x). \tag{10.52}$$

For the remainder we have the estimate

$$\| u_q \|_{s+m,(b)} \leq c_q e^{qt} \| \varphi(x) \|_{s,a}. \tag{10.53}$$

The relations (10.52), (10.53) hold for $t \geq t_0 \gg 1$ so that formula (10.52) represents the asymptotic expansion of solution u for $t \to +\infty$.

Notice that the sum on the right-hand side of formula (10.52) consists of a finite number of terms.

Similar results can be derived for spaces of even dimensions.

Theorem 10.9 is deduced from Theorem 10.8 in the following way. Applying the formula for the inverse Laplace transformation we obtain

$$u(t,x) = \frac{1}{2\pi} \int_{c-i\infty}^{c+i\infty} e^{-ikt} R_k \varphi(x)\, dk,$$

where $c > 0$ is such that operator $\hat{R}_k$ has no poles for $\text{Im } k \geq c$. By virtue

of the estimate given in Theorem 10.9, the integration path can be shifted
parallel to the real axis so that function u is equal to the integral along
the straight line $\operatorname{Im} k = q < c$ plus the sum of the residues of the integrand
for the poles lying in the strip $c < \operatorname{Im} k < q$. The rigorous implementation of
these arguments yields Theorem 10.9.

Similar results were obtained in [84] for the mixed problem concerning
the non-stationary Equation (10.26) and the corresponding stationary
one. Thus consider the problem (10.26), (10.29) for $t > 0, x \in \Omega$, where Ω is
the complement of a bounded domain in $\mathbf{R}^n$ with boundary $\partial\Omega$ of class C^∞
The following boundary conditions are put on $\partial\Omega$:

$$B(x, D_t, D_x)u(t, x) = 0, \qquad (t > 0, x \in \partial\Omega); \tag{10.54}$$

here B is an $((mN/2) \times N)$-matrix of differential operators of orders
$\leq m - 1$ with the coefficients of class $C^\infty(\partial\Omega)$. It is assumed that the
boundary-value problem for stationary equation

$$\mathscr{L}_k v = f, \quad x \in \Omega; \qquad B(x, k, D_x)v = \varphi, \quad x \in \partial\Omega,$$

represents a coercive problem with a parameter for one ray $k = \rho e^{i\varphi_0}$
$(0 < \rho < \infty, 0 < \varphi_0 < \pi)$ at least; for details see [83]. The conditions
L10.1–L10.4 are assumed satisfied as before. Instead of assumption
L10.5 (which represents a requirement on the behaviour of trajectories of
the Hamilton system associated with operator $\mathscr{L}$) one introduces a
condition with the following content. Let $G(t, x, x^0)$ be the Green matrix
of the mixed problem (i.e., G satisfies the homogeneous equation, the
homogeneous boundary conditions and the Cauchy data (10.31)); it is
required that the discontinuities of Green matrix G go to infinity for
$t \to +\infty$.

In the work [84], the asymptotic solution of the scattering problem
within the semi-classical approximation was also derived for the stationary
Schrödinger equation with a compactly supported potential.

§ 11. Matrix Hamiltonians

This paragraph is devoted to the Cauchy problem with rapidly oscillating
initial data for the systems of equations. It is assumed that the
characteristics of the considered operators are real and with a constant
multiplicity. The transport equations are derived and the global asymptotic solution of the Cauchy problem is constructed.

1. *The Hamilton–Jacobi Equation*

Let u be an N-vector (a column) $u(x) = (u_1(x), \ldots, u_N(x))$, $x \in \mathbf{R}^n$, and L be an $(N \times N)$-matrix with elements $L_{jk}(x, p; (i\lambda)^{-1})$. Consider the system of N equations

$$L(\overset{2}{x}, \lambda^{-1}\overset{1}{D}_x; (i\lambda)^{-1})u = 0. \tag{11.1}$$

Everywhere in this section we assume that the following assumption holds.

L11.1. The symbol of operator L belongs to class T_+^m (see Definition 2.5), i.e. all elements L_{jk} of matrix $L(x, p; (i\lambda)^{-1})$ belong to class T_+^m with one and the same m.

We shall look for a f.a. solution of Equation (11.1) in the form of the formal series

$$u(x, \lambda) = \exp\left[i\lambda S(x)\right] \sum_{m=0}^{\infty} (i\lambda)^{-m} \varphi^m(x). \tag{11.2}$$

By applying Theorem 2.6 to operators L_{jk}, we obtain

$$\exp(-i\lambda S)\, L(\varphi \exp(i\lambda S)) = R_0\, \varphi + (i\lambda)^{-1} R_1\, \varphi + O_{-2}(x, \lambda). \tag{11.3}$$

The operators R_0, R_1 are given by

$$(R_0\, \varphi)(x) = L\left(x, \frac{\partial S(x)}{\partial x}; 0\right)\varphi(x),$$

$$(R_1\, \varphi)(x) = \frac{\partial L}{\partial p}\frac{\partial \varphi(x)}{\partial x} + \tfrac{1}{2}\,\mathrm{Sp}\,(S''_{xx}(x)L''_{pp})\varphi(x) + \frac{\partial L}{\partial \varepsilon}\bigg|_{\varepsilon=0}\varphi(x), \tag{11.4}$$

where the value of matrix L is taken at the point $(x, \partial S(x)/\partial x; 0)$. We remind that $R_0\, \varphi, R_1\, \varphi$ are N-vectors (columns). For continuity of our exposition we retain all notations used for the scalar Hamiltonians. Componentwise we have

$$(R_1\, \varphi)_m = \sum_{j=1}^{N} \sum_{k=1}^{n} \frac{\partial L_{mj}}{\partial p_k}\frac{\partial \varphi_j}{\partial x_k} +$$

$$+ \frac{1}{2} \sum_{j=1}^{N} \sum_{k,m=1}^{n} \frac{\partial^2 L_{mj}}{\partial p_k \partial p_m}\frac{\partial^2 S}{\partial x_k \partial x_m}\varphi_j + \frac{\partial L_{mj}}{\partial \varepsilon}\bigg|_{\varepsilon=0}\varphi_j. \tag{11.4'}$$

Let $L_m = (\dot{L}_{m1}, \ldots, L_{mN})$ be the mth row of matrix L; then formula (11.4')
takes the more compact form

$$(R_1 \varphi)_m = \frac{\partial L_m}{\partial p} \frac{\partial \varphi}{\partial x} + \tfrac{1}{2} \operatorname{Sp} \left(S''_{xx} \frac{\partial^2 L_m}{\partial p^2} \right) \varphi + \frac{\partial L_m}{\partial \varepsilon} \bigg|_{\varepsilon = 0} \varphi. \qquad (11.4'')$$

Here $\partial \varphi / \partial x$ is a Jacobi matrix, $\partial^2 L / \partial p^2$ is a tensor, i.e.

$$\frac{\partial \varphi}{\partial x} = \left(\frac{\partial \varphi_j}{\partial x_k} \right), 1 \leq j, k \leq n, \quad \text{and} \quad \left(\frac{\partial^2 L}{\partial p^2} \right)_{jk} = \frac{\partial^2 L}{\partial p_j \, \partial p_k} .$$

We introduce the notation: $f(x, p)$ and $f^*(x, p)$ are the right and left
null-vectors, respectively, of the matrix $L^0 = L(x, p; 0)$ (f being a column,
f^* a row), i.e.

$$L^0 f(x, p) = 0, \qquad f^*(x, p) L^0 = 0. \qquad (11.5)$$

Let Ω be a bounded domain in $\mathbf{R}^n_x$. We are interested in f.a. solutions
of Equation (11.1) for $\lambda \to + \infty, x \in \Omega$. We look for a solution of the form

$$u(x, \lambda) = \exp \left[i\lambda S(x) \right] \varphi(x). \qquad (11.6)$$

Inserting u in Equation (11.1) and setting the coefficient of λ^0 in the
expression $e^{-i\lambda S} L e^{i\lambda S} \varphi$ equal to zero, we get, by virtue of (11.3) and (11.4),
the equation

$$L \left(x, \frac{\partial S(x)}{\partial x}; 0 \right) \varphi(x) = 0. \qquad (11.7)$$

Consequently, we have

$$\Delta \left(x, \frac{\partial S(x)}{\partial x} \right) = 0 \qquad (11.8)$$

for $x \in \Omega$, where $\Delta(x, p) = \det L(x, p; 0)$. This equation is the *Hamilton–
Jacobi equation* associated with operator L.

PROPOSITION 11.1. *Let function* $S(x) \in C^\infty(\Omega)$ *be real-valued and
satisfy Hamilton–Jacobi equation* (11.8) *in a domain* Ω. *Let the vector
function* $\varphi(x) \in C^\infty(\Omega)$ *be a right null-vector of the matrix* $L(x, \partial S(x)/\partial x; 0)$
for $x \in \Omega$. *Then the vector function* (11.6) *is a solution of Equation* (11.1).

The null-vector $\varphi(x)$ is obviously not determined uniquely; for instance,
if $\sigma(x)$ is an arbitrary scalar function, then $\sigma(x) \varphi(x)$ is a null-vector of
matrix $L(x, \partial S(x)/\partial x; 0)$. The normalization of the null-vector $\varphi(x)$ (more

precisely, of the basis in the subspace of null-vectors) is found when the next term of the asymptotic expansion is evaluated.

We shall discuss similarly the evolution system of N equations

$$Lu \equiv [I\lambda^{-1}D_t + H(t, \overset{2}{x}, \lambda^{-1}\overset{1}{D}_x; (i\lambda)^{-1})]u(t, x) = 0. \tag{11.9}$$

Here $x \in \mathbf{R}^n, t \in \mathbf{R}, u = (u_1, \ldots, u_N)$ (a column vector), H is an $(N \times N)$-matrix with elements H_{jk}, I is the unit $(N \times N)$-matrix. Let H_{jk} satisfy the condition L11.1. We formulate the Cauchy problem

$$u\big|_{t=0} = \exp[i\lambda S_0(x)]u^0(x), \tag{11.10}$$

where $S_0(x) \in C^\infty(\mathbf{R}^n), u^0(x) \in C_0^\infty(\mathbf{R}^n)$, and function S_0 is real-valued. The f.a. solution of the Cauchy problem (11.9), (11.10) will be sought in the form

$$u(t, x, \lambda) = \exp[i\lambda S(t, x)][\varphi^0(t, x) + (i\lambda)^{-1}\varphi^1(t, x) + \ldots]. \tag{11.11}$$

It follows from (11.8) that function S must satisfy the Hamilton–Jacobi equation

$$\det\left(I\frac{\partial S}{\partial t} + H\left(t, x, \frac{\partial S}{\partial x}; 0\right)\right) = 0. \tag{11.12}$$

We shall study the case when the eigenvalues $h_\alpha(t, x, p)$ of the matrix $H(t, x, p; 0)$ are distinct or have a constant multiplicity. Then the phase function S will satisfy one of the equations

$$\frac{\partial S}{\partial t} + h_\alpha\left(t, x, \frac{\partial S}{\partial x}\right) = 0, \tag{11.13}$$

which will also be called the *Hamilton–Jacobi equations*. Further, let $f^\alpha(t, x, p)$ and $f^{*\alpha}(t, x, p)$ be the right and left eigenvectors of matrix $H(t, x, p; 0)$, respectively. If S fulfils Equation (11.13), then the vector function $\varphi^0(t, x)$ will be the right eigenvector of matrix $H(t, x, \partial S/\partial x; 0)$, i.e.

$$H\left(t, x, \frac{\partial S}{\partial x}; 0\right)\varphi^0 = h_\alpha\varphi^0. \tag{11.14}$$

2. *The Transport Equations*

Consider Equation (11.1). We suppose in this section that the symbol $L(x, p; \varepsilon)$ is real for real x, p, ε, and that the equation

$$\Delta(x, p) = 0 \tag{11.15}$$

determines a C^∞-manifold of dimension $2n - 1$ in the phase space. All constructions have a local character (i.e. in a small neighbourhood of some point (x^0, p^0)). All null-vectors are also taken to be real.

The Hamilton system corresponding to Equation (11.15) has the form

$$\frac{dx}{d\tau} = \frac{\partial \Delta}{\partial p}, \qquad \frac{dp}{d\tau} = -\frac{\partial \Delta}{\partial x}. \tag{11.16}$$

The vector function $R_1 \varphi$ contains the term $(\partial L/\partial p)(\partial \varphi/\partial x)$ (see (11.4)), while the derivative of vector function $\varphi(x)$ with respect to the Hamilton system (11.16) is equal to $(\partial \Delta/\partial p)(\partial \varphi/\partial x) = \dot{\varphi}$. We shall establish the connection between the derivatives $\partial L/\partial p_j$ and $\partial \Delta/\partial p_j$. First we consider the case in which matrix $L(x, p; 0)$ has a simple null-vector.

LEMMA 11.2 [11]. *Let*

$$\Delta(x^0, p^0) = 0, \qquad \frac{\partial \Delta(x^0, p^0)}{\partial p} \neq 0, \tag{11.17}$$

and rank $L(x, p; 0) = N - 1$ *in a neighbourhood of the point* (x^0, p^0). *Then there exists a constant C such that*

$$f^* \frac{\partial L}{\partial p_j} f = C \frac{\partial \Delta}{\partial p_j}, \qquad 1 \leq j \leq n, \tag{11.18}$$

in the point (x^0, p^0).

Proof. We fix $x = x^0$. By virtue of condition (11.17), Equation (11.15) defines an $(n - 1)$-dimensional C^∞-manifold M in a neighbourhood of the point p^0. Let $A_{ij}(p)$ be the algebraic complement of the element $L_{ij}(x^0, p; 0)$. By assumption, at least one of the numbers $A_{ij}(p^0)$ is different from zero; thus let $A_{i_0 j_0}(p^0) \neq 0$. Then vector f with the components $A_{i_0 1}(p), \ldots, A_{i_0 N}(p)$ is a right null-vector of matrix $L(x^0, p; 0)$ for p close to $p^0, p \in M$, it belongs to class C^∞ and is different from zero. The right null-vector is unique up to a factor by hypothesis; analogous statements are true for the left null-vector $f^*(p)$. Differentiating the identities $\Delta = 0$, $Lf = 0$ with respect to p (for $p \in M$) and then multiplying the latter from the left by f^*, we obtain the relations

$$\sum_{j=1}^{n} \frac{\partial \Delta}{\partial p_j} dp_j = 0, \qquad \sum_{j=1}^{n} f^* \frac{\partial L}{\partial p_j} f \, dp_j = 0.$$

The second relation holds for all dp_j constrained by the first one, provided $p = p^0$. Since, due to condition (11.17), among the differentials dp_j

there are exactly $n - 1$ independent, the coefficients $f^*(\partial L/\partial p_j)f$ of differentials dp_j are proportional to the coefficients $\partial \Delta/\partial p_j$.

The lemma can be extended to the case of the characteristics with constant multiplicities.

LEMMA 11.3. [11]. *Suppose that in a neighbourhood of the point* (x^0, p^0):

(1) *Equation* (11.17) *is equivalent to the equation* $\Delta_0(x, p) = 0$, *where* $\partial \Delta_0(x^0, p^0)/\partial p \neq 0$;

(2) *there exists a basis* $\{f^1(x, p), \ldots, f^k(x, p)\}$ *of class* C^τ *of the right null-space of matrix* $L(x, p; 0)$.

Then there exists a constant C *such that*

$$f^{*\alpha} \frac{\partial L}{\partial p_j} f^\beta = C \frac{\partial \Delta_0}{\partial p_j} \tag{11.19}$$

for all $\alpha, \beta \in \{1, \ldots, k\}$, $1 \leq j \leq n$, *where* $\{f^{*1}, \ldots, f^{*k}\}$ *is a basis of the left null-space of the matrix* $L(x, p; 0)$.

The proof is completely analogous to that of the previous lemma.

Inserting (11.2) into Equation (11.1) we obtain the system of recursion relations

$$R_0 \varphi^0 = 0, \ldots, R_0 \varphi^k = -R_1 \varphi^{k-1} - R_2 \varphi^{k-2} - \ldots - R_k \varphi^0. \tag{11.20}$$

The first of these equations has already been analyzed, hence we shall study the others. Namely, we show that these equations lead to the transport equations along trajectories of the Hamilton system (11.16). These equations turn out to be more complex than those for the scalar Hamiltonians. The next considerations will have a local character.

Let function $S(x)$ satisfy the Hamilton–Jacobi equation in a neighbourhood of the point x^0 and let the assumptions of Lemma 11.2 be fulfilled in a neighbourhood of the point (x^0, p^0), $p^0 = \partial S(x^0)/\partial x$, i.e. S is a simple characteristic. Let $f(x)$ and $f^*(x)$ be the right and the left null-vectors, respectively, of the matrix $L^0 = L(x, \partial S(x)/\partial x; 0)$ of class C^∞ for x close to x^0. Then

$$\varphi^0(x) = \sigma(x)f(x), \tag{11.21}$$

where σ is a scalar function. Consider the equation

$$R_0 \varphi^1(x) = -R_1 \varphi^0(x). \tag{11.22}$$

For x fixed we get the system of n linear algebraic equations in the unknowns $\varphi_1, \ldots, \varphi_N$, with the determinant of the system equal to zero. The necessary and sufficient condition for solvability of the system is the identity

$$f^*(x) R_1 \varphi^0(x) = 0 \tag{11.23}$$

(in a neighbourhood of point x^0). Consequently, function $\sigma(x)$ has to satisfy the equation

$$\sum_{k=1}^{n} \frac{\partial \sigma}{\partial x_k} f^* \frac{\partial L}{\partial p_k} f + M_1 \sigma = 0, \tag{11.24}$$

where

$$M_1 = \sum_{k=1}^{n} f^* \frac{\partial L}{\partial p_k} \frac{\partial f}{\partial x_k} + \tfrac{1}{2} f^* \operatorname{Sp}\left(S''_{xx} L''_{pp}\right) f + f^* \frac{\partial L}{\partial \varepsilon}\bigg|_{\varepsilon=0} f \tag{11.25}$$

(in the same notation as in (11.4)). Let $(x(\tau), p(\tau))$ be a solution of the system (11.16); then

$$\frac{d}{d\tau}(\sigma \circ x)(\tau) = \sum_{k=1}^{n} \frac{\partial \sigma}{\partial x_k} \frac{\partial \Delta}{\partial p_k}.$$

By virtue of Lemma 11.2 we have

$$f^* \frac{\partial L}{\partial p_k} f = C(x) \frac{\partial \Delta}{\partial p_k} \tag{11.26}$$

(the values of f^*, f are taken at point x, and the values of the derivatives of L and Δ at the point $(x, \partial S(x)/\partial x; 0)$). As a consequence, Equation (11.24) along the trajectory $x = x(\tau), p = p(\tau)$ has the form

$$C \frac{d\sigma}{d\tau} + M_1 \sigma = 0. \tag{11.27}$$

We shall write down the leading term of the f.a. solution of Equation (11.1). Suppose compatible values of S and ∇S are given on a smooth $(n-1)$-dimensional manifold M in $\mathbf{R}^n_x$ (i.e. Equation (11.8) is satisfied in a neighbourhood of M). Consider the Lagrangian Cauchy problem

$$x\big|_{\tau=0} = x^0, \qquad p\big|_{\tau=0} = p^0 = \frac{\partial S(x^0)}{\partial x}, \qquad x^0 \in M,$$

for the Hamilton system (11.16), and let $x(\tau; x^0), p(\tau; x^0)$ be its solution.

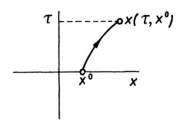

Fig. 14.

We take an arbitrary right null-vector of matrix $L(x, \partial S(x)/\partial x; 0)$ as $f(x)$. Then the leading term of the f.a. solution of Equation (11.1) (near M) has the form

$$u(x, \lambda) = \exp\left[i\lambda S(x) - \int_0^\tau C^{-1} M_1 \, d\tilde{\tau}\right] f(x) \qquad (11.28)$$

(see (11.6)). Here C and M_1 are determined by (11.18), (11.25), and the integral is taken along a phase trajectory such that $x(\tau; x^0) = x$ (Figure 14).

3. *The Cauchy Problem* (11.9), (11.10) *in the Small*

The main problem we are interested in, is the Cauchy problem (11.9), (11.10). In this case the formulae are much simpler, since operator L is a first order operator with respect to the variable t. We introduce the assumption

L11.2. The matrix $H^0 = H(t, x, p; 0)$ is Hermitian for $t \geq 0$, $(x, p) \in \mathbf{R}^{2n}$.

The discussion which follows has a *local character* – it concerns a small neighbourhood U of a fixed point $Q_0 = (t_0, x^0, p^0)$. We introduce the assumption

L11.3. The eigenvalues $h_\alpha(Q_0)$, $1 \leq \alpha \leq N$, are distinct.

Then eigenvalues h_α are infinitely differentiable in the neighbourhood U and there exists an orthonormal basis $\{f^\alpha(t, x, p)\}$, $1 \leq \alpha \leq N$, formed by the eigenvectors of matrix H^0.

The Hamilton system associated with the Hamilton–Jacobi equation (11.13) is given by

$$\frac{dx}{dt} = \frac{\partial h_\alpha}{\partial p}, \qquad \frac{dp}{dt} = -\frac{\partial h_\alpha}{\partial x} \qquad (11.29)$$

Let us derive the transport equations. Inserting (11.11) into evolution Equation (11.9) and setting the coefficient of λ^{-1} equal to zero, we obtain the equation

$$R_0 \varphi^1 = -R_1 \varphi^0, \tag{11.30}$$

where

$$R_0 = I \frac{\partial S}{\partial t} + H\left(t, x, \frac{\partial S}{\partial x}; 0\right).$$

If function S satisfies Equation (11.13), then det $R_0 = 0$, and φ^0 is the eigenvector corresponding to the eigenvalue h_α. Let us fix a (smooth) eigenvector f^α. Then

$$\varphi^0 = \sigma f^\alpha,$$

where σ is a scalar function, since h_α is a simple eigenvalue. The necessary and sufficient condition of solvability of Equation (11.30) is the condition

$$\langle f^\alpha, R_1 \sigma f^\alpha \rangle = 0. \tag{11.31}$$

The operator R_1 has the form

$$R_1 \varphi = \left[\frac{\partial \varphi}{\partial t} + \sum_{j=1}^{n} \frac{\partial H}{\partial p_j} \frac{\partial \varphi}{\partial x_j} \right] +$$

$$+ \left[\frac{1}{2} \sum_{j,k=1}^{n} \frac{\partial^2 S}{\partial x_j \partial x_k} \frac{\partial^2 H}{\partial p_j \partial p_k} + \frac{\partial H}{\partial \varepsilon} \right] \varphi = \tilde{R}_1 \varphi + \tilde{\tilde{R}}_1 \varphi, \tag{11.32}$$

where $H = H(t, x, p; 0)$, $p = \partial S(x)/\partial x$. We shall transform Equation (11.31). We introduce the determinant

$$J_\alpha(t, x) = \det \frac{\partial x^\alpha(t, y)}{\partial y}. \tag{11.33}$$

Here $\{x^\alpha(t, y), p^\alpha(t, y)\}$ is the solution of the Hamilton system (11.29) with the Cauchy data

$$x\big|_{t=0} = y, \qquad p\big|_{t=0} = \frac{\partial S_0(y)}{\partial y}. \tag{11.34}$$

Now we shall prove the analogue of Lemma 3.14 for Equation (11.31).

LEMMA 11.4. *Let the assumptions* L11.1–L11.3 *be satisfied,* $\sigma = \sigma(t, x)$

be a scalar function, and a vector function f^α be such that $\langle f^\alpha, f^\alpha \rangle \equiv 1$. Then

$$\langle f^\alpha, R_1(\sigma f^\alpha) \rangle = \frac{d\sigma}{dt} + \frac{\sigma}{2}\frac{d}{dt} \ln J_\alpha + M_\alpha \sigma. \qquad (11.35)$$

Here d/dt *is the derivative with respect to the Hamilton system* (11.29),

$$M_\alpha = \sum_{j=1}^{n} \left\langle f^\alpha, \left(\frac{\partial H}{\partial p_j} - \frac{\partial h_\alpha}{\partial p_j}\right)\frac{\partial f^\alpha}{\partial x_j}\right\rangle -$$

$$- \frac{1}{2}\sum_{j=1}^{n} \frac{\partial^2 h_\alpha}{\partial x_j \partial p_j} + \left\langle f^\alpha, \frac{\partial H}{\partial \varepsilon}\bigg|_{\varepsilon=0} f^\alpha \right\rangle + \left\langle f^\alpha, \frac{df^\alpha}{dt} \right\rangle,$$

$$\qquad\qquad (11.36)$$

$$H = H\left(t, x, \frac{\partial S}{\partial x}; 0\right).$$

Proof. Vector function f^α has the form $f^\alpha = f^\alpha(t, x, p)$, where $p = \partial S(t, x)/\partial x$, i.e. it is a composite function of x. The symbol D_{x_j} denotes the total derivative with respect to the variable x_j, i.e.

$$D_{x_j} f^\alpha = \frac{\partial f^\alpha}{\partial x_j} + \sum_{k=1}^{n} \frac{\partial f^\alpha}{\partial p_k}\frac{\partial^2 S}{\partial x_j \partial x_k}.$$

Notice that total derivatives appear in formula (11.32). Furthermore, if $\varphi = \varphi(t, x, p)$ is a scalar or a vector function, its derivative with respect to the Hamilton system (11.29) has the form

$$\frac{d\varphi}{dt} = \frac{\partial \varphi}{\partial t} + \sum_{j=1}^{n} \frac{\partial h_\alpha}{\partial p_j} D_{x_j}\varphi$$

(here $p = \partial S/\partial x$).

We derive some algebraic identities. By differentiating the identity

$$H(t, x, p; 0)f^\alpha(t, x, p) = h_\alpha(t, x, p)f^\alpha(t, x, p)$$

with respect to the variable p_j, we obtain

$$\left(\frac{\partial H}{\partial p_j} - \frac{\partial h_\alpha}{\partial p_j}\right)f^\alpha = (h_\alpha - H)\frac{\partial f^\alpha}{\partial p_j}. \qquad (11.37)$$

In this and the subsequent identities, t, x, and p are independent variables. Forming the scalar product of identity (11.37) with f^β, we obtain

$$\left\langle f^\beta, \frac{\partial H}{\partial p_j}f^\alpha \right\rangle = \frac{\partial h_\alpha}{\partial p_j}\delta_{\alpha\beta} + (h_\alpha - h_\beta)\left\langle f^\beta, \frac{\partial f^\alpha}{\partial p_j} \right\rangle. \qquad (11.38)$$

By differentiating the initial identity with respect to the variables p_j, p_k and forming the scalar product with f^β, we get

$$\left\langle f^\beta, \frac{\partial^2 H}{\partial p_j \partial p_k} f^\alpha \right\rangle = \frac{\partial^2 h_\alpha}{\partial p_j \partial p_k} \delta_{\alpha\beta} + \left\langle f^\beta, \left(\frac{\partial h_\alpha}{\partial p_j} - \frac{\partial H}{\partial p_j} \right) \frac{\partial f^\alpha}{\partial p_k} \right\rangle +$$

$$+ \left\langle f^\beta, \left(\frac{\partial h_\alpha}{\partial p_k} - \frac{\partial H}{\partial p_k} \right) \frac{\partial f^\alpha}{\partial p_j} \right\rangle. \tag{11.39}$$

We have

$$R_1(\sigma f^\alpha) = A(\sigma) f^\alpha + B\sigma$$

with the notation:

$$A(\sigma) = \frac{\partial \sigma}{\partial t} + \sum_{j=1}^{n} \frac{\partial H}{\partial p_j} D_{x_j} \sigma,$$

$$B = \frac{\partial f^\alpha}{\partial t} + \sum_{j=1}^{n} \frac{\partial H}{\partial p_j} D_{x_j} f^\alpha + \frac{1}{2} \sum_{j,k=1}^{n} \frac{\partial^2 H}{\partial p_j \partial p_k} f^\alpha + \frac{\partial H}{\partial \varepsilon} \bigg|_{\varepsilon=0} f^\alpha.$$

Taking identity (11.38) into account we get

$$\langle f^\alpha, A(\sigma) f^\alpha \rangle = \frac{d\sigma}{dt}.$$

To evaluate the expression $\langle f^\alpha, B \rangle$, we use the Liouville formula (§ 3)

$$\frac{d}{dt} \ln J_\alpha = \sum_{j,k=1}^{n} \frac{\partial^2 S}{\partial x_j \partial x_k} \frac{\partial^2 h_\alpha}{\partial p_j \partial p_k} + \sum_{j=1}^{n} \frac{\partial^2 h_\alpha}{\partial x_j \partial p_j},$$

the identity

$$\sum_{j=1}^{n} \frac{\partial f^\alpha}{\partial p_j} \frac{\partial^2 S}{\partial x_j \partial x_k} = D_{x_k} f^\alpha - \frac{\partial f^\alpha}{\partial x_k}$$

and identity (11.39). This leads to

$$\langle f^\alpha, B \rangle = \left\langle f^\alpha, \frac{\partial f^\alpha}{\partial t} + \sum_{j=1}^{n} \frac{\partial H}{\partial p_j} D_{x_j} f^\alpha + \right.$$

$$+ \frac{1}{2} \sum_{j,k=1}^{n} \frac{\partial^2 h_\alpha}{\partial p_j \partial p_k} \frac{\partial^2 S}{\partial x_j \partial x_k} + \sum_{j,k=1}^{n} \frac{\partial h_\alpha}{\partial p_j} \frac{\partial f^\alpha}{\partial p_k} \frac{\partial^2 S}{\partial x_j \partial x_k} -$$

$$- \sum_{j,k=1}^{n} \frac{\partial H}{\partial p_j} \frac{\partial f^\alpha}{\partial p_k} \frac{\partial^2 S}{\partial x_j \partial x_k} + \frac{\partial H}{\partial \varepsilon} \bigg|_{\varepsilon=0} f^\alpha \right\rangle =$$

$$= \left\langle f^{\alpha}, \frac{\partial f^{\alpha}}{\partial t} + \sum_{j=1}^{n} \frac{\partial h_{\alpha}}{\partial p_{j}} D_{x_{j}} f^{\alpha} \right\rangle +$$

$$+ \frac{1}{2} \left[\frac{d}{dt} \ln J_{\alpha} - \sum_{j=1}^{n} \frac{\partial^{2} h_{\alpha}}{\partial x_{j} \partial p_{j}} \right] +$$

$$+ \sum_{j=1}^{n} \left\langle f^{\alpha}, \left(\frac{\partial H}{\partial p_{j}} - \frac{\partial h_{\alpha}}{\partial p_{j}} \right) \frac{\partial f^{\alpha}}{\partial x_{j}} \right\rangle + \left\langle f^{\alpha}, \frac{\partial H}{\partial \varepsilon} \bigg|_{\varepsilon=0} f^{\alpha} \right\rangle.$$

Applying the formula for the derivative df^{α}/dt with respect to the Hamilton system, we obtain (11.35) and (11.36).

Suppose the assumptions L11.1 and L11.2 are fulfilled as before. Instead of assumption L11.3' introduce the assumption.

L11.3″. In neighbourhood U of point Q_0 the matrix H has an eigenvalue $h(t, x, p)$ with a constant multiplicity κ, and there exists an orthonormal basis $\{f^{\alpha}\}$, $1 \leq \alpha \leq \kappa$, consisting of the eigenvectors of class $C^{\infty}(U)$ corresponding to eigenvalue h.

This condition is satisfied, e.g., for the Dirac equation (§ 14).

Then any eigenvector f corresponding to eigenvalue h can be represented in the form

$$f = \sum_{\alpha=1}^{\kappa} \sigma_{\alpha} f^{\alpha},$$

where σ_{α} are scalar functions. In this case we obtain the system of equations for functions σ_{α} instead of the scalar transport equation. Namely, we have

LEMMA 11.5. *Let the assumptions L11.1, L11.2, L11.3' be satisfied. Then*

$$\langle f^{\beta}, R_{1}(\sigma f^{\alpha}) \rangle = \delta_{\alpha\beta} \left(\frac{d}{dt} + \frac{1}{2} \frac{d}{dt} \ln J \right) \sigma + M_{\alpha\beta} \sigma,$$

$$1 \leq \alpha, \beta \leq \kappa, \qquad (11.40)$$

$$M_{\alpha\beta} = \left\langle f^{\beta}, \frac{df^{\alpha}}{dt} \right\rangle + \sum_{j=1}^{n} \left\langle f^{\beta}, \left(\frac{\partial H}{\partial p_{j}} - \frac{\partial h}{\partial p_{j}} \right) f^{\alpha} \right\rangle -$$

$$- \frac{1}{2} \delta_{\alpha\beta} \sum_{j=1}^{n} \frac{\partial^{2} h}{\partial x_{j} \partial p_{j}} + \left\langle f^{\beta}, \frac{\partial H}{\partial \varepsilon} \bigg|_{\varepsilon=0} f^{\alpha} \right\rangle. \qquad (11.41)$$

The proof of the lemma follows from that of Lemma 11.4. The system of transport equations has the form

$$\left(\frac{d}{dt} + \frac{1}{2}\frac{d}{dt}\ln J\right)\sigma_\beta + \sum_{\alpha=1}^{\kappa} M_{\alpha\beta}\sigma_\alpha = 0, \qquad 1 \leq \beta \leq \kappa.$$

We look for a particular f.a. solution of the system (11.9) in the form

$$u^\alpha(t, x, \lambda) = \exp\left[i\lambda S_\alpha(t, x)\right] \sum_{k=0}^{m} (i\lambda)^{-k}\, \varphi^{k\alpha}(t, x), \tag{11.42}$$

where S_α is a solution of Equation (11.13). Then the f.a. solution of the Cauchy problem (11.9), (11.10) will be written as a superposition of the particular f.a. solutions. For our systems, two new features appear which were absent in the scalar case.

(1) The first approximation is found with the help of the equation for the second approximation.

Actually, we have seen above that to evaluate φ^0 Equation (11.30) had to be used.

(2) In order to obtain a f.a. solution with accuracy to $\dot{O}(\lambda^{-2})$, one has to know the first two approximations.

Indeed, if we consider only the leading term, then

$$e^{-i\lambda S_\alpha} L(e^{i\lambda S_\alpha}\varphi^{0\alpha}) = \frac{1}{i\lambda} R_1\,\varphi^{0\alpha} + O(\lambda^{-2})$$

and $R_1\,\varphi^{0\alpha} \neq 0$. Hence the f.a. solution with accuracy to $O(\lambda^{-2})$ should be taken in the form

$$u^\alpha = \exp\left[i\lambda S_\alpha\right]\left(\varphi^{0\alpha} + \frac{1}{i\lambda}\varphi^{1\alpha}\right). \tag{11.43}$$

For the scalar equation one can set $\varphi^{1\alpha} \equiv 0$.

We introduce further assumptions.

The assumption L11.3′ will be replaced by a stronger one:

L11.3. The eigenvalues of the matrix $H(t, x, p; 0)$ are distinct for $0 \leq t \leq T, (x, p) \in \mathbf{R}^{2n}$.

L11.4. The Cauchy problem (11.29), (11.34) has a unique solution

$$\{x^\alpha(t, y), p^\alpha(t, y)\} \in C^\infty([0, T] \times \mathbf{R}^n_y).$$

L11.5. The map $(t, y) \to x^\alpha(t, y), t \in [0, T], y \in \mathbf{R}^n$, is a diffeomorphism.

The symbol $O_{-M,T}$ will denote the class of vector functions $\psi(t, x; (i\lambda)^{-1})$ such that:

(1) $\psi \in C^\infty([0, T] \times \mathbf{R}^n_x \times \mathbf{R}^+_\lambda)$;

(2) for arbitrary multi-indices α, β and for arbitrary integer q the estimates

$$\| D_x^\alpha D_t^\beta \psi(t, x; (i\lambda)^{-1}) \| \leq C_{\alpha, \beta, q} \lambda^{-M + |\alpha| + \beta} (1 + |x|)^{-q} \qquad (11.44)$$

hold for $\lambda \geq 1, 0 \leq t \leq T, x \in \mathbf{R}^n$, where the constants C do not depend on t, x, λ.

The estimate (11.44) represents an improved analogue of estimate (3.54). As an example of a vector function $\psi \in O_{-M,T}$ one may consider

$$\psi = \lambda^{-M} \exp [i\lambda S(t, x)] \varphi(t, x),$$

where S is a real-valued C^∞ function for $0 \leq t \leq T, x \in \mathbf{R}^n$, φ is a vector function and also infinitely differentiable, and $\varphi(t, x) \equiv 0$ for $|x| \geq a$, $0 \leq t \leq T$.

If moreover, all derivatives of function S are bounded for $0 \leq t \leq T$, $x \in \mathbf{R}^n$, then we can take as φ a vector function such that

$$\| D_x^\alpha D_t^\beta \varphi(t, x) \| \leq C_{\alpha, \beta, M} (1 + |x|)^{-M}$$

for $0 \leq t \leq T, x \in \mathbf{R}^n$ and for arbitrary α, β, M.

We shall briefly say that the vector function $u(t, x, \lambda)$ is a f.a. solution of the system (11.9) with accuracy to $O(\lambda^{-M})$, if

$$[\lambda^{-1} D_t I + H(t, \overset{2}{x}, \lambda^{-1} \overset{1}{D}_x ; (i\lambda)^{-1})] u = \psi^M \in O_{-M,T}. \qquad (11.45)$$

Now we shall construct particular f.a. solutions of the system (11.9) under the assumptions L11.1–L11.5.

First we shall describe the construction procedure for f.a. solutions, and then we shall formulate the corresponding theorem.

(1) Function $S_\alpha(t, x)$ entering in formula (11.42) is defined as the solution of Equation (11.13) with the Cauchy data

$$S_\alpha(0, x) = S(x). \qquad (11.46)$$

By assumption, $S_\alpha \in C^\infty$ for $0 \leq t \leq T, x \in \mathbf{R}^n$.

(2) Now we show how the vector function $\varphi^{0\alpha}$ can be evaluated (see (11.42)).

Let $\{ f^\alpha(t, x, p) \}$, $1 \leq \alpha \leq N$, be an orthonormal basis in $\mathbf{R}^n$ (for any fixed (t, x, p)) of class C^∞ for $0 \leq t \leq T, (x, p) \in \mathbf{R}^{2n}$, consisting of eigenvectors of matrix $H(t, x, p; 0)$. We choose these vector functions in such a way that all vector functions $f^\alpha(0, x, \partial S(x)/\partial x)$ have compact support. We take $\varphi^{0\alpha}$ in the form

$$\varphi^{0\alpha}(t, x) = \sigma_\alpha(t, x) f^\alpha(t, x, p),$$

where $p = \partial S_\alpha(t, x)/\partial x$, σ_α is an unknown scalar function. Then $R_0 \, \varphi^{0\alpha} \equiv 0$. From (11.31) and Lemma 11.4 we obtain the *transport equation* for σ_α:

$$\frac{d\sigma_\alpha}{dt} + \frac{\sigma_\alpha}{2} \frac{d \ln J_\alpha}{dt} + M_\alpha \sigma_\alpha = 0. \tag{11.35'}$$

By taking some Cauchy data for $t = 0$ and solving this equation, we get

$$\sigma_\alpha = \sqrt{\frac{J_\alpha(0, y)}{J_\alpha(t, y)}} \exp\left[-\int_0^t M_\alpha \, d\tau \right] \sigma_\alpha(0, y).$$

Thus the leading term of the asymptotics is given by

$$u^{\alpha 0}(t, x, \lambda) =$$

$$= \exp\left[i\lambda S_\alpha(t, x) \right] \sqrt{\frac{J_\alpha(0, y)}{J_\alpha(t, y)}} \exp\left[-\int_0^t M_\alpha \, d\tau \right] \sigma_\alpha \bigg|_{t=0} f^\alpha(t, x, p).$$

$$\tag{11.47}$$

In this formula, $x = x^\alpha(t, y)$, $p = p^\alpha(t, y)$ (this is the solution of the Cauchy problem (11.34) for the Hamilton system (11.29); Jacobian J_α and the function M_α are determined from (11.33) and (11.36), respectively).

Notice that $u^{\alpha 0}(t, x, \lambda) \equiv 0$ outside the set

$$x = x^\alpha(t, y), \ 0 \le t \le T, \ y \in \operatorname{supp} f^\alpha(0, y, \partial S(y)/\partial y) \tag{11.48}$$

i.e. outside the set filled by the trajectories originating from the support of the vector function $f^\alpha|_{t=0}$. It is evident that vector function $u^{\alpha 0}$ is smooth for $0 \le t \le T, x \in \mathbf{R}^n, \lambda \ge 1$.

(3) Now we evaluate the next terms of expansion (11.42). By inserting (11.42) into (11.9) and putting the coefficients at powers of λ^{-1} equal to zero, we obtain the system of recursion relations

$$R_0 \varphi^{1\alpha} = -R_1 \varphi^{0\alpha},$$
$$R_0 \varphi^{2\alpha} = -R_1 \varphi^{1\alpha} - R_2 \varphi^{2\alpha}, \tag{11.49}$$

$$\dotsc\dotsc\dotsc\dotsc\dotsc\dotsc\dotsc\dotsc\dotsc\dotsc\dotsc$$

In the scalar case, R_0 is just the operator of multiplication by a function, and $R_0 \equiv 0$ provided S_α is a solution of the Hamilton–Jacobi equation; in the vector case $R_0 \not\equiv 0$.

The evaluation of $\varphi^{1\alpha}$. We have

$$\varphi^{1\alpha} = \sum_{\beta=1}^{N} C_{\alpha\beta}^1 f^\beta \qquad (11.50)$$

and $R_0 f^\beta = (h_\beta - h_\alpha) f^\beta$. Taking the scalar product of the first Equation (11.49) with f^β we get

$$C_{\alpha\beta}^1 = \frac{\langle f^\beta, R_1 \varphi^{0\alpha} \rangle}{h_\alpha - h_\beta}, \qquad \beta \neq \alpha. \qquad (11.50')$$

Here, as in (11.47), the values of functions and vector functions are taken in the point $(t, x^\alpha(t, y), p^\alpha(t, y))$. Since the vector functions $f^\beta(0, x, \partial S(x)/\partial x)$ have compact supports, we have $|h_\alpha - h_\beta| \geq C > 0$ on the set determined by conditions (11.48), so that $C_{\alpha\beta}^1$ are smooth functions.

(4) The value of the coefficient $C_{\alpha\alpha}^1$ remains unknown. If we would like to consider a f.a. solution with accuracy to $O(\lambda^{-2})$ only, then we may set $C_{\alpha\alpha}^1 \equiv 0$. For the construction of a more accurate f.a. solution the coefficient $C_{\alpha\alpha}^1$ has to be found from the next approximation.

The necessary and sufficient condition for solvability of the second equation (11.49) is

$$\langle f^\alpha, R_1 \varphi^{1\alpha} \rangle + \langle f^\alpha, R_2 \varphi^{0\alpha} \rangle = 0,$$

so that

$$\langle f^\alpha, R_1 (C_{\alpha\alpha}^1 f^\alpha) \rangle = g^{1\alpha},$$

where

$$g^{1\alpha} = - R_2 \varphi^{0\alpha} - R_1 \sum_{\beta \neq \alpha} C_{\alpha\beta}^1 f^\beta$$

is a known vector function and $g^{1\alpha} \equiv 0$ outside the set (11.48), i.e. outside the projection on $\mathbf{R}_{t,x}^{n+1}$ of the tube of trajectories originating from supp f^α. By virtue of Lemma 11.4, we obtain the transport equation (an ordinary differential equation along the trajectory):

$$\frac{dC_{\alpha\alpha}^1}{dt} + \tfrac{1}{2} C_{\alpha\alpha}^1 \frac{d \ln J_\alpha}{dt} + M_\alpha C_{\alpha\alpha}^1 = g^{1\alpha}.$$

By specifying the compactly supported Cauchy data $C_{\alpha\alpha}^1|_{t=0}$ we find $C_{\alpha\alpha}^1$ along the tracjectory.

Thus the procedure for deriving the terms of the f.a. solution can be summarized as follows. We have

$$\varphi^{k\alpha} = \sum_{\beta=1}^{N} C_{\alpha\beta}^{k} f^{\beta}.$$

The coefficients $C_{\alpha\beta}^{k}$, $\alpha \neq \beta$, are found from the algebraic relations (see (11.50′)), and the coefficients $C_{\alpha\alpha}^{k}$ from the transport equations. In order to evaluate $C_{\alpha\alpha}^{k}$ we need the equation for the $(k+1)$-th approximation. We proved

THEOREM 11.6. *Let the assumptions* L11.1–L11.5 *be satisfied,* $S(x) \in C^{\infty}$, *and let the eigenvectors* $\{ f^{\alpha}(0, x, \partial S(x)/\partial x \}$, $1 \leq \alpha \leq N$, *of the matrix* $H(0, x, \partial S(x)/\partial x; 0)$ *be compactly supported functions of class* C^{∞}.

Then for any integer $M \geq 1$ *there exists a f.a. solution of the system* (11.9) *in the form* (11.42) *with accuracy to* $O(\lambda^{-M})$.

This means that relation (11.45) is true for u^{α}.

Notice that all vector functions $\varphi^{k\alpha} \in C^{\infty}$ for $x \in \mathbf{R}^{n}$, $0 \leq t \leq T$, and are identically zero outside the set specified by relations (11.48).

The leading term of the asymptotics is given by (11.47) and the f.a. solution with accuracy to $O(\lambda^{-2})$ has the form (11.43).

Consider the Cauchy problem (11.9) and (11.10).

THEOREM 11.7 *Let the assumption* L11.1–L11.5 *be fulfilled,* $S(x) \in C^{\infty}(\mathbf{R}^{n})$, $u^{0}(x) \in C_{0}^{\infty}(\mathbf{R}^{n})$. *Then for any integer* $M \geq 1$, *there exists a vector function* u^{M} *which is a solution of the system* (11.9) *with accuracy to* $O(\lambda^{-M})$ *and has the form*

$$u^{M}(t, x, \lambda) = \sum_{\alpha=1}^{N} u^{\alpha M}(t, x, \lambda), \tag{11.51}$$

where solutions $u^{\alpha M}$ *have the form* (11.42). *Vector function* u^{M} *satisfies the Cauchy data* (11.10):

$$u^{M}\big|_{t=0} = \exp[i\lambda S(x)] u^{0}(x).$$

Vector functions $u^{\alpha M}$ *have the same properties as in Theorem* 11.6. *Moreover,* $u^{M}(t, x, \lambda) \equiv 0$ *outside the set*

$$x = x^{\alpha}(t, y), 0 \leq t \leq T, y \in \text{supp } u^{0}(y).$$

Proof. Vector function u^{M} of the form (11.51), by virtue of Theorem 11.6, is a f.a. solution of the system (11.9) with accuracy to $O(\lambda^{-M})$. It remains to satisfy the Cauchy data by exploiting an arbitrariness in the choice of f.a. solutions $u^{\alpha M}$. We have

$$u^M\big|_{t=0} = \exp[i\lambda S(x)] \sum_{k=0}^{M} (i\lambda)^{-k} \psi^k(x),$$

$$\psi^0(x) = \sum_{\alpha=1}^{N} \sigma_\alpha f^\alpha\big|_{t=0},$$

$$\psi^k(x) = \sum_{\alpha,\beta=1}^{N} (C_{\alpha\beta}^k f^\beta)\big|_{t=0}, \qquad (k \geq 1).$$

Since the vectors $\{f^\alpha\big|_{t=0}\}$ form a smooth orthonormal basis in $\mathbf{R}^N$ for any x, then by putting

$$\sigma_\alpha\big|_{t=0} = \langle u^0(x), f^\alpha\big|_{t=0} \rangle$$

for $t = 0$, we obtain $\psi^0(x) = u^0(x)$. Furthermore $\sigma_\alpha\big|_{t=0} \in C_0^\infty(\mathbf{R}^n)$. The Cauchy data will be satisfied, if $\psi^k(x) \equiv 0$ is valid for $k \geq 1$. From the equation $\psi^1(x) \equiv 0$ we obtain the relations

$$\sum_{\alpha=1}^{N} C_{\alpha\beta}^1\big|_{t=0} \equiv 0, \qquad 1 \leq \beta \leq N.$$

We recall that the coefficients $C_{\alpha\beta}^1, \alpha \neq \beta$, are uniquely determined from the known σ_α, f^α, and the coefficients $C_{\alpha\alpha}^1$ fulfil the transport equations, but the Cauchy data can be specified arbitrarily. Setting

$$C_{\beta\beta}^1\big|_{t=0} = -\sum_{\alpha \neq \beta} C_{\alpha\beta}^1\big|_{t=0}, \qquad 1 \leq \beta \leq N,$$

we obtain $\psi^1(x) \equiv 0$. It is essential that the coefficient $C_{\beta\beta}^1\big|_{t=0} \equiv 0$ outside the support of the vector function $u^0(x)$. Following this reasoning we get the f.a. solution for which the Cauchy data are fulfilled exactly.

EXAMPLE 11.1   We consider the first order system which does not depend on t:

$$\left[\lambda^{-1} I D_t + \lambda^{-1} \sum_{j=1}^{n} A_j(x) D_{x_j} + B(x) \right] u = 0. \tag{11.52}$$

Here $A_j(x), B(x)$ are real symmetric matrices.

In this case, formula (11.47) for the leading term of the asymptotics is simple, since vectors f^α can be chosen to be independent of t and since

$H_{p_j p_k} \equiv 0$. Thus the leading term of the asymptotics is given by

$$u^{\alpha 0} = \exp\left[i\lambda S_\alpha\right] \sqrt{\frac{J_\alpha(0, y)}{J_\alpha(t, y)}} \sigma_\alpha \Bigg|_{t=0} \times$$

$$\times \exp\left[\int_0^t \left[\frac{1}{2} \sum_{j=1}^n \frac{\partial^2 h_\alpha}{\partial x_j \partial p_j} - \sum_{j=1}^n \left\langle f^\alpha, A_j(x) \frac{\partial f^\alpha}{\partial x_j} \right\rangle \right] dt' \right] f^\alpha .$$

$$(11.53)$$

The arguments of all functions and vector functions are the same as in (11.47).

We shall give the analogue of Lemma 11.4 for operator L in which the order of the actions of the operators x_j, D_{x_j} is opposite.

LEMMA 11.8.   *Let the assumptions of Lemma* 11.4 *be satisfied and operator L has the form*

$$L = \lambda^{-1} D_t + H(t, \overset{1}{x}, \lambda^{-1} \overset{2}{D}_x ; (i\lambda)^{-1}).$$

$$(11.52')$$

Then formula (11.35) *holds, where function M_α is given by*

$$M_\alpha = \left\langle f^\alpha, \frac{df^\alpha}{dt} \right\rangle + \sum_{j=1}^n \left\langle f^\alpha, \left(\frac{\partial h_\alpha}{\partial x_j} - \frac{\partial H}{\partial x_j}\right) \frac{\partial f^\alpha}{\partial p_j} \right\rangle +$$

$$+ \frac{1}{2} \sum_{j=1}^n \frac{\partial^2 h_\alpha}{\partial x_j \partial p_j} + \left\langle f^\alpha, \frac{\partial H^0}{\partial \varepsilon} \Bigg|_{\varepsilon=0} f^\alpha \right\rangle.$$

$$(11.53')$$

Proof. Let R_1^0 be the operator (11.32), and R_1 be an analogous operator corresponding to the Hamiltonian (11.52'). Then, by virtue of (2.11), (2.15),

$$R_1 \varphi = R_1^0 \varphi + \sum_{j=1}^n \frac{\partial^2 H}{\partial x_j \partial p_j} \varphi.$$

Differentiating identity (11.38) with respect to x_k, we obtain

$$\frac{\partial^2 H}{\partial p_j \partial x_k} f^\alpha + \frac{\partial H}{\partial p_j} \frac{\partial f^\alpha}{\partial x_k} + \frac{\partial H}{\partial x_k} \frac{\partial f^\alpha}{\partial p_j} + H \frac{\partial^2 f^\alpha}{\partial p_j \partial x_k} =$$

$$= \frac{\partial^2 h_\alpha}{\partial p_j \partial x_k} f^\alpha + \frac{\partial h_\alpha}{\partial p_j} \frac{\partial f^\alpha}{\partial x_k} + \frac{\partial h_\alpha}{\partial x_k} \frac{\partial f^\alpha}{\partial p_j} + h_\alpha \frac{\partial^2 f^\alpha}{\partial p_j \partial x_k}.$$

Taking the scalar product of this identity with f^α, we get the identity

$$\left\langle f^\alpha, \frac{\partial^2 H}{\partial p_j \partial x_k} f^\alpha \right\rangle + \left\langle f^\alpha, \frac{\partial(H-h_\alpha)\partial f^\alpha}{\partial p_j \partial x_k} \right\rangle +$$

$$+ \left\langle f^\alpha, \frac{\partial(H-h_\alpha)\partial f^\alpha}{\partial x_k \partial p_j} \right\rangle = \frac{\partial^2 h_\alpha}{\partial p_j \partial x_k}. \tag{11.54}$$

This formula and Lemma 11.4 yield (11.53'), q.e.d.

4. *The global asymptotics of the Cauchy problem (11.9), (11.10)*

This Cauchy problem generates N Lagrangian manifolds of dimension n:

$$\Lambda_{t,\alpha}^n = \{(x,p): x = x^\alpha(t,y), p = p^\alpha(t,y), y \in \mathbf{R}^n\}, \quad 1 \leq \alpha \leq N,$$

in phase space $\mathbf{R}^{2n}$, and N Lagrangian manifolds of dimension $n+1$

$$\Lambda_\alpha^{n+1} = \bigcup_{0 \leq t \leq T} \Lambda_{t,\alpha}^n$$

in phase space $\mathbf{R}^{2n+2}$ with the coordinates (t, x, E, p) (E is the variable conjugate to t). We supplement the notation of § 10 by adding the index α everywhere (the manifold label). Let G_α^t be the displacement operator along trajectories of the Hamilton system

$$\frac{dx}{d\tau} = \frac{\partial h_\alpha}{\partial p}, \quad \frac{dp}{d\tau} = -\frac{\partial h_\alpha}{\partial x}, \quad \frac{dt}{d\tau} = 1, \quad \frac{dE}{d\tau} = -\frac{\partial h_\alpha}{\partial t}. \tag{11.55}$$

Consider the Lagrangian manifold Λ_α^{n+1}; it is constructed by means of the Hamiltonian $h_\alpha(t, x, p)$ exactly as in § 8, Section 3.

We derive the commutation formula for operator L and the canonical operator $K_\alpha = K_{\Lambda_\alpha^{n+1}}^{r_0}$. Let manifolds Λ_β^{n+1} and the eigenvalues h_β, $1 \leq \beta \leq N$, satisfy the assumptions of Theorem 8.4; and let the assumptions L11.1–L11.4 be fulfilled.

THEOREM 11.9   *Suppose $f: \Lambda_\alpha^{n+1} \to \mathbf{R}^N$ is a vector function of class C^∞ and the projection of supp f on $\mathbf{R}_{x,p}^{2n}$ is contained in some compact set for all $t \in [0, T]$. Then the commutation formula*

$$LK_\alpha f = K_\alpha \sum_{j=0}^m (i\lambda)^{-j} R_j f + O_{-m-1,T} \tag{11.56}$$

holds for arbitrary integer $m \geq 0$. Here R_j are differential operators on

Λ_α^{n+1} of order $\leq j$, with the coefficients of class $C^\infty(\Lambda_\alpha^{n+1})$ which are independent of λ. Furthermore

$$R_0 f = (H^0 - h_\alpha I) f. \tag{11.57}$$

If $f = \sigma_\alpha f^\alpha$, where σ_α is a scalar function of class $C^\infty(\Lambda_\alpha^{n+1})$, $\langle f^\alpha, f^\alpha \rangle \equiv 1$, then

$$R_0(\sigma_\alpha f^\alpha) = 0,$$

$$\langle f^\alpha, R_1(\sigma_\alpha f^\alpha) \rangle = \frac{d\sigma_\alpha}{dt} + M_\alpha \sigma_\alpha, \tag{11.58}$$

where function M_α is given by (11.36).

In all these formulae $f = f(t, x, p)$ where the point $(x, p) \in \Lambda_{t,\alpha}^n$. The theorem plays exactly the same rôle for the systems as Theorem 8.4 does for scalar equations.

The relations (11.56)–(11.58) form the basic content of this paragraph.

Proof of Theorem 11.9. We present the proof for precanonical operators; the passage to the canonical operator will then be done in the same manner as in part 3 of the proof of Theorem 8.4.

Taking into account the invariance of manifold Λ_α^{n+1} with respect to the displacements along trajectories of the dynamical system (11.55), we can make a special choice of the canonical atlas. Namely, let Ω be a chart on $\Lambda_{t_0,\alpha}^n$ which admits a diffeomorphic projection on the Lagrangian plane $p_{(\gamma)}, x_{(\delta)}$ (here $(\gamma) \cup (\delta) = \{1, 2, \ldots, n\}$ and the sets $(\gamma), (\delta)$ are disjoint). Then the charts $g^t \tilde{\Omega}$ have the same property, if $|t - t_0| < \varepsilon, \varepsilon > 0$ is sufficiently small, and $\tilde{\Omega}$ is a chart on $\Lambda_{t_0,\alpha}^n$ which is contained in Ω together with its closure. We take the atlas of charts, the projections of which on $\mathbf{R}_{x,p}^{2n}$ have the form $\bigcup_{|t - t_0| < \varepsilon} g^t \tilde{\Omega}$, as the canonical atlas on Λ_α^{n+1}.

(1) We prove the commutation formula in a non-singular chart. Let Ω be a non-singular chart, i.e. the variables (t, x) can be chosen as coordinates on Ω. In accordance with the above mentioned facts, Ω can be taken in the form

$$x = x^\alpha(t, y), \qquad p = p^\alpha(t, y),$$

$$E = E_0 - \int_0^t \frac{\partial h_\alpha}{\partial \tau} d\tau, \qquad t_0 - \varepsilon < t < t_0 + \varepsilon, \qquad y \in \tilde{\Omega}.$$

Here $\tilde{\Omega}$ is a domain in R_y^n, $\{x^\alpha, p^\alpha\}$ is a solution of the system (11.29) with the Cauchy data $x|_{t=0} = x^0(y)$, $p|_{t=0} = p^0(y)$ where the point $(x^0(y),$

$p^0(y)) \in \Lambda_{t_0,\alpha}^n$. The points $r \in \Omega$ are parametrized by the variables (t, y): $r = r(t, y)$. The precanonical operator K in chart Ω, with accuracy to an inessential numerical factor, has the form

$$(K\varphi(r))(t, x) = \exp(i\lambda S) \sqrt{\frac{J_\alpha(t_0, y)}{J_\alpha(t, y)}} (\varphi \circ r)(t, x)$$

(for simplicity, all indices of operator K are omitted). Here

$$S = \int_{t_0}^{t} (\langle p^\alpha, dx^\alpha \rangle - h_\alpha \, dt'),$$

the integral is taken along the bicharacteristic, and S in the point

$$r = (t, x^\alpha(t, y), E, p^\alpha(t, y)).$$

We have

$$\frac{\partial S_\alpha}{\partial t} = -h_\alpha(t, x, p)$$

in chart Ω. Let vector function $f \in C_0^\infty(\Omega)$. Then, due to (11.3), (11.4),

$$LKf = \exp(i\lambda S) \sqrt{\frac{J_\alpha(t_0, y)}{J_\alpha(t, y)}} (R_0 f + \frac{1}{i\lambda} R_1 f) + O_{-2, T},$$

$$R_0 f = \left(I \frac{\partial S}{\partial t} + H\right) f = (H - h_\alpha I) f.$$

Here and later on, $H = H(t, x, p; 0)$, the point $(x, p) \in \Lambda_{t,\alpha}^n$. If $f = \sigma_\alpha f^\alpha$, then Lemma 11.4 implies (11.58). We are considering here only the first two terms of expansion (11.56); the existence of this expansion with any number of terms follows from § 9.

(2) Suppose that Ω is a singular chart, and that the variables (t, p) can be taken as coordinates in Ω. The relations (11.56), (11.57) can be proved exactly as before. Relation (11.57) is obtained by combining Lemma 11.8 and the formulae for the Fourier transformation of a λ-p.d. operator.

We have

$$K\varphi = F^{-1}\left(\sqrt{\frac{\tilde{J}_\alpha(t_0, \tilde{y})}{\tilde{J}_\alpha(t, \tilde{y})}} \exp(i\lambda \tilde{S}) \varphi\right), \tag{11.59}$$

where $F^{-1} = F_{\lambda, p \to x}^{-1}$; function $\tilde{S}$ and Jacobian $\tilde{J}_\alpha$ will be specified below.

Due to (5.3) we obtain

$$LF^{-1} = F^{-1}\tilde{L},$$
$$\tilde{L} = \lambda^{-1}D_t + H(t, -\lambda^{-1}\overset{1}{D}_p, \overset{2}{p}; (i\lambda)^{-1}).$$

Notice that here the operators of differentiation and multiplication by a function act in the reverse order than in the original operator

$$L = \lambda^{-1}D_t + H(t, \overset{1}{x}, \lambda^{-1}\overset{2}{D}_x; (i\lambda)^{-1}).$$

Function $\tilde{S}$ depends on the variables (t, p). Its explicit form is not essential at the moment. It is only important that this function fulfils the Hamilton–Jacobi equation

$$\frac{\partial\tilde{S}}{\partial t} + h_\alpha\left(t, -\frac{\partial\tilde{S}}{\partial p}, p\right) = 0.$$

We make the substitution

$$\tilde{x} = p, \qquad \tilde{p} = -x, \tag{11.60}$$

in the Hamilton system (11.29) and set

$$\tilde{h}_\alpha(t, \tilde{x}, \tilde{p}) = h_\alpha(t, x, p). \tag{11.61}$$

The system (11.29) corresponds to the Hamilton–Jacobi equation (11.13), and $p = \partial S/\partial x$ holds on the trajectories of this system (see (3.10)). Variable t plays the rôle of a parameter.

The transformation (11.60) is canonical, since, according to Theorem 4.16 it transforms the Hamilton system (11.29) into the Hamilton system with Hamiltonian $\tilde{h}_\alpha$. For clarity, we shall write down the corresponding equations side by side; here $S = S(t, x)$, $\tilde{S} = \tilde{S}(t, p)$ and, by virtue of (11.60), (11.61),

$$\tilde{h}_\alpha(t, x, p) = h_\alpha(t, p, -x). \tag{11.61'}$$

The following table demonstrates the duality of x- and p-representations.

x-representation	p-representation
$\dfrac{\partial S}{\partial t} + h_\alpha\left(t, x, \dfrac{\partial S}{\partial x}\right) = 0$	$\dfrac{\partial\tilde{S}}{\partial t} + \tilde{h}_\alpha\left(t, -\dfrac{\partial S}{\partial p}, p\right) = 0$
$\dfrac{dx}{dt} = \dfrac{\partial h_\alpha}{\partial p}, \quad \dfrac{dp}{dt} = -\dfrac{\partial h_\alpha}{\partial x}$	$\dfrac{d\tilde{x}}{dt} = \dfrac{\partial\tilde{h}_\alpha}{\partial\tilde{p}}, \quad \dfrac{d\tilde{p}}{dt} = -\dfrac{\partial\tilde{h}_\alpha}{\partial\tilde{x}}$

on a trajectory	on a trajectory
$$p = \frac{\partial S}{\partial x}$$	$$\tilde{p} = \frac{\partial \tilde{S}}{\partial \tilde{x}}$$

In order to exhibit the symmetry, we write down the Hamilton–Jacobi equations and the operators L, $\tilde{L}$ in the form

$$\frac{\partial S}{\partial t} + h_\alpha(t, x, p) = 0$$	$$\frac{\partial \tilde{S}}{\partial t} + \tilde{h}_\alpha(t, \tilde{x}, \tilde{p}) = 0$$
$$L = \lambda^{-1} D_t + \hat{H}(t, \overset{2}{\tilde{x}}, \lambda^{-1}\overset{1}{p} ; \varepsilon)$$	$$\tilde{L} = \lambda^{-1} D_t + \hat{\tilde{H}}(t, \overset{1}{\tilde{x}}, \lambda^{-1}\overset{2}{\tilde{p}} ; \varepsilon)$$
$$p = D_x$$	$$\tilde{p} = D_{\tilde{x}}$$

where the symbols $H, \tilde{H}$ are related exactly as $h_\alpha, \tilde{h}_\alpha$:

$$\tilde{H}(t, \tilde{x}, \tilde{p} ; \varepsilon) = H(t, x, p ; \varepsilon). \tag{11.61''}$$

By applying Lemma 11.8 (in the considered case we have $\tilde{J}_\alpha = \det[\partial \tilde{x}^\alpha(t, \tilde{y})/\partial \tilde{y}]$, analogously to (11.33)), and by using (11.53), we obtain the following expression for operator M_α:

$$M_\alpha = -\sum_{j=1}^{n} \left\langle \tilde{f}^\alpha, \frac{\partial(h_\alpha - \tilde{H})\partial \tilde{f}^\alpha}{\partial \tilde{x}_j \; \partial \tilde{p}_j} \right\rangle +$$
$$+ \frac{1}{2}\sum_{j=1}^{n} \frac{\partial^2 \tilde{h}_\alpha}{\partial \tilde{x}_j \partial \tilde{p}_j} + \left\langle \tilde{f}^\alpha, \frac{\partial \tilde{H}^0}{\partial \varepsilon}\Big|_{\varepsilon=0} \tilde{f}^\alpha \right\rangle.$$

Furthermore, due to (11.60), (11.61'), (11.61''), we have

$$\frac{\partial^2 \tilde{h}_\alpha(t, \tilde{x}, \tilde{p})}{\partial \tilde{x}_j \; \partial \tilde{p}_j} = -\frac{\partial^2 h_\alpha(t, x, p)}{\partial x_j \; \partial p_j},$$

$$-\frac{\partial \tilde{H}}{\partial \tilde{x}_j}\frac{\partial \tilde{f}^\alpha}{\partial \tilde{p}_j} = \frac{\partial H}{\partial p_j}\frac{\partial f^\alpha}{\partial x_j}.$$

We recall that in a neighbourhood of each non-singular point of chart $\Omega \subset \Lambda_\alpha^{n+1}$ one can choose either (t, x) or (t, p) as the local coordinates, and that the points

$$(t, x, E, p), \qquad (t, \tilde{x}, \tilde{E}, \tilde{p})$$

determine one and the same point on Λ_α^{n+1}, if

$$p = \frac{\partial S_\alpha(t, x)}{\partial x}, \qquad \tilde{x} = p, \qquad \tilde{p} = \frac{\partial \tilde{S}_\alpha(t, p)}{\partial p}.$$

(3) In the same way we can investigate the case when the variable t, a part of the variables x and a part of the variables p can be chosen as local coordinates in chart Ω. Thus let $\{1, 2, ..., n\} = (\delta) \cup (\gamma)$, the sets $(\delta), (\gamma)$ be disjoint, and $(t, p_{(\delta)}, x_{(\gamma)})$ be the local coordinates in chart Ω, $F^{-1} = F^{-1}_{\lambda, \, p_{(\delta)} \to x_{(\delta)}}$, and let the operator K have the form (11.59). We perform the canonical transformation

$$\tilde{x}_k = p_k, \qquad \tilde{p}_k = - x_k, \qquad k \in (\delta),$$
$$\tilde{x}_k = x_k, \qquad \tilde{p}_k = p_k, \qquad k \in (\gamma),$$

so that

$$\tilde{x} = (p_{(\delta)}, x_{(\gamma)}), \qquad \tilde{p} = (- x_{(\delta)}, p_{(\gamma)}),$$

and set

$$\tilde{h}_\alpha(t, \tilde{x}, \tilde{p}) = h_\alpha(t, x, p),$$
$$\tilde{H}(t, \tilde{x}, \tilde{p}; 0) = H(t, x, p; 0),$$

as in (11.61), (11.61'). Then we obtain the formulae on p. 221 with the only difference that

$$\tilde{L} = \lambda^{-1} D_t + L(\lambda^{-1} \overset{2}{\tilde{D}}_{\tilde{x}_{(\delta)}}, \overset{2}{\tilde{x}}_{(\gamma)}, \overset{1}{\tilde{p}}_{(\gamma)}, \lambda^{-1} \overset{1}{\tilde{D}}_{\tilde{p}_{(\delta)}}; (i\lambda)^{-1}).$$

Namely, the operators of multiplication by a function and of differentiation with respect to the variables $\tilde{x}_{(\delta)}$ and $\tilde{x}_{(\gamma)}$ act in reverse order; thus in order to evaluate M_α it is necessary to apply both Lemma 11.4 and Lemma 11.8. However, as we have already seen in Sections 1 and 2, the summands composing function M_α do not depend on the order in which the considered operators act, and hence formula (11.36) remains true.

Now consider the Cauchy problem

$$u|_{t=0} = u^0(x) \exp[i\lambda S_0(x)] \tag{11.62}$$

for system (11.9) where, as usual, function $S_0(x) \in C^\infty(\mathbf{R}^n)$ and is real-valued, vector function $u^0(x) \in C_0^\infty(\mathbf{R}^n)$. Let $\{f^\alpha(t, x, p)\}$ be an orthonormal basis consisting of the eigenvectors of matrix $H(t, x, p; 0)$ of class C^∞ for $0 \leq t \leq T, (x, p) \in \mathbf{R}^{2n}$. Then, as shown in the proof of Theorem 11.7, vector function $u^0(x)$ can be expanded in terms of the eigenvectors $f^\alpha|_{t=0} = f^\alpha(0, x, \partial S_0(x)/\partial x)$:

$$u^0(x) = \sum_{\alpha=1}^{N} \sigma_\alpha(x) f^\alpha|_{t=0}, \tag{11.63}$$

where $\sigma_\alpha(x) \in C_0^\infty(\mathbf{R}^n)$. Thus the Cauchy problem reduces to N Cauchy problems

$$u^{0\alpha}\big|_{t=0} = \sigma_\alpha(x) f^\alpha\big|_{t=0}. \tag{11.64}$$

The solution of the problem (11.9), (11.62) is equal to the sum of the solutions of problems (11.9), (11.64). Recall that the assumptions L11.1–L11.4 are satisfied and that all Lagrangian manifolds $\Lambda_\alpha^{n+1}, 1 \leq \alpha \leq N$, fulfil the assumptions of Theorem 8.4.

THEOREM 11.10. *Let the just mentioned assumptions be fulfilled. Then for any $M \geq 1$ there exists a vector function $u^{\alpha M}(t, x, \lambda)$ which exactly satisfies the Cauchy data (11.64) and fulfils the Equation (11.9) with accuracy to $O(\lambda^{-M})$, i.e.*

$$Lu^{\alpha M} \in O_{-M,T}. \tag{11.65}$$

This f.a. solution has the form

$$u^{\alpha M}(t, x, \lambda) = \exp[i\lambda S_0(y^0)] K_{\Lambda_\alpha^{\widetilde{n}}+1}^{\widetilde{n}} \sum_{j=0}^{M} (i\lambda)^{-j} \varphi^{\alpha j} \tag{11.66}$$

with the notation as in (10.12).

We write down the leading term of the asymptotics

$$\varphi^{\alpha 0}(t, x, p) = \exp\left(-\int_0^t M_\alpha \, d\tau\right) \sigma_\alpha(y) f^\alpha(t, x, p). \tag{11.67}$$

In this formula $x = x^\alpha(t, y), p = p^\alpha(t, y)$ (it is the solution of the Cauchy problem (11.34) for the Hamilton system (11.29)) and function M_α has the form (11.36).

We remark that all vector functions $\varphi^{\alpha j}$ identically vanish on Λ_α^{n+1} outside the 'strip' cut out by the bicharacteristics, i.e. $\varphi^{\alpha j}$, for each $t \in [0, T]$, are different from zero only in the points which under the projection on $\mathbf{R}_{x,p}^{2n}$ are mapped into the set

$$\{(x, p): x = x^\alpha(t, y), p = p^\alpha(t, y), y \in \operatorname{supp} \sigma_\alpha\}.$$

Proof. We look for a f.a. solution of the form (11.67). According to commutation formula (11.56) we have

$$LK_\alpha u^{\alpha M} = K_\alpha \sum_{j=0}^{M} \varepsilon^j \left(\sum_{k=0}^{M-j} \varepsilon^k R_k \varphi^{\alpha M}\right) + O_{-S-1,T}, \quad \varepsilon = (i\lambda)^{-1}.$$

(Here $K_\alpha = K_{\Lambda_\alpha^n + 1}^{r_\alpha^0}$.) Setting the coefficients at powers of ε consecutively equal to zero, we get $R_0 \varphi^{\alpha 0} = 0$, i.e. $\varphi^{\alpha 0} = \sigma_\alpha f^\alpha$. For function σ_α we have the transport equation

$$\frac{d\sigma_\alpha}{dt} + M_\alpha \sigma_\alpha = 0,$$

and taking the Cauchy data

$$\sigma_\alpha|_{t=0} = \sigma_\alpha^0,$$

we obtain (11.67). The next approximations are found by repeating the procedure used in the proof of Theorem 11.7.

The leading term of the asymptotics can be written also in the form (10.17) in which H and Λ_t^n should be replaced by h_α and $\Lambda_{t,\alpha}^n$, respectively.

5. The Characteristics of Constant Multiplicity

We shall derive the transport equations in the case when Equation (11.15) has roots with constant multiplicities. We consider Equation (11.1) with the symbol $L(x, p;(i\lambda)^{-1})$ which satisfies assumption L11.1. Let $h_j(x, p)$ be the eigenvalues of matrix $L(x, p; 0)$. Further we assume:

(1) The multiplicity of the eigenvalue $h_1(x, p)$ is equal to k and does not depend on $(x, p) \in \mathbf{R}^{2n}$.

(2) The set $M_1 : h_1(x, p) = 0$ is a $(2n-1)$-dimensional C^∞-manifold in $\mathbf{R}_{x,p}^{2n}$. Function $h_1(x, p)$ is real-valued in a neighbourhood of this set.

(3) Eigenvalue h_1 is isolated in the following sense: there exist constants $\varepsilon_0 > 0, \delta_0 > 0$ such that

$$|h_j(x, p)| \geq \delta_0$$

for all (x, p) such that $|h_1(x, p)| \leq \varepsilon_0$, and for all $j \neq 1$.

(4) The resolvent $(\mu I - L(x, p; 0))^{-1}$ has (for fixed (x, p)) a simple pole at the point $h_1(x, p)$ for $|h_1(x, p)| < \varepsilon_0$.

The last condition means that the normal Jordan form of matrix $L(x, p; 0)$ has no Jordan cells corresponding to an eigenvalue in a neighbourhood of the manifold $h_1 = 0$.

LEMMA 11.11 *Let the assumptions (1)–(4) be satisfied; then there exists a matrix $\tilde{L}(x, p)$ of class $C^\infty (\mathbf{R}^{2n})$ such that*

$$\tilde{L}(x, p)L(x, p; 0) = L(x, p; 0)\tilde{L}(x, p) = h_1(x, p)I \qquad (11.67')$$

$$\tilde{L}^2 = L.$$

Here I is the unit $(N \times N)$-matrix.

Proof. We fix a point (x^0, p^0) such that $h_1(x^0, p^0) = 0$ and choose its closed neighbourhood U, in which $|h_1(x, p)| \leq \varepsilon$, where $\varepsilon \leq \varepsilon_0$ will be specified below. Due to assumption (4) we have

$$(\mu I - L(x, p; 0))^{-1} = \frac{A(x, p)}{\mu - h_1(x, p)} + B(\mu, x, p)$$

for $(x, p) \in U$. Matrix B is holomorphic with respect to μ for

$$|\mu - h_1(x, p)| \leq \underset{j \neq 1}{\text{Min}} |h_1(x, p) - h_j(x, p)|,$$

since the poles of the resolvent coincide with the eigenvalues of matrix $L(x, p; 0)$. We set $\varepsilon = \delta_0/4$ and choose ρ such that $\varepsilon < \rho < \delta_0 - \varepsilon$. Then the resolvent is holomorphic on the circle $|\mu| = \rho$ for arbitrary fixed $(x, p) \in U$. Indeed, we have $|\mu - h_1(x, p)| \geq \rho - \varepsilon > 0$ on that circle, and for $j \neq 1$ we get

$$|\mu - h_j(x, p)| \geq |h_j(x, p) - h_1(x, p)| + |\mu| - |h_1(x, p)| \geq$$

$$\geq \delta_0 + \rho - \varepsilon > 0.$$

We have

$$A(x, p) = \underset{\mu = h_1(x, p)}{\text{res}} \ (\mu I - L(x, p; 0))^{-1} =$$

$$= \frac{1}{2\pi i} \int_{|\mu| = \rho} (\mu I - L(x, p; 0))^{-1} \, d\mu,$$

which implies that $A(x, p) \in C^\infty(U)$. In fact, due to the choice of ρ, U, the matrix $(\mu I - L)^{-1}$ is infinitely differentiable with respect to the variables (μ, x, p) for $|\mu| = \rho, (x, p) \in U$. As follows from the known properties of the residue of a resolvent [14], we have

$$L(x, p; 0) A(x, p) = A(x, p) L(x, p; 0), \quad A^2 = A.$$

We put $\tilde{L}(x, p) = A(x, p)$ in the domain $|h_1(x, p)| < \varepsilon$ and then extend this matrix function on $\mathbf{R}_{x, p}^{2n}$ preserving its smoothness.

Notice that later we shall need the existence of matrix $\tilde{L}(x, p)$ of class C_0^∞ only in a small neighbourhood of some point $(x^0, p^0), h_1(x^0, p^0) = 0$.

Let the assumptions of Lemma 11.11 be satisfied. We construct the Lagrangian manifold Λ^n from the Hamiltonian $h_1(x, p)$ (the construction of and the assumptions on the Hamiltonian are as in Theorem 8.4), and the canonical operator $K = K_{\Lambda^n}^{r^0}$.

THEOREM 11.12 *Let the assumptions formulated above be fulfilled, and $f: \Lambda^n \to \mathbf{R}^N$ be a matrix function of class $C_0^\infty(\Lambda^n)$. Then the commutation formula (11.56) holds with the remainder of class $O_{-m-1}^+(\mathbf{R}_x^n)$. Furthermore,*

$$R_0 = L^0 = L(x, p; 0),$$

$$\tilde{L}R_1\tilde{L} = \left(\frac{d}{d\tau} - \frac{1}{2}\sum_{j=1}^n \frac{\partial^2 h^1(x,p)}{\partial x_j \partial p_j} + \sum_{j=1}^n \frac{\partial \tilde{L}}{\partial p_j}\frac{\partial L^0}{\partial x_j} + \frac{\partial L^0}{\partial \varepsilon}\bigg|_{\varepsilon=0} \right)\tilde{L}.$$

$$(11.68)$$

Here $d/d\tau$ is the derivative with respect to the Hamilton system

$$\frac{dx}{d\tau} = \frac{\partial h_1}{\partial p}, \quad \frac{dp}{d\tau} = -\frac{\partial h_1}{\partial x},$$

and $\tilde{L}(x, p)$ is the matrix constructed in Lemma 11.11.

Proof of the theorem reduces to the proof of formula (11.68). Since the estimates for the remainders have already been obtained, we shall simply write $O(\varepsilon^k)$, $\varepsilon = (i\lambda)^{-1}$, without specifying the exact meaning of the symbol O. Furthermore, we choose a matrix $\tilde{L}(x, p) \in C_0^\infty$ with the support concentrated in a neighbourhood of a point (x^0, p^0), in which $h_1(x^0, p^0) = 0$. We recall that $h_1(x, p) \equiv 0$ on Λ^n. We have

$$LK = \varepsilon KR_1 + O(\varepsilon^2),$$

so that

$$LK\tilde{L} = \varepsilon KR_1\tilde{L} + O(\varepsilon^2),$$

where $\tilde{L}$ is the operator of multiplication by the matrix $\tilde{L}(x, p)$. If $\hat{\tilde{L}}$ is the operator $\tilde{L}(\overset{2}{x}, \lambda^{-1}\overset{1}{D}_x)$, then

$$\hat{\tilde{L}}LK\tilde{L} = \varepsilon K\tilde{L}R_1\tilde{L} + O(\varepsilon^2).$$

$$(11.69)$$

Now we shall group the operators in another way:

$$\hat{\tilde{L}}LK\tilde{L} = (\hat{\tilde{L}}L)(K\tilde{L}).$$

Due to (2.33)–(2.35), we have

$$\hat{\tilde{L}}L = \hat{h}_1 I + \varepsilon \hat{Q} + O(\varepsilon^2),$$

where

$$\hat{Q} = Q(\overset{2}{x}, \lambda^{-1}\overset{1}{D}_x)$$

is the λ-p.d. operator with the symbol

$$Q(x, p) = \sum_{j=1}^{n} \frac{\partial \tilde{L}(x, p)}{\partial p_j} \frac{\partial L(x, p\,; 0)}{\partial x_j} + \frac{\partial L}{\partial \varepsilon}\bigg|_{\varepsilon = 0}. \tag{11.70}$$

By using the commutation formula for the scalar operator $\hat{h}_1$ (see Theorem 8.4) we obtain

$$\hat{h}_1 K = \varepsilon K\left(I\left(\frac{d}{d\tau} - \frac{1}{2}\sum_{j=1}^{n} \frac{\partial^2 h_1}{\partial x_j \partial p_j}\right)\right) + O(\varepsilon^2).$$

Consequently,

$$\hat{\tilde{L}}LK\tilde{L} = \varepsilon K\left(I\left(\frac{d}{d\tau} - \frac{1}{2}\sum_{j=1}^{n} \frac{\partial^2 h_1}{\partial x_j \partial p_j}\right) + Q\right)\tilde{L} + O(\varepsilon^2). \tag{11.71}$$

The comparison of the right-hand sides of the formulae (11.69) and (11.71) leads to

$$\tilde{L}R_1\tilde{L} = \left(\frac{d}{d\tau} - \frac{1}{2}\sum_{j=1}^{n} \frac{\partial^2 h_1}{\partial x_j \partial p_j}\right)\tilde{L} + Q\tilde{L},$$

which proves formula (11.68).

Yet another variant of the transport equations corresponding to the system (11.1) will be discussed in § 14.

§ 12. The Semi-Classical Asymptotics of the Cauchy Problem for the Schrödinger Equation

1. Formulation of the Problem

Consider the Schrödinger equation

$$ih\frac{\partial \psi}{\partial t} = -\frac{h^2}{2m}\Delta\psi + V(x)\psi \tag{12.1}$$

and set up the Cauchy problem

$$\psi\big|_{t=0} = \psi_0(x)e^{(i/h)S_0(x)}. \tag{12.2}$$

Here $x \in \mathbf{R}^n$, and $S_0(x)$ is a real-valued function.

Equation (12.1) describes the motion of a non-relativistic quantum particle of mass m in a potential field with a potential energy $V(x)$. The potential $V(x)$ is assumed real.

Consider the operator $L : L_2(\mathbf{R}^n) \to L_2(\mathbf{R}^n)$ with the domain of definition $D(L) = C_0^\infty(\mathbf{R}^n)$ acting according to the formula

$$L\psi = -\frac{h^2}{2m}\Delta\psi + V(x)\psi.$$

Operator L is symmetric:

$$(L\psi_1, \psi_2) = (\psi_1, L\psi_2)$$

for any $\psi_{1,2} \in C_0^\infty(\mathbf{R}^3)$, where the scalar product $(.,.)$ is defined by

$$(\psi, \varphi) = \int_{\mathbf{R}_3} \overline{\psi(x)}\varphi(x)\mathrm{d}x.$$

Under certain assumptions on potential $V(x)$ the minimal symmetric operator L can be extended to a self-adjoint operator $A : L_2(\mathbf{R}^n) \to L_2(\mathbf{R}^n)$ with a domain of definition $D(A)$ dense in $L_2(\mathbf{R}^n)$. For this it is sufficient that, e.g., [25]:

$$V(x) \in C^\infty(\mathbf{R}^n), \qquad \inf_{x \in \mathbf{R}^n} V(x) > -\infty.$$

Then the Schrödinger equation can be expressed in the form

$$ih\frac{\mathrm{d}\psi}{\mathrm{d}t} = A\psi. \tag{12.3}$$

We shall restrict ourselves to potentials $V(x) \in \mathscr{S}(\mathbf{R}^n)$, where $\mathscr{S}$ is the Schwartz space. We are interested in the asymptotic solution of the Cauchy problem (12.1), (12.2) for $h \to 0$ in the domain $x \in \mathbf{R}^n, 0 \leq t \leq T$. Equation (12.1) can be written in the form

$$ih\frac{\partial\psi}{\partial t} + H(\overset{2}{x}, h\overset{1}{D}_x)\psi = 0, \tag{12.1'}$$

$$H(x, p) = \frac{p^2}{2m} + V(x), \tag{12.4}$$

where $p^2 = \langle p, p \rangle$. The left-hand side of Equation (12.1') is an h^{-1}-pseudo-differential operator with the symbol $-E + H(x, p)$, where E is the variable conjugate to t.

Now we apply the procedure developed in § 10 to the Schrödinger equation. The Hamilton–Jacobi equation associated with the Schrödinger equation is the Hamilton–Jacobi equation of classical mechanics,

$$\frac{\partial S}{\partial t} + \frac{1}{2m}(\nabla_x S)^2 + V(x) = 0. \tag{12.5}$$

Function S is the *classical action*.

The truncated Hamilton system corresponding to the Schrödinger equation has the form

$$m\frac{dx}{dt} = p, \qquad \frac{dp}{dt} = -\nabla V(x). \tag{12.6}$$

This system is equivalent to the Newton system

$$m\frac{d^2x}{dt^2} = -\nabla V(x). \tag{12.7}$$

The complete Hamilton system is obtained by joining the system

$$\frac{dt}{dt} = 1, \qquad \frac{dE}{dt} = 0, \tag{12.8}$$

to the system (12.6), with function S satisfying the equation

$$\frac{dS}{dt} = \frac{p^2}{2m} - V(x).$$

If $x = x(\tau), p = p(\tau), 0 \leq \tau \leq T$, is a solution of Hamilton system (12.6), then the action along the trajectory is evaluated according to the formula

$$S(x(t)) = S(x(0)) + \int_0^t \left[\frac{1}{2m}p^2(\tau)\,d\tau - V(x(\tau))\right]d\tau \tag{12.9}$$

or

$$S(x(t)) = S(x(0)) + \int_0^t \left[\frac{m}{2}\dot{x}^2(\tau) - V(x(\tau))\right]d\tau, \tag{12.9'}$$

where $x(\tau)$ is the solution of Newton system (12.7).

The Lagrangian Cauchy problems

$$x\big|_{t=0} = y, \qquad p\big|_{t=0} = \frac{\partial S_0(y)}{\partial y}, \quad y \in \mathbf{R}^n, \tag{12.10}$$

for the truncated Hamilton system (12.6), and

$$x|_{t=0} = y, \quad p|_{t=0} = \frac{\partial S_0(y)}{\partial y}, \quad t|_{t=0} = 0 \quad E|_{t=0} = E_0, \quad y \in \mathbf{R}^n,$$

$$(12.11)$$

for the Hamilton system (12.6), (12.8), correspond to the Cauchy problem (12.1) (12.2).

The set given by Equation (12.10) is an n-dimensional Lagrangian C^∞-manifold Λ_0^n in the phase space $\mathbf{R}_{x,p}^{2n}$; the set (12.11) is an n-dimensional Lagrangian manifold $\tilde{\Lambda}_0^n$ in the $(2n+2)$-dimensional phase space $\mathbf{R}^{2n+2}$.

Since the potential $V(x) \in \mathscr{S}(\mathbf{R}^n)$, the solution of the above stated Cauchy problems exists globally, i.e. for $-\infty < t < \infty$.

2. The Asymptotic Solution of the Cauchy Problem in the Small.

We put, as usual,

$$J(t, y) = \det \frac{\partial x(t, y)}{\partial y},$$

$$(12.11')$$

where $(x(t, y), p(t, y))$ is the solution of the Cauchy problem (12.6), (12.10).

The asymptotic solution of the Cauchy problem (11.1), (11.2) immediately follows from the results of § 10. However, for some particular concrete cases, it is possible to derive simpler formulae for both the leading term of the asymptotics and the next approximations. We shall always assume that

S12.1.   Functions $S_0(x)$, $V(x)$ are real-valued, and

$$S_0(x) \in C^\infty(\mathbf{R}^n), \quad \psi_0(x) \in C_0^\infty \mathbf{R}^n).$$

Under this assumption, the solution of the Cauchy problem (12.1), (12.2) exists, is unique and belongs to $C^\infty(\mathbf{R}_x^n \times \mathbf{R}_t)$ for each fixed $h \neq 0$ [25].

Now we are going to derive the characteristic representation of the Schrödinger equation in the small (formula (12.14)). We introduce the sets

$$\Pi(M) = \{(t, x, p) : x = x(t, y), p = p(t, y), -\infty < t < \infty, y \in M\},$$

$$\Pi_T(M) = \Pi(M) \cap \{0 \le t \le T\},$$

where M is some set in $\mathbf{R}_y^n$ (Figure 4).

The set $\Pi(M)$ is a strip filled by the bicharacteristics for which $x|_{t=0} \in M$

(a tube of trajectories). We denote the projections of the strips $\Pi(M)$, $\Pi_T(M)$ on $\mathbf{R}^{n+1}_{t,x}$ by $\Pi_x(M)$, $\Pi_{T,x}(M)$. If the set M is compact, then the sets $\Pi_T(M)$, $\Pi_{T,x}(M)$ are also compact in the spaces $\mathbf{R}^{2n+1}$, $\mathbf{R}^{n+1}$, respectively.

LEMMA 12.1   *Let the assumption* S12.1 *be fulfilled and* M *be the ball* $|y| \leq R$. *Then there exists* $T = T(R) > 0$ *such that*

(1) $J(t, y) \in C^\infty$, $|J(t, y)| \geq \delta > 0$ *for* $0 \leq t \leq T$, $y \in M$, *where* J *is Jacobian* (12.11');

(2) *the solution* $S(t, x)$ *of the Cauchy problem* $S(0, x) = S_0(x)$ *for the Hamilton–Jacobi equation* (12.5) *exists, is unique and infinitely differentiable for* $(t, x) \in \Pi_{T,x}(M)$.

Proof. Due to assumption S12.1, the vector function

$$x = x(t, y) \in C^\infty(\mathbf{R}_t \times \mathbf{R}^n_y),$$

hence the Jacobian $J(t, y)$ is infinitely differentiable for all t, y. Since $J(0, y) \equiv 1$, the last statement follows from the compactness of the ball M. By the implicit function theorem, the equation $x = x(t, y)$ is uniquely solvable with respect to y and has the unique solution $y = \varphi(t, x)$ of class C^∞ in a neighbourhood of any point (t_0, y^0), where $0 \leq t_0 \leq T, |y^0| \leq R$. Since M is compact, then, by applying the Heine–Borel lemma, we obtain that $y = \varphi(t, x) \in C^\infty$ for $(t, x) \in \Pi_{T_1,x}(M), 0 < T_1 \leq T$. Consequently,

$$p = p(t, y) = p(t, \varphi(t, x)) = \tilde{p}(t, x) \in C^\infty$$

for the same t, x, and function $S(t, x)$ given by formula (12.9) is infinitely differentiable in the strip $\Pi_{T_1,x}(M)$ (Figure 4). This function solves the Cauchy problem as shown in § 3.

Thus it is possible to introduce curvilinear coordinates (t, y) (Figure 14) in the strip $\Pi_{T,x}(M)$ instead of the coordinates t, x. We set

$$\psi(t, x) = \frac{1}{\sqrt{J(t, y)}} e^{(i/h)S(t,x)} \varphi(t, x) \tag{12.12}$$

(S is determined by (12.9), where $x = x(t, y)$). Further, we put

$$\varphi(t, x(t, y)) = \tilde{\varphi}(t, y). \tag{12.13}$$

LEMMA 12.2.   *Let* T, M *be the same as in Lemma* 12.1. *Then the Schrödinger equation has the form*

$$\frac{\partial \tilde{\varphi}(t, y)}{\partial t} = -\frac{ih}{2m} \sqrt{J(t, y)} \Delta_x \frac{\tilde{\varphi}(t, y)}{\sqrt{J(t, y)}} \tag{12.14}$$

in the strip $\Pi_{T,x}(M)$.

Notice that the Laplace operator Δ_x with respect to the (curvilinear) ray coordinates y is given by

$$\Delta_x \tilde{\varphi}(t, y) = \sum_{j=1}^{n} \left(\sum_{k=1}^{n} \frac{\partial y_k}{\partial x_j} \frac{\partial}{\partial y_k} \right)^2 \tilde{\varphi}.$$

Proof. Inserting (12.12) into the last expression and using the formula

$$\Delta(\varphi_1 \varphi_2) = \varphi_2 \Delta\varphi_1 + 2\langle \nabla\varphi_1, \nabla\varphi_2 \rangle + \varphi_1 \Delta\varphi_2,$$

we get, after dividing by $\exp((i/h)S)$, the equation

$$
\begin{aligned}
\frac{ih}{\sqrt{J}} \left[\frac{\partial\varphi}{\partial t} + \frac{1}{m} \langle \nabla\varphi, \nabla S \rangle \right] + \frac{h^2}{2m} \Delta\left(\frac{\varphi}{\sqrt{J}} \right) &= \\
= \frac{\varphi}{\sqrt{J}} \left[\frac{\partial S}{\partial t} + \frac{1}{m} (\nabla S)^2 + V \right] &+ ih\varphi \left[-\frac{\partial}{\partial t}\left(\frac{1}{\sqrt{J}} \right) - \right. \\
&\left. - \frac{1}{m} \left\langle \nabla\left(\frac{1}{\sqrt{J}} \right), \nabla S \right\rangle - \frac{1}{2m} \frac{1}{\sqrt{J}} \Delta S \right],
\end{aligned}
\tag{12.15}
$$

where $\nabla = \partial/\partial x$. The right-hand side of this equation vanishes: the coefficient at h^0 is equal to zero, since the action S satisfies the Hamilton–Jacobi equation, and the coefficient at h is zero since J satisfies the transport equation.

Notice that it is by no means necessary to write down the explicit form of the right-hand side of Equation (12.15); it suffices to realise that it has the form $a\varphi + hb\varphi$, where a, b are functions independent of φ. Indeed, function (12.12) for $\varphi \equiv 1$ is a solution of the Schrödinger equation with accuracy to $O(h^2)$, as shown in § 3, Theorem 3.15, so that $a \equiv 0$, $b \equiv 0$. Finally,

$$\frac{\partial\varphi}{\partial t} + \frac{1}{m} \langle \nabla\varphi, \nabla S \rangle = \frac{d\tilde{\varphi}(t, y)}{dt},$$

where d/dt is the derivative with respect to the Hamilton system (12.6). Hence the lemma is proved.

Now we introduce the space $C_T(L_2)$ which is the closure with respect to the norm

$$\|\varphi(t, x)\|_{C_T(L_2)} = \underset{|t| \leq T}{\mathrm{Max}} \|\varphi(t, x)\|_{L_2(\mathbf{R}^n)} = \underset{|t| \leq T}{\mathrm{Max}} \left(\int_{\mathbf{R}^n} |\varphi(t, x)|^2 \, dx \right)^{1/2}$$

of the functions which are infinitely differentiable for $|t| \leq T$ and have compact supports for each fixed $t \in [-T, T]$.

THEOREM 12.3. *Let assumption S12.1 be fulfilled and let $T > 0$ be sufficiently small. Then the solution of the Cauchy problem (12.1), (12.2) has the form*

$$\psi(t, x, h) = \frac{1}{\sqrt{J(t, y)}} \exp\left[\frac{i}{h} S(t, x)\right] \sum_{k=0}^{N} (ih)^k \psi_k(t, x) +$$
$$+ R_{N+1}(t, x, h). \tag{12.16}$$

Here $N \geq 1$ is an arbitrary integer; the remainder satisfies the estimate

$$\|R_{N+1}(t, x, h)\|_{C_T(L_2)} \leq C_N h^{N+1} \tag{12.17}$$

for $0 < h \leq h_0$, where $h_0 > 0$ is sufficiently small.
   The explicit expressions for functions ψ_k have the form:

$$\psi_0(t, x) = \psi_0(y), \tag{12.18}$$

$$\psi_{k+1}(t, x) = \int_0^t \sqrt{J(t, y)} \Delta_x\left(\frac{\psi_k(t, x(t, y))}{\sqrt{J(t, y)}}\right) dt;$$

(in these formulae $x = x(t, y)$ and the integrals are taken along the trajectories of the Hamilton system (12.9)).
   All functions $\psi_k(t, x)$ are infinitely differentiable in the strip $\Pi_{T,x}(\mathrm{supp}\,\psi_0)$ and vanish outside the strip.
   Proof. By virtue of assumption S12.1, Lemma 12.1 and Theorem 3.15, there exists the function

$$\psi_{N+1}(t, x, h) = \sum_{k=0}^{N+1} (i h)^k \tilde{\psi}_k(t, x),$$

satisfying the Cauchy data (12.2) for any $N \geq 1$, and such that

$$L\psi_{N+1} = f_N \in O_{-N-2, T}, \qquad L = ih\frac{\partial}{\partial t} + \frac{h^2}{2m}\Delta - V.$$

If function $f \in O_{-k, T}$, then

$$\|f\|_{C_T(L_2)} \leq C h^k,$$

which follows from the definition of class $O_{-k, T}$ (§ 10).

If the Schrödinger equation is written in the form (12.3), then

$$ih\frac{d\psi_{N+1}}{dt} - A\psi_{N+1} = f_N.$$

It follows from Proposition 10.3 that

$$\|\psi - \psi_{N+1}\|_{C_T(L_2)} \le h^{-1} C_N \|f_N\|_{C_T(L_2)} \le C'_N h^{N+1},$$
$$0 < h \le h_0,$$

since $f_N \in O_{-N-2,T}$. And finally since

$$\|\tilde{\psi}_{N+1}(t,x)\|_{C_T(L_2)} < \infty,$$

we have

$$\|\psi - \psi_N\|_{C_T(L_2)} \le \tilde{C}_N h^{N+1}.$$

In this way the estimate (12.17) for the remainder is established.

It remains to prove that $\tilde{\psi}_k \equiv \psi_k J^{-1/2}$, i.e. that $\tilde{\psi}_k J^{1/2}$ has the form (12.18). Since the existence of the asymptotic expansion has already been established and the asymptotic expansion in powers of h is unique, the proof of formula (12.18) can be reduced to formal arguments. We set

$$\psi = J^{-1/2} e^{(i/h)S} \sum_{k=0}^{N} (ih)^k \psi_k,$$

insert this into the Schrödinger equation, and after dividing by $\exp((i/h)S)$ we put the coefficients at powers of h equal to zero. Using the characteristic representation (12.12) for the Schrödinger equation, we obtain the system of recursion relations

$$\frac{\partial\tilde{\psi}_0}{\partial t} = 0, \quad \frac{\partial\tilde{\psi}_1}{\partial t} + Q\tilde{\psi}_0 = 0, \dots, \quad \frac{\partial\tilde{\psi}_{j+1}}{\partial t} + Q\tilde{\psi}_j = 0, \dots,$$

$$Q = \sqrt{J}\Delta_x \frac{1}{\sqrt{J}}.$$

Here $\tilde{\psi}_j(t,y) = \psi_j(t, x(t,y))$. The Cauchy data are given by

$$\tilde{\psi}_0(0,y) = \psi_0(y), \quad \tilde{\psi}_j(0,y) = 0, \quad j \ge 1.$$

By solving these Cauchy problems step by step, we get (12.18). The last formula implies that functions ψ_k are smooth in the strip $\Pi_{T,x}$ and vanish outside the strip.

Theorem 12.3 is just a rigorous demonstration of the known physical fact: *the semi-classical wave function of a quantum mechanical particle is concentrated near its classical trajectory.* Indeed, if ψ is the solution of the Cauchy problem (12.1), (12.2), then it follows from Theorem 12.3 that $|\psi|^2$ is a quantity of order 1 in the ray tube $\Pi_{T,x}$ filled by the classical trajectories, and that, outside the ray tube, $|\psi|^2$ decreases faster than any power of h for $h \to 0$. This means that probability to find a particle outside the ray tube is arbitrarily small for small h.

3. *The Asymptotic Solution of the Cauchy Problem in the Large*

Below we suppose condition S12.1 as well as

S12.2. The sets $\Lambda^n_t = g^t \Lambda^n_0$ (for $0 \le t \le T_0$) and $\Lambda^{n+1}_{T_0} = \bigcup_{0 < t < T_0} \Lambda^n_t$ are C^∞-manifolds.

Proposition 4.19 implies that $\Lambda^{n+1}_{T_0}$ is a Lagrangian manifold of dimension $n + 1$.

THEOREM 12.4. *Let the assumptions* S12.1, S12.2 *be satisfied for* $0 < T < T_0$; *then the solution of the Cauchy problem* (12.1), (12.2) *has the form*

$$\psi(t, x, h) = K_{\Lambda^{n+1}_T} \sum_{k=0}^{N} (ih)^k \psi_k + R_{N+1}(t, x, h). \qquad (12.19)$$

The estimate (12.17) *of the remainder is valid. Functions* $\psi_k \in C^\infty(\Lambda^{n+1}_T)$ *and vanish outside the strip* $\Pi_T (\operatorname{supp} \psi_0)$.

Proof of the theorem follows from the existence of a f.a. solution $\tilde{\psi}$

$$ih \frac{d\tilde{\psi}}{dt} - A\tilde{\psi} = f_N \in O_{-N-2,T},$$

proved in § 10, Theorem 10.1, and from the estimate of the inverse operator (Proposition 10.3).

We shall give the asymptotic formulae for solutions of the Schrödinger equation at non-focal points. Let $x = x(t, y)$ be a trajectory of the system (12.6). Point t_0 is called a *focus* on the trajectory $x = x(t, y)$, if $J(t_0, y) = 0$.

Now we shall write down the asymptotic solution of the Cauchy problem (12.1), (12.2) in non-focal points. We fix t_0, the point $x^0 \in \mathbf{R}^n$, and suppose that $y^j \in \mathbf{R}^n, j = 1, 2, \ldots, N = N(x^0)$, are all points for which

$$x(t_0, y^j) = x^0. \qquad (12.20)$$

The number of such points y^j is finite, provided the point (t_0, x^0) is non-

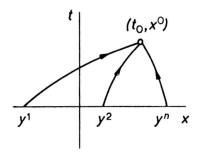

Fig. 15.

focal [69] (Figure 15). We notice (see § 7) that the Morse index μ of a trajectory (with non-focal end-points) is equal to the number of the focal points on the trajectory counted with their multiplicity.

Theorems 12.3 and 10.2 yield

THEOREM 12.5. *Let the point* (t_0, x^0) *be non-focal. Then*

$$\psi(t_0, x^0, h) = \sum_{j=1}^{N} \psi_0(y^j) \left| \det \frac{\partial x(t_0, y^j)}{\partial y} \right|^{-1/2} \times$$

$$\times \exp\left[\frac{i}{h} S_j(t_0, x^0) - \frac{i\pi}{2} \mu_j \right] + O(h), \quad (h \to 0).$$

$$(12.21)$$

Here $S_j(t_0, x^0)$ is the action along the classical trajectory joining the points y^j, x^0, i.e.

$$S_j(t_0, x^0) = S_0(y^j) + \int_0^{t_0} \left(m \frac{\dot{x}^2}{2} - V(x) \right) dt,$$

where the integral is taken along a trajectory such that

$$x\big|_{t=0} = y^j, \quad x\big|_{t=t_0} = x^0,$$

and μ_j is the Morse index of the trajectory.

As an example consider the quantum mechanical oscillator:

$$ih \frac{\partial \psi}{\partial t} = -\frac{h^2}{2} \Delta \psi + \tfrac{1}{2} \langle x, x \rangle \psi,$$

$$(12.22)$$

where $x \in \mathbf{R}^n$. The fundamental solution $G(t, x, y)$ of the Cauchy problem for Equation (12.22), i.e. the solution with the Cauchy data

$$G|_{t=0} = \delta(x - y),$$

can be evaluated explicitly [21]:

$$G(t, x, y) = (2\pi \, hi \sin t)^{-n/2} \exp\left[\frac{i}{h} S(t, x; 0, y)\right],$$

$$S(t, x; 0, y) = \frac{1}{2 \sin t}[(\langle x, x \rangle + \langle y, y \rangle)\cos t - 2\langle x, y \rangle]. \quad (12.23)$$

Recall that S is a two-point characteristic function, i.e. the action along the ray $x = X(\tau)$ such that $X(0) = y$, $X(t) = x$ (see (3.72)). Let us also remark that formula (12.23) only holds for $0 < t < \pi/2$.

Consequently, for $0 < t < \pi/2$ the following representation of the solution of the Cauchy problem is valid:

$$\psi(t, x) = \int_{\mathbf{R}_y^n} G(t, x, y) \, \psi(0, y) \, dy, \quad (12.24)$$

provided $\psi(0, y) \in C_0^\infty(\mathbf{R}^n)$.

Now consider the Cauchy problem

$$\psi|_{t=0} = \psi_0(x) \exp\left(\frac{i}{h}\langle p^0, x \rangle\right), \quad (12.25)$$

where $\psi_0 \in C_0^\infty(\mathbf{R}^n)$, $p^0 \neq 0$ is a fixed vector, and write down the asymptotic formulae for the solution of the Cauchy problem (12.22) and (12.25). The Hamilton system has the form

$$\frac{dx}{dt} = p, \qquad \frac{dp}{dt} = -x;$$

the Cauchy data for this system induced by the Cauchy data (12.25) are

$$x|_{t=0} = y, \qquad p|_{t=0} = p^0, \qquad y \in \mathbf{R}^n.$$

The solution of the Cauchy problem for the Hamilton system is given by (see Example 3.22, § 3)

$$x = p^0 \sin t + y \cos t, \quad p = p^0 \cos t - y \sin t. \quad (12.26)$$

The Jacobian $J(t, y) = \partial x(t, y)/\partial y$ is equal to

$$J(t, y) = (\cos t)^n$$

and vanishes for $t = t_k = k\pi + (\pi/2)$, $k = 0, \pm 1, \pm 2, \ldots$. Formula (12.26), where y runs through $\mathbf{R}^n$, determines a Lagrangian C^∞-manifold Λ_t^n in phase space $\mathbf{R}_{x,p}^{2n}$. This manifold admits a diffeomorphic projection on $\mathbf{R}_x^n$ for $t \neq t_k$. For $t = t_k$, we have

$$\Lambda_{t_k}^n = \{(x, p): x = (-1)^k p^0, p = (-1)^{k+1} y, y \in \mathbf{R}^n\},$$

i.e. the manifold $\Lambda_{t_k}^n$ is an n-plane parallel to the coordinate Lagrangian plane p. This manifold is diffeomorphically projected on $\mathbf{R}_p^n$.

Let us investigate the family of rays $x = x(t, y)$, $y \in \mathbf{R}^n$. If $t = t_k$, then

$$x(t_k, y) = (-1)^k p^0$$

for all y, i.e. all rays meet at one point – the ray focusation takes place (Figure 7). If $t \neq t_k$, then exactly one ray $x = x(t, y)$ arrives in an arbitrary point $\tilde{x} \in \mathbf{R}^n$ after time t; here y is determined from the relation

$$y = \frac{\tilde{x} - p^0 \sin t}{\cos t} \qquad (12.27)$$

(see also (3.69)). We write down the asymptotic formulae for the solution of the Cauchy problem (12.22), (12.25) for $t \neq t_k$. The action S is equal to $S(t, x) = S(t, x; 0, y)$, where y is determined by (12.27), so that

$$S(t, x) = \tfrac{1}{2} \tan t (\langle p^0, p^0 \rangle - \langle x, x \rangle). \qquad (12.28)$$

(1) $0 < t < \pi/2$. After time t exactly one ray arrives in the given point x; since there are no focal points on that ray, its index equals zero. Consequently, from (12.21) we obtain

$$\psi(t, x) = (\cos t)^{-n/2} \exp\left[\frac{i}{h} S(t, x)\right] \psi\left(\frac{x - p^0 \sin t}{\cos t}\right) + O(h)$$

$$(0 < t < \pi/2), \qquad (12.29)$$

where S is given by (12.28).

(2) $\pi/2 < t < 3\pi/2$. In this case, there is again exactly one ray arriving in each point after time t, but there exists a focal point on the ray $x = x(\tau, y)$ for $\tau = \pi/2$. Since the multiplicity of the zero of the Jacobian $J = (\cos \tau)^n$

is n, the Morse index for that ray is equal to n (§ 7). From (12.21) we obtain

$$\psi(t, x) = |\cos t|^{-n/2} e^{-i\pi n/2} \exp\left[\frac{i}{h} S(t, x)\right] \times$$

$$\times \psi_0\left(\frac{x - p^0 \sin t}{\cos t}\right) + O(h), \left(\frac{\pi}{2} < t < \frac{3\pi}{2}\right). \qquad (12.30)$$

(3) $(\pi/2) + k\pi < t < (\pi/2) + (k + 1)\pi, k \geq 0$.

It follows from the additivity of the Morse index that the index of the ray reaching point x after time t is equal to $n(k + 1)$, and thus

$$\psi(t, x) = |\cos t|^{-n/2} \exp\left[\frac{i}{h} S(t, x)\right] e^{-i\pi n(k+1)/2} \times$$

$$\times \psi_0\left(\frac{x - p^0 \sin t}{\cos t}\right) + O(h) \qquad (12.31)$$

$$\left(\frac{\pi}{2} + k\pi < t < \frac{\pi}{2} + (k + 1)\pi\right).$$

Remainder $O(h)$ belongs to class $O_{-1,T}(\mathbf{R}^n)$ in all formulae (12.29)–(12.31).

This example is remarkable, since we can directly verify the asymptotic formulae on it. Namely, let $0 < t < \pi/2$; then the solution of the Cauchy problem (12.22), (12.25) has the form

$$\psi(t, x) = (2\pi hi \sin t)^{-n/2} \int_{\mathbf{R}^n} \exp\left[\frac{i}{h} \tilde{S}(t, x, y)\right] \psi_0(y)\, dy,$$

$$\tilde{S} = S(t, x; 0, y) + \langle p^0, y \rangle.$$

We evaluate the asymptotics of this integral for fixed t and for $h \to 0$ by using the stationary phase method (§ 1). The stationary points of $\tilde{S}$ (as a function of y) are found from the equation $\partial\tilde{S}/\partial y = 0$; one concludes that the stationary point $y = y(t, x)$ is unique and of the form

$$y = \frac{x - p^0 \sin t}{\cos t}$$

(cf. (12.27)). The value of function $\tilde{S}$ at the stationary point is equal to

$$\tilde{S}_{st} = S(t, x),$$

where $S(t, x)$ has the form (12.28). Now we evaluate $\partial^2 \tilde{S}/\partial y^2$. We have

$$\frac{\partial^2 \tilde{S}}{\partial y^2} = I \cot t,$$

hence

$$\det \frac{\partial^2 \tilde{S}}{\partial y^2} = (\operatorname{ctg} t)^n, \qquad \operatorname{sgn} \frac{\partial^2 \tilde{S}}{\partial y^2} = n$$

at the stationary point. From formula (1.4) we obtain the expression

$$\psi(t, x) = (2\pi i h \sin t)^{-n/2} \left| \det \frac{\partial^2 \tilde{S}}{\partial y^2} \right|^{-1/2} \times$$

$$\times \exp\left[\frac{i}{h}\tilde{S} + \frac{i\pi}{4} \operatorname{sgn} \frac{\partial^2 \tilde{S}}{\partial y^2} \right] \psi_0 [1 + O(h)]$$

(the values of all functions are taken at the stationary point); it coincides with the right-hand side of formula (12.20). The Cauchy data (12.2) can be verified in exactly the same way.

This example has still one interesting application. As follows from formulae (12.23), (12.24),

$$\psi\left(\frac{\pi}{2}, 0\right) = (2\pi h i)^{-n/2} \int_{\mathbf{R}^n} \exp\left[\frac{i}{h} S_0(x) \right] \psi_0(y)\, dy, \qquad (12.32)$$

provided $\psi(t, x)$ represents the solution of the Cauchy problem (12.22), (12.2).

However, this integral is an integral of the form (1.1), i.e. an integral of an arbitrary rapidly oscillating function. Thus the asymptotic expansion of integral (12.32) and explicit expressions for the coefficients at powers of h can be obtained by means of the semi-classical asymptotic solution of the Cauchy problem (12.22), (12.2). This is worked out in [64].

Now we shall present the asymptotic formulae for the Green function $G(t, x, y)$. We consider the boundary-value problem for the classical Newton equation

$$m \frac{d^2 X}{dt^2} = -\frac{\partial V(X)}{\partial X}, \qquad (12.33)$$

$$X(0) = y, \qquad X(t) = x.$$

Let its solution $X(\tau; t, x, y)$ be unique. The action along the classical trajectory is equal to

$$S(t, x, y) = \int_0^t \left[\frac{m\dot{X}^2}{2} - V(X(\tau; t, x, y)) \right] d\tau. \tag{12.34}$$

We fix $y \in \mathbf{R}^n$, and suppose the solution of the problem (12.33) to be unique for all $x \in \mathbf{R}^n$ and for all $t \in [0, T]$. The potential is restricted by the same conditions as in Section 1. Then the semi-classical asymptotics of the Green function is given by [56], [57]

$$G(t, x, y) = (2\pi i h)^{-n/2} \sqrt{\left| \det \frac{\partial^2 S}{\partial x \, \partial y} \right|} \times$$

$$\times \exp \left[\frac{i}{h} S(t, x, y) \right] [1 + hz]. \tag{12.35}$$

For the function $z = z(t, x, y; h)$ the estimate

$$\| z \|_{L_2(\mathbf{R}^n)} \leq C$$

is true for $0 \leq t \leq T, x \in \mathbf{R}^n, 0 < h \leq 1$.

If potential $V(x)$ is given by a quadratic function, then the asymptotic formula (12.35) (for $z \equiv 0$) leads to the exact Green function (for $0 \leq t \leq T$, where T is determined by the potential V).

The asymptotics of the Green function can be extended to a larger time interval by using the standard convolution formula

$$G(t_1 + t_2, x, y) = \int_{\mathbf{R}^n} G(t_2, x, \tilde{y}) \, G(t_1, \tilde{y}, y) \, d\tilde{y}.$$

The asymptotics of the last integral can be evaluated with the help of the stationary phase method.

Let the boundary-value problem (12.33) have a finite number of distinct solutions $X_k(\tau; t, x, y)$ ($k = 1, 2, \ldots, N$) for fixed t, y, and let $S_k(t, x, y)$ be the action along the kth solution. Then the asymptotics of the Green function has the form

$$G(t, x, y) = (2\pi i h)^{-n/2} \left[\sum_{k=1}^{N} \left| \det \frac{\partial^2 S_k}{\partial x \, \partial y} \right|^{1/2} \times \right.$$

$$\left. \times \exp \left(\frac{i}{h} S_k(t, x, y) - \frac{i\pi}{2} \mu_k \right) + O(h) \right], \tag{12.36}$$

where μ_k is the Morse index of the kth trajectory. It is assumed that the end-points of each trajectory are non-focal.

4. The Semi-Classical Asymptotics of the Green Function for the Stationary Schrödinger Equation

Consider the stationary Schrödinger equation

$$h^2 \Delta_x G + (E - V(x))G = \delta(x - x^0). \tag{12.37}$$

Here $x, x^0 \in \mathbf{R}^n$, the point x^0 is fixed, the potential $V(x)$ is a real-valued function of class $C^\infty(\mathbf{R}^n)$, $E > 0$ is a constant,

$$E - V(x) \geqq a > 0, \quad x \in \mathbf{R}^n. \tag{12.38}$$

Suppose potential $V(x)$ decreases sufficiently fast at infinity, namely, for any multi-index α the estimate

$$|D_x^\alpha V(x)| \leqq C_\alpha (1 + |x|)^{-|\alpha| - 2}, \quad x \in \mathbf{R}^n, \tag{12.39}$$

is true.

Moreover, we suppose that the Sommerfield radiation conditions

$$G(x, x^0) = O(|x|^{(1-n)/2}),$$

$$\frac{\partial G}{\partial |x|} + i \sqrt{\frac{E}{h}} G = o(|x|^{(1-n)/2}), \quad (|x| \to \infty), \tag{12.40}$$

are valid at infinity.

As is well known, the solution of the problem (12.37), (12.40) exists and is unique, provided the assumptions on the potential formulated above are satisfied. This solution is a ψ-function of a beam of particles emitted with energy E from a source placed in point x^0 and moving in a potential field with potential energy $V(x)$. This problem can be also interpreted as the problem of determining the field of a point light source in an inhomogeneous medium with the refraction index $n(x) = \sqrt{E - V(x)}$, and with the wave number $k = h^{-1}$. The condition (12.38) expresses in this case the fact that the square of the refraction index is positive everywhere, so that there is no absorption in the medium.

The scattering problem in the one-dimensional case is well understood. Even such fine questions as the quasi-stationary level problem, or the above-barrier reflection problem have been studied (see [17], [18], [58], [73]).

The scattering problem for $n \geq 2$ is considerably more complicated.

In this section we shall briefly summarize the rigorous results obtained in [42] by the method of the canonical operator.

The Hamilton–Jacobi equation

$$(\nabla S(x))^2 = E - V(x) \tag{12.41}$$

and the Hamilton system

$$\frac{dx}{d\tau} = 2p, \qquad \frac{dp}{d\tau} = -\nabla V(x) \tag{12.42}$$

are associated with Equation (12.37).

We set up the Cauchy problem for the Hamilton system:

$$p^2\big|_{\tau=0} = E - V(x^0), \qquad x\big|_{\tau=0} = x^0. \tag{12.43}$$

We keep in mind that the Hamiltonian $H(x, p) = p^2 + V(x)$ is a first integral of the Hamilton system (see § 3), hence

$$p^2 + V(x) \equiv E \tag{12.44}$$

on any solution $\{x(\tau), p(\tau)\}$ of Cauchy problem (12.42), (12.43).

The physical interpretation of the Cauchy data (12.43) is: at 'time' $\tau = 0$, all rays are emitted from the point x^0, in all possible directions, each having energy E. We shall determine the Lagrangian manifold associated with the problem (12.37), (12.40). The initial momenta $p\big|_{\tau=0}$ fill up the sphere $S^{n-1} = \{p \in \mathbf{R}^n : p^2 = E - V(x^0)\}$. Let $\theta = (\theta_1, \theta_2, \ldots, \theta_{n-1})$ be angular coordinates on the sphere; by $\{x(\tau, \theta), p(\tau, \theta)\}$ we denote the solution of the problem (12.42), (12.43) for which the initial momentum $p(0, \theta)$ has angular coordinates θ. In the rest of this paragraph we assume the validity of the assumption:

S12.3. Potential $V(x)$ satisfies the conditions (12.38), (12.39), and system (12.42) does not admit *finite motions*.

The assumption means that

$$\lim_{\tau \to +\infty} |x(\tau, \theta)| = \infty$$

for any $x^0 \in \mathbf{R}^n$ and all $\theta \in S^{n-1}$. Then the set

$$\Lambda^n = \{(x, p) \in \mathbf{R}^{2n} : x = x(\tau, \theta), p = p(\tau, \theta), \theta \in S^{n-1}, 0 < \tau < \infty\} \tag{12.45}$$

is a Lagrangian C^∞-manifold of dimension n with the boundary

$$\partial \Lambda^n = \{(x, p) \in \mathbf{R}^{2n} : x = x^0, p^2 = E - V(x^0)\}.$$

The asymptotics of the Green function $G(x, x^0)$ for $h \to 0$ is constructed with the help of the canonical operator K_{Λ^n} corresponding to manifold Λ^n.

In constructing the asymptotics of the Green function a difficulty arises, connected with the fact that function $G(x, x^0)$ has a singularity in point $x = x^0$, and therefore the asymptotics of G behaves differently in the neighbourhood $U = \{x \in \mathbf{R}^n : |x - x^0| < \varepsilon\}$ of point x^0 and outside this neighbourhood. In the following $\varepsilon > 0$ is fixed and sufficiently small.

The approximate Green function $G_N(x, x^0)$ will be sought in the form

$$G_N(x, x^0) = G_N^0(x, x^0) + G_N^1(x, x^0), \tag{12.46}$$

where $G_N^1 \equiv 0$ for $x \in U$, and function G_N^0 is concentrated in a neighbourhood point x^0.

(1) The construction of the *singular part* G_N^0 of the Green function.

Function G_N^0 is constructed with the help of the non-stationary Schrödinger equation. Consider the auxiliary Cauchy problem

$$-ih\frac{\partial \psi}{\partial t} = h^2 \Delta \psi + n^2(x)\psi, \tag{12.47}$$

$$\psi|_{t=0} = \frac{-ih^{n-1}}{(2\pi)^n} \exp\left[\frac{i}{h}\langle p, x\rangle\right], \tag{12.48}$$

where $n^2(x) = E - V(x)$, $p \in \mathbf{R}^n$. Let $\psi(t, x, p)$ be the solution of the problem. Since

$$\delta(x - x^0) = (2\pi h)^{-n} \int_{\mathbf{R}^n} \exp\left[\frac{i}{h}\langle p, x - x^0\rangle\right] dp,$$

the formal application of the Fourier transformation leads to the formula:

$$G(x, x^0) = \int_{\mathbf{R}^n} dp \int_0^\infty \exp\left[-\frac{i}{h}\langle p, x^0\rangle\right]\psi(t, x, p)\, dp.$$

We have already discussed the diffculties connected with such formal approach in § 10, Section 5. However, we are going to construct the asymptotics of the Green function only in a small neighbourhood U of point x^0. In this neighbourhood the rays are 'almost straight' lines so that the variables (τ, θ) are the natural coordinates in U, provided $0 \leq \tau \leq t_0$ and $t_0 \ll 1$.

We introduce a cut-off function $\eta(t) \in C^\infty$ at $t = 0$, which equals 1 for small t and vanishes for $t \geq t_0$. Denoting the f.a. solution of the Cauchy problem (12.48) with accuracy to $O(|p|^{-N})$ by ψ_N (the construction of

ψ_N see below), we set

$$G_N^0(x, x^0) = \int_{\mathbf{R}^n} dp \int_0^{\infty} dt\, \eta(t)\psi_N(t, x, p) \exp\left[-\frac{i}{h}\langle p, x^0\rangle \right]. \quad (12.49)$$

This integral is formally divergent; it can be regularized by the standard methods used for oscillatory integrals.

Function $\hat{G}_N^0 \in C^{\infty}$ in a small 'holey' neighbourhood of point x^0.

It remains to derive the explicit formulae for the functions ψ_N. This task reduces to the construction of the asymptotics (for $h \to 0, |p| \geq 1$) of the Cauchy problem (12.49) in the small. As follows from the results of § 10, we have

$$\psi_N(t, x, p) = \exp\left[\frac{i}{h}S(t, x, p) \right] |J|^{1/2} \sum_{j=0}^{N} (-ih)^j \varphi_j(t, x, p).$$

$$(12.50)$$

Action S is the solution of the Cauchy problem

$$\frac{\partial S}{\partial t} + (\nabla S)^2 + n^2(x) = 0, \qquad S|_{t=0} = \langle x, p\rangle,$$

and appears to be a homogeneous function of degree 1 in the variables p. Further, J denotes the Jacobian,

$$J = \det \frac{\partial x(t, y, p)}{\partial y}, \quad (12.51)$$

where $x = x(t, y, p)$ is the solution of the Cauchy problem for the Newton system

$$\frac{d^2 x}{dt^2} = -\nabla n^2(x), \qquad x|_{t=0} = y, \qquad \frac{dx}{dt}\bigg|_{t=0} = p. \quad (12.52)$$

Thus we have completed the construction of a singular part of the Green function.

As is apparent from formulae (12.49)–(12.52) function G_N^0 can be expressed in terms of the solutions of the dynamical system (12.52). This function satisfies the equation

$$[h^2\Delta + n^2(x)]G_N^0 = \delta(x - x^0) + h^{N+1} f_N(x, h),$$

in a domain U, where $f_N \in C^\infty$, and fulfils the estimates

$$|D_x^\alpha f_N| \leq C_\alpha h^{-|\alpha|}$$

for $x \in U$.

(2) The asymptotics of function G_N^0 outside singularities.

Integral (12.49) has a singularity for $x = x^0$ and its behaviour in a small neighbourhood of point x^0 is pretty complicated. However, going away from this point, G_N^0 becomes a rapidly oscillating function. This can be shown by applying the stationary phase method to integral (12.49). The asymptotics is expressed in terms of the Lagrangian manifold Λ^n (see (12.45)) and of the dynamical system (12.42). We introduce the action

$$S(x, x^0) = \int_{r^0}^{r} \langle p, dx \rangle. \tag{12.53}$$

The integral is taken along a path lying on Λ^n and connecting the points $r^0 = (x^0, p^0)$, $r = (x, p)$; p^0 is any point such that $(p^0)^2 = n^2(x^0)$ (i.e., x^0, x are projections of the points r^0, r on x-space). Function S is single-valued, provided x belongs to a small neighbourhood U of the point x^0. Further, we introduce the Jacobian

$$J(x, x^0) = \frac{dx}{d\omega\, d\tau}\bigg|_{\tau=t}. \tag{12.54}$$

Here $x = x(\tau, \theta)$ (a ray),

$$d\omega = \sin^{n-2}\theta_1 \ldots \sin\theta_{n-2}\, d\theta_1 \ldots d\theta_{n-1}$$

and $x(t, \theta) = x$ (i.e. the ray goes through the point x). Time $t \in (0, t_0)$ and point $x \in U$, so the ray is unique for small t_0, θ.

Finally, we obtain the asymptotic formula $(h \to +0)$

$$G_N^0(x, x^0) = h^{-(n+1)/2}(n(x^0))^{(n-2)/2} e^{(i\pi/4)(n+1)}|J(x, x^0)|^{-1/2} \times$$

$$\times \exp\left[\frac{i}{h}S(x, x^0)\right]\left[1 + \sum_{j=1}^{m} \varphi_j(x, x^0)h^j + O(h^{m+1})\right]. \tag{12.55}$$

This formula holds in the domain $U \setminus U_0$, where U_0 is a neighbourhood of point x^0, smaller than U; function $\varphi_j \in C^\infty(U \setminus U_0)$.

(3) The construction of function G_N^1.

This function is expressible via canonical operator K_{Λ^n} corresponding to Lagrangian manifold Λ^n (see (12.45)).

Let Λ_1^n be a part of manifold Λ^n lying over the domain $\mathbf{R}_x^n \setminus V$, V being a neighbourhood of the point x^0 lying strictly inside the neighbourhood U chosen above. Manifold Λ_1^n is unbounded. Nevertheless, it can be covered by a finite number of canonical charts, because finite motions are absent (assumption S12.3) and potential $V(x)$ converges rapidly to zero for $|x| \to \infty$. This permits to introduce the canonical operator $K_{\Lambda_1^n}$ by the same formulae as in § 8. The volume element on Λ_1^n is chosen as follows:

$$d\sigma^n = d\omega \, d\tau.$$

This volume element is invariant under the displacements along trajectories of the system (12.35).

Function G_N^1 is given by the formula

$$G_N^1(x, x^0) = K_{\Lambda_1^n}\left[\rho\left(g_0 + \frac{h}{i}g_1 + \ldots + \left(\frac{h}{i}\right)^N g_N \right)\right]. \tag{12.56}$$

Here ρ is a cut-off function equal to 1 on Λ_1^n and to 0 in a neighbourhood of the boundary $\partial\Lambda_n$. The functions g_j are constructed in terms of functions φ_j found above (see (12.50)).

THEOREM 12.6.   *The estimate*

$$\left|G(x, x^0) - G_N(x, x^0)\right| \leq Ch^{N-n}(1 + |x|)^{(1-m)\,2}. \tag{12.57}$$

holds for $x \in \mathbf{R}^n$ and $0 < h \leq 1$.

Finally, we present the asymptotic formula for the Green function in a non-focal point. Let $\tilde{x} \neq x^0$ be the non-focal point. Then there exists a finite number of rays

$$x = x(\tau, \theta^j), \quad 1 \leq j \leq k,$$

passing through the point $\tilde{x}$, i.e.

$$x(\tau_j, \theta^j) = \tilde{x}$$

for some τ_j. We denote by $\tilde{S}_j(\tilde{x}, x^0)$ the action along the jth ray, by $\mu_j(\tilde{x})$ the index of the jth ray, and by J_j the Jacobian (12.54) evaluated for

$\tau = \tau_j$, $\theta = \theta^j$. Function G is given by

$$G(x, x^0) = d_n h^{(n-7)/2} \sum_{j=1}^{k} |J_j|^{-1/2} \times$$

$$\times \exp\left[-\frac{i}{h} S_j(\tilde{x}, x^0) + \frac{i\pi}{2}\mu_j(\tilde{x}) \right] + O(h), \qquad (12.58)$$

where d_n is the constant

$$d_n = 2\sqrt{\pi} \left(\frac{E - V(x^0)}{2\pi} \right)^{(n/2)-1} \exp\left[i(n+1)\frac{\pi}{4} \right].$$

5. The Stationary Phase Method and Asymptotic Solution of the Cauchy Problem in the Small

We shall review here the method suggested in [89] for deriving the successive terms of the expansion of the integral (1.1)

$$I(h) = \frac{1}{(2\pi h)^{n/2}} \int_{-\infty}^{+\infty} e^{(i/h)\Phi(x)} \varphi(x)\, dx, \qquad x = (x_1, \ldots, x_n),$$

in powers of $h = \lambda^{-1}$, where $\Phi(x) \in C^{\infty}(\mathbf{R}^n)$, $\varphi(x) \in C_0^{\infty}(\mathbf{R}^n)$.

The method is based on such careful choice of the coefficients a_{ij} in the free Schrödinger equation

$$ih\frac{\partial\psi(x, t)}{\partial t} = \sum_{i,j} a_{ij} \frac{\partial^2\psi(x, t)}{\partial x_i \partial x_j} \qquad (12.59)$$

that for some initial condition the solution of this equation at the point $x = 0$, $t = 1$ would coincide with the investigated integral and, for $0 \leq t \leq 1$ the WKB method would be applicable to construct the asymptotic solution of Equation (12.59), i.e. the bicharacteristics should have no focal points.

Let $x = \xi^k$ be an isolated simple stationary point of the function $\Phi(x)$:

$$\operatorname{grad} \Phi(\xi^k) = 0,\ \det C(\xi^k) \neq 0,$$

where $C(x) = (\Phi_{x_i x_j}(x))$. The following lemma is important for further discussion.

LEMMA 12.7.   Let Ω_{ε}^k be an ε-neighbourhood of the point ξ^k and

$$X^k(x_0, t) \stackrel{\text{def}}{=} 2tC^{-1}(\xi^k)\nabla\Phi(x_0) + (1 - t)x_0.$$

Then for sufficiently small ε and $t \geq 0$,

$$J(x_0, t) \overset{\text{def}}{=} \det \left(\frac{\partial X_i^k}{\partial x_{0j}} \right) \neq 0,$$

and there exists a unique smooth solution $x_0 = x_0^k(x, t)$ of the equation $x = X^k(x_0, t)$.

Proof. By explicit calculation we can show that

$$J(\xi^k, t) = \det \left(\frac{\partial X_i^k}{\partial x_{0j}} \right) \Bigg|_{x_0 = \xi^k} = (t + 1)^n,$$

from which the statement follows.

We introduce the notation:

$$\left(\frac{1}{1 - A} \right)_N = \sum_{k=0}^{N} A^k,$$

$$R_k \psi(x, t) = \int_0^t \sqrt{J(x_0^k(x, \tau), \tau)} \, \langle \nabla, C^{-1}(\xi^k) \nabla \rangle \frac{\psi(x, \tau)}{\sqrt{J(x_0^k(x, \tau), \tau)}} \, d\tau,$$

$$\nabla = \left(\frac{\partial}{\partial x_1}, \dots, \frac{\partial}{\partial x_n} \right),$$

where $J(x_0, t)$ and $x_0^k(x, t)$ are defined in the lemma.

THEOREM 12.8. *Let there exist s isolated simple stationary points $\xi^1, \xi^2, \dots, \xi^s$ of function $\Phi(x)$ in the domain supp $\varphi(x)$ (i.e. $\nabla \Phi(\xi^k) = 0$, $\det C(\xi^k) \neq 0$, $k = 1, \dots, s$, $C(x) = (\Phi_{x_i x_j}(x))$). Then for any $N \geq 1$ the following relation is true:*

$$I(h) = \sum_{k=1}^{s} \frac{e^{i(\pi/4)n} \exp \left\{ -i\frac{\pi}{2} \text{inerdex } C(\xi^k) \right\} e^{(i/h)\Phi(\xi^k)}}{\sqrt{|\det C(\xi^k)|}} \times$$

$$\times \left(\varphi(\xi^k) + ih \left[\left(\frac{1}{1 - ihR_k} \right)_N R_k e_k(x_0^k(x, t)) \varphi(x_0^k(x, t)) \right] \Bigg|_{\substack{x=0 \\ t=1}} \right) +$$

$$+ Z_h + O(h^\infty), \tag{12.60}$$

where $e_k(x)$ belong to C_0^∞ and are such that $e_k(x) \equiv 1$ for $x = \xi^k$ and equal zero for x lying outside a sufficiently small neighbourhood of the point ξ^k; the *remainder is given by*

$$Z_h = \sum_{k=1}^{s} h^{N+1} \int_0^1 \frac{1}{(\pi h)^{n/2}(1-\tau)^{n/2} 2^{n/2}} \times$$

$$\times \int_{-\infty}^{+\infty} \exp\left\{\frac{i}{4h(1-\tau)}\langle -\eta, C(\xi^k)(-\eta)\rangle\right\} \times$$

$$\times \exp\left\{\frac{i}{h}\left[\langle \nabla\Phi(x_0^k(\eta,\tau)), C^{-1}(\xi^k)\nabla\Phi(x_0^k(\eta,\tau))\rangle\tau + \right.\right.$$

$$\left.\left. + \tfrac{1}{4}(\tau-1)\langle x_0^k(\eta,\tau), C(\xi^k)x_0^k(\eta,\tau)\rangle\right]\right\}\exp\left\{\frac{i}{h}\Phi(x_0^k(\eta,\tau))\right\} \times$$

$$\times \langle \nabla, \dot{C}^{-1}(\xi^k)\nabla\rangle \frac{R_k^N \varphi(x_0^k(\eta,\tau))e_k(x_0^k(\eta,\tau))}{\sqrt{J(x_0^k(\eta,\tau),\tau)}}\, d\eta\, d\tau, \quad (12.61)$$

where $|Z_h| \leqq Ch^{N-[n/2]}$, $N > [n/2]$.

Proof. It follows from Lemma 12.7 that $J(x_0^k, t) \neq 0$ for $x_0^k \in \operatorname{supp} e_k(x)$. We can always find such partition of unity that

$$\varphi(x) = \sum_{k=1}^{s} \varphi(x)\, e_k(x) + \sum_{j=s+1}^{s_1} \varphi(x)e_j(x),$$

where $\xi^k \in \operatorname{supp} e_j(x)$; $k = 1, \ldots, s$; $j = s+1, \ldots, s_1$. Consequently, the problem reduces to the derivation of the asymptotics of the integrals

$$I_k(h) = \frac{1}{(2\pi h)^{n/2}} \int_{-\infty}^{+\infty} e^{(i/h)\Phi(x)} \varphi(x)\, e_k(x)dx, \qquad x = (x_1, \ldots, x_n).$$

Integrating by parts we obtain that integral $I_k(h)$ is equal to $o(h^\infty)$ for $k > s$. Consider the case $k \leqq s$. In order to deduce the asymptotics we use the semi-classical asymptotic solution of the Cauchy problem for the free Schrödinger equation

$$i\frac{\partial\psi(x,t)}{\partial t} = -h\langle \nabla, C^{-1}(\xi^k)\nabla\rangle\psi(x,t). \qquad (12.62)$$

It is not hard to check (by direct substitution) that the exact solution of this problem is given by

$$\psi(x, t) = \frac{\exp\left\{-i\frac{\pi}{4}\sum_{j=1}^{n}\text{sgn }\lambda_j^k\right\}}{(\pi t h)^{n/2}\, 2^n}\sqrt{|\det C(\xi^k)|}\times$$

$$\times \int_{-\infty}^{+\infty} \exp\left\{\frac{i}{4ht}\langle x - \eta, C(\xi^k)(x - \eta)\rangle\right\}\times$$

$$\times \exp\left\{\frac{i}{h}S_0^k(\eta)\right\}\varphi(\eta)e_k(\eta)\, d\eta, \, \eta = (\eta_1, \ldots, \eta_n), \quad (12.63)$$

(here λ_j^k are the eigenvalues of $C(\xi^k)$).

We put $S_0^k(x) = \Phi(x) - \frac{1}{4}\langle x, C(\xi^k)x\rangle$. One easily sees that the wanted integral is expressible in terms of the solution $\psi(x, t)$:

$$I_k(h) = |\det C(\xi^k)|^{-1/2}\exp\left\{i\frac{\pi}{4}\sum_{j=1}^{n}\text{sgn }\lambda_j^k\right\}2^{n/2}\psi(0, 1). \quad (12.64)$$

It remains to find the asymptotic solution of problem (12.62) for $t = 1$, $x = 0$. Due to Lemma 12.7, this asymptotic solution can be derived by applying the standard WKB method. Indeed, after the substitution

$$\psi(x, t, h) = \exp\left\{\frac{i}{h}S^k(x, t)\right\}\varphi(x, t, h), \quad (12.65)$$

where $S^k(x, t)$ is the solution of the problem

$$\frac{\partial S^k(x, t)}{\partial t} + \langle \nabla S^k(x, t), C^{-1}(\xi^k)\nabla S^k(x, t)\rangle = 0, \quad (12.66)$$

$$S^k|_{t=0} = S_0^k(x) = \Phi(x) - \frac{1}{4}\langle x, C(\xi^k)x\rangle,$$

we obtain the equation

$$i\frac{\partial \varphi}{\partial t} + 2i\langle \nabla S, C^{-1}(\xi^k)\nabla\varphi\rangle + i\varphi\langle \nabla, C^{-1}(\xi^k)\nabla S\rangle +$$

$$+ h\langle \nabla, C^{-1}(\xi^k)\nabla\varphi\rangle = 0. \quad (12.67)$$

The solution of problem (12.66) is given by

$$S^k(x, t) = \langle \nabla\Phi(x_0^k), C^{-1}(\xi^k)\nabla\Phi(x_0^k)\rangle t -$$

$$- \frac{1}{4}(t - 1)\langle x_0^k, C(\xi^k)x_0^k\rangle + \Phi(x_0^k(x, t)), \quad (12.68)$$

where $x_0^k = x_0^k(x, t)$ is the solution of the equation

$$x = 2tC^{-1}(\xi^k)\,\text{grad}\,\Phi(x_0^k) + (1 - t)x. \tag{12.69}$$

By virtue of the lemma, $J(x_0^k, t) = \det(\partial X_i^k/\partial x_0^j) \neq 0$ for $t \geq 0$, $x_0^k \in \text{supp}$ $e_k(x)$; thus this solution exists, and, as a consequence, the standard WKB method is applicable.

The solution of Equation (12.67) is sought in the form

$$\varphi(x, t, h) = \frac{1}{\sqrt{J(x_0^k(x, t), t)}} \sum_{v=1}^{N} h^v \varphi_v(x, t). \tag{12.70}$$

This leads to the following system of equations for determining $\varphi_v(x, t)$:

$$\frac{d\varphi_0}{d\tau} = 0, \qquad \varphi_0|_{\tau=0} = \varphi(x_0^k)e_k(x_0^k),$$

$$\cdots \cdots \cdots \cdots \cdots \cdots \cdots \cdots \tag{12.71}$$

$$i\frac{d\varphi_v}{d\tau} = -\sqrt{J(x_0^k(x, t), t)}\,\langle \nabla, C^{-1}(\xi^k)\nabla \rangle \frac{\varphi_{v-1}}{\sqrt{J(x_0^k(x, t), t)}},$$

where

$$\frac{d}{d\tau} \overset{\text{def}}{=} \frac{\partial}{\partial t} + 2\langle \nabla S, C^{-1}(\xi^k)\nabla \rangle.$$

From here we have $\varphi_0(x, t) = \varphi(x_0^k(x, t))\,e_k(x_0^k(x, t))$, and for any $N \geq 1$, φ_v are given by the formula

$$\varphi_v(x, t) = i^v R_k^v \varphi(x_0^k(x, t))\,e_k(x_0^k(x, t)), \tag{12.72}$$

thus

$$\psi_N(x, t, h) = \frac{\exp\left\{\dfrac{i}{h}S^k(x, t)\right\}}{\sqrt{J(x_0^k(x, t), t)}}\left[e_k(x_0^k(x, t))\,\varphi(x_0^k(x, t)) + \right.$$

$$\left. + \sum_{v=1}^{N} ihR_k(x_0^k)^v\,e_k(x_0^k(x, t))\,\varphi(x, t) \right] =$$

$$= \frac{\exp\left\{\dfrac{i}{h}S^k(x, t)\right\}}{\sqrt{J(x_0^k(x, t), t)}}\left[e_k(x_0^k(x, t))\,\varphi(x_0^k(x, t)) + \right.$$

$$\left. + ih\left(\frac{1}{1 - ihR_k}\right)_N R_k e_k(x_0^k(x, t))\,\varphi(x_0^k(x, t)) \right]. \tag{12.73}$$

Now we prove that function $\psi_N(x, t, h)$ really determines the asymptotic solution of problem (12.62). This function satisfies the equation

$$i\frac{\partial \psi_N}{\partial t} = -h\langle \nabla, C^{-1}(\xi^k)\nabla\rangle\psi_N - h^{N+1}e^{(i/h)S^k(x,t)}\langle \nabla, C^{-1}(\xi^K)\nabla\rangle \times$$

$$\times \frac{\varphi_N}{\sqrt{J(x_0^k(x,t),t)}},$$

$$\psi_N|_{t=0} = e^{(i/h)S^k(x)}\varphi(x);$$

consequently, $\tilde{\psi}(x, t, h) = \psi_N(x, t, h) - \psi(x, t)$ is the solution of the problem

$$i\frac{\partial\tilde{\psi}(x,t,h)}{\partial t} = -h\langle \nabla, C^{-1}(\xi^k)\nabla\rangle\tilde{\psi}(x,t,h) -$$

$$- h^{N+1}e^{(i/h)S^k(x,t)}\langle \nabla, C^{-1}(\xi^k)\nabla\rangle\frac{\varphi_N}{\sqrt{J}}, \quad \tilde{\psi}|_{t=o} = 0.$$

$$(12.74)$$

The solution of this problem has the form

$$\tilde{\psi}(x,t,h) = h^{N+1}\int_0^t \frac{\exp\left\{-i\frac{\pi}{4}\sum_{j=1}^n \mathrm{sgn}\lambda_j^k\right\}|\det C(\xi^k)|^{1/2}}{(\pi h)^{n/2}(t-\tau)^{n/2}2^n} \times$$

$$\times \int_{-\infty}^{+\infty}\exp\left\{\frac{i}{4h(t-\tau)}\langle x-\eta, C(\xi^k)(x-\eta)\rangle\right\} \times$$

$$\times \exp\left\{\frac{i}{h}[\langle \nabla\Phi(x_0^k(\eta,\tau)), C^{-1}(\xi^k)\nabla\Phi(x_0^k(\eta,\tau))\rangle\tau +\right.$$

$$\left. + \tfrac{1}{4}(\tau-1)\langle x_0^k(\eta,\tau), C(\xi^k)x_0^k(\eta,\tau)\rangle]\right\} \times$$

$$\times \exp\left\{\frac{i}{h}\Phi(x_0^k(\eta,\tau))\right\}\langle \nabla, C^{-1}(\xi^k)\nabla\rangle \times$$

$$\times \frac{R_k^N\varphi(x_0^k(\eta,\tau))e_k(x_0^k(\eta,\tau))}{\sqrt{J(x_0^k(\eta,\tau),\tau)}}\,d\eta\,d\tau. \qquad (12.75)$$

This solution can also be expressed in the form

$$\tilde{\psi}(x, t, h) = h^{N+1} \int_0^t e^{i\langle \nabla, C^{-1}(\xi^k)\nabla \rangle(t-\tau)} e^{(i/h)S^k(x,\tau)} \times$$

$$\times \langle \nabla, C^{-1}(\xi^k)\nabla \rangle \frac{\varphi_N(x_0^k(x, \tau))}{\sqrt{J(x_0^k(x, \tau), \tau)}} \, d\tau.$$

And since $\| e^{iHt} \| = \| e^{i\langle \nabla, C^{-1}(\xi^k)\nabla \rangle t} \| = 1$, then

$$\| \tilde{\psi}(x, t, h) \| \leq$$

$$\leq h^{N+1} \int_0^t \left\| \langle \nabla, C^{-1}(\xi^k)\nabla \rangle \frac{\varphi_N(x_0^k(x,\tau))}{\sqrt{J(x_0^k(x, \tau), \tau)}} \right\| d\tau \leq \text{const } t \, h^{N+1}.$$

As

$$\frac{\partial \tilde{\psi}}{\partial x} = h^{N+1} \int_0^t e^{i\langle \nabla, C^{-1}(\xi^k)\nabla \rangle(t-\tau)} \times$$

$$\times \frac{\partial}{\partial x} \left(e^{(i/h)S^k(x,\tau)} \langle \nabla, C^{-1}(\xi^k)\nabla \rangle \frac{\varphi_N(x_0^k(x, \tau))}{\sqrt{J(x_0^k(x, \tau), \tau)}} \right) d\tau =$$

$$= ih^N \int_0^t e^{i\langle \nabla, C^{-1}(\xi^k)\nabla \rangle(t-\tau)} \frac{\partial S^k(x, \tau)}{\partial x} \times$$

$$\times \left(e^{(i/h)S^k(x,\tau)} \langle \nabla, C^{-1}(\xi^k)\nabla \rangle \frac{\varphi_N(x_0^k(x, \tau))}{\sqrt{J(x_0^t(x, \tau), \tau)}} \right) d\tau +$$

$$+ h^{N+1} \int_0^t e^{i\langle \nabla, C^{-1}(\xi^k)\nabla \rangle(t-\tau)} e^{(i/h)S^k(x,\tau)} \times$$

$$\times \frac{\partial}{\partial x} \left(\langle \nabla, C^{-1}(\xi^k)\nabla \rangle \frac{\varphi_N(x_0^k(x, \tau))}{\sqrt{J(x_0^k(x, \tau), \tau)}} \right) d\tau,$$

we get

$$\left\| \frac{\partial \tilde{\psi}}{\partial x} \right\| \leq \text{const } t \, h^N.$$

Similarly

$$\left\| \frac{\partial^j \tilde{\psi}}{\partial x^j} \right\| \leq \text{const } t \, h^{N+1-j}.$$

Therefore

$$\| \tilde{\psi} \|_{W_2^j} \leq \text{const } t \, h^{N+1-j}.$$

Due to the embedding theorem

$$Max \, |\tilde{\psi}| \leq c \, \| \tilde{\psi} \|_{W_2^{[n/2]+1}} \leq c_1 \, t h^{N-[n/2]}.$$

The remainder Z_h can be obtained from (12.75) and (12.64). From (12.64) and (12.73) we get the expansion of the integral

$$I_k(h) = \sum_{k=1}^{s} \frac{e^{-i(\pi/4)n} \exp\left\{ i\frac{\pi}{4}\left(n - \sum_{j=1}^{n} \text{sgn}\,\lambda_j^k \right) \right\} 2^{n/2}}{\sqrt{|\det C(\xi^k)|} \sqrt{J(x_0^k(0,1),1)}} \, e^{(i/h)S(0,1)} \times$$

$$\times \left[\varphi(x_0^k(x,t))e_k(x_0^k(x,t)) + ih\left(\frac{1}{1-ihR_k} \right)_N \times \right.$$

$$\left. \times R_k \, e(x_0^k(x,t)) \, \varphi(x_0^k(x,t)) \right] \Bigg|_{\substack{x=0 \\ t=1}} + Z_h.$$

From (12.69) at the point $x = 0, t = 1$ we obtain $\nabla\Phi(x_0^k) = 0$, but for $x \in \text{supp } e_k(x)$, $\nabla\Phi(x) = 0$ holds at one and only one point $x = \xi^k$. Hence $x_0^k(0,1) = \xi^k$.

From Lemma 12.7 we get

$$J(x_0^k(x,t),t)\big|_{\substack{x=0 \\ t=1}} = J(\xi^k,1) = 2^n,$$

and moreover

$$S(0,1) = \left[\tfrac{1}{4}\langle C(\xi^k)x_0^k, C^{-1}(\xi^k)C(\xi^k)x_0^k \rangle - \right.$$

$$\left. - \tfrac{1}{4}\langle x_0^k, C(\xi^k)x_0^k \rangle + \Phi(x_0^k) \right]\Big|_{\substack{x=0 \\ t=1}} = \Phi(\xi^k),$$

whence the required expansions follow.

§ 13. The Asymptotic Series for the Eigenvalues
(Bohr's Quantization Rule)

Special asymptotic series for the eigenvalues and approximate eigen-functions of self-adjoint differential operators in $L_2(\mathbf{R}^n)$ are constructed with the help of the canonical operator.

1. *Formulation of the Problem*

Let H be a separable Hilbert space, $A : H \to H$ be a linear operator. We recall the definition of the spectrum $\sigma(A)$ of operator A. A point λ_0 is called the point of the *spectrum* of operator A, if there exists a sequence of vectors $f_m \in D(A)$, $m = 1, 2, \ldots$, such that

$$\|f_m\| \geq C > 0, \qquad \|(A - \lambda_0)f_m\| = \varepsilon_m \to 0, \qquad (m \to \infty).$$

The vectors f_m are called *approximate eigenvectors* of operator A. If A is a self-adjoint operator and its approximate eigenvector is known, then the following information on the spectrum of operator A can be derived:

LEMMA 13.1.   *Let* $A : H \to H$ *be a self-adjoint operator,* $f \in D(A)$ *and*

$$\|f\| \geq C > 0, \quad \|(A - \lambda)f\| \leq \varepsilon \tag{13.1}$$

Then

$$d(\lambda) \leq \varepsilon/C, \tag{13.2}$$

where $d(\lambda)$ is the distance of the point λ from the spectrum $\sigma(A)$.
    Proof. Since operator A is self-adjoint, we have

$$\|(A - \lambda)^{-1}\| \leq 1/d(\lambda).$$

If $\lambda \in \sigma(A)$, then $d(\lambda) = 0$ and inequality (13.2) is trivial. If $\lambda \notin \sigma(A)$, then

$$C \leq \|f\| = \|(A - \lambda)^{-1}(A - \lambda)f\| \leq$$

$$\leq \|(A - \lambda)^{-1}\| \|(A - \lambda)f\| \leq \frac{\varepsilon}{d(\lambda)},$$

which proves estimate (13.2).
    It follows from inequality (13.2) that there are points of the spectrum of operator A on the segment $[\lambda - (\varepsilon/C), \lambda + (\varepsilon/C)]$.
    In this paragraph we shall investigate the behaviour of the discrete spectrum of a family of self-adjoint operators $A(h)$ which act in the Hilbert space H, for $h \to +0$. Here $h \in (0, h_0)$, $h_0 > 0$; the domains of definition of

operators $A(h)$ do not depend on h. It is assumed that the spectrum $\sigma(A(h))$ of operator $A(h)$ is purely discrete for each fixed $h \in (0, h_0)$, and consists of the eigenvalues $\{E_j(h)\}$. We are interested in the asymptotic behaviour of the spectrum $\sigma(A(h))$ for $h \to + 0$.

In order to elucidate the formulation of the problem, consider the example of the one-dimensional Schrödinger equation

$$- h^2 \psi'' + V(x)\psi = E\psi.$$

Here

$$H = L_2(R), \qquad A(h) = - h^2 \frac{\mathrm{d}^2}{\mathrm{d}x^2} + V(x),$$

function $V(x)$ is real-valued, continuous, and tends to $+ \infty$ for $|x| \to \infty$. Then for each fixed $h > 0$, the spectrum $\sigma(A(h))$ is purely discrete and consists of the simple eigenvalues

$$E_0(h) < E_1(h) < \cdots < E_m(h) < \cdots \to + \infty.$$

The spectrum of operator $A(0)$ (i.e. of the operator of multiplication by function $V(x)$) is continuous and consists of the half-axis $[V_0, + \infty)$, $V_0 = \mathrm{Min}_{x \in R} V(x)$.

We look for that part of the spectrum $\sigma(A(h))$ which belongs to a fixed interval $I = [E_0, E_1]$, $V_0 < E_0 < E_1$. For simplicity, let us consider the spectrum of the oscillator $V(x) = x^2/4$. Then the spectrum can be exactly calculated and has the form

$$E_m(h) = (m + \tfrac{1}{2})h, \qquad m = 0, 1, 2, \ldots.$$

This example clearly demonstrates the following two facts.

(1) If h decreases, the number $N(h)$ of the points of the spectrum $\sigma(A(h))$ belonging to interval I increases indefinitely: $N(h) \sim (E_1 - E_0)/h$.

(2) The indices m of the eigenvalues belonging to I increase indefinitely for $h \to 0$: $m > E_0/h$. If m is fixed, then $E_m(h) \to 0$ with $h \to 0$, and the eigenvalue $E_m(h)$ runs out of the interval I for $h \to 0$.

For this reason we shall not investigate the behaviour of an individual eigenvalue $E_m(h)$ for $h \to 0$, but we shall study the structure of the set $\sigma(A(h)) \cap I$ for $h \to 0$. Namely, we shall derive the asymptotics for a certain series of eigenvalues (for $h \to 0$) for the family $\{A(h)\}$ of self-adjoint differential operators in $L_2(\mathbf{R}^n)$, and, more precisely, for h^{-1}-p.d. operators.

Another variant of the problem of the asymptotics of the spectrum consists in the following. We consider h as a spectral parameter, i.e., we consider a family of operators $A(h)$; the eigenvalues h_m and the eigen-vectors f_m are determined by the equation $A(h_m)f_m = 0$. In the case of the oscillator we have

$$h_m = \frac{E}{m + \frac{1}{2}}, \quad m = 0, 1, 2, \ldots,$$

where E is a fixed number. Thus in this problem we investigate the asymptotics of the spectrum of operator family $\{A(h)\}$.

2. The Canonical Operator on a Closed Lagrangian Manifold

In order to investigate the asymptotics of the spectrum of a family of h^{-1}-p.d. operators $\{A(h)\}$, the approximate eigenfunctions of operators will be constructed, i.e. functions $\varphi_h \in D(A)$ such that

$$A(h)\varphi_h = (E(h) + \varepsilon(h))\varphi_h,$$

where $\varepsilon(h) = o(E(h))(h \to +0)$. In other words, an approximate eigenfunction is an approximate solution of the equation

$$A(h)\varphi = E(h)\varphi.$$

Such a solution is locally given by the precanonical operator (§§ 3, 5, and 6). The construction of the global solution is realized by patching the local solutions, i.e. by means of the construction of the canonical operator. Here new difficulties arise, since the corresponding Lagrangian manifolds Λ^n have a non-trivial fundamental group $\pi_1(\Lambda^n)$. For instance, with the oscillator eigenvalue problem

$$-h^2\psi'' + x^2\psi = E\psi$$

a family of Lagrangian manifolds in the plane is associated:

$$\{\Lambda^1(E)\} : x^2 + p^2 = E, \quad E_0 < E < E_1.$$

In this connection we shall reanalyse the construction of the canonical operator. Let Λ^n be a smooth n-dimensional Lagrangian manifold in phase space $\mathbf{R}^{2n}$, $\{\Omega_j\}$ be a canonical atlas on Λ^n, $\{e_j\}$ be a partition of unity on Λ^n (the notation as in § 8) and r^0 a distinguished point on Λ^n. For simplicity, the dependence on r^0 will be omitted; we shall, however, use the notation $K_{\Lambda^n}(h)$, to emphasize the dependence on the parameter $h = \lambda^{-1}$. The canonical operator is defined by formula (8.3):

$$(K_{\Lambda^n}(h)\varphi)(x) = \sum_j K(\Omega_j, h)(e_j\varphi)(x) \tag{13.3}$$

on functions $\varphi \in C_0^\infty(\Lambda^n)$ for $0 < h < h_0$. For each fixed h we have a mapping

$$K_{\Lambda^n}(h) : C_0^\infty(\Lambda^n) \to C^\infty(\mathbf{R}_x^n)$$

and $K_{\Lambda^n}(h)$ is, from this point of view, a family of mappings. The basic properties of this family are:

(1) If x lies outside some ε-neighbourhood of the set π_x supp φ (π_x is the projection on $\mathbf{R}_x^n$), then for any multi-index α and for any integer $N \geq 0$

$$|D_x^\alpha(K_{\Lambda^n}(h)\varphi)(x)| \leq C_{\alpha,N} h^N(1 + |x|)^{-N}, \tag{13.4}$$

where $C_{\alpha,N}$ are constants (depending on φ).

(2) If $\hat{L} = L(\overset{2}{x}, h\overset{1}{D}_x ; h)$ is a h^{-1}-p.d. operator with the symbol of class $T_+^m(\mathbf{R}_x^n)$, then the first commutation formula

$$(\hat{L}K_{\Lambda^n}(h)\varphi)(x) = (K_{\Lambda^n}(h)L(x, p; 0)\varphi)(x) + h\chi(x, h) \tag{13.5}$$

is true, where $\chi \in O_0(\mathbf{R}_x^n)$ (see § 1).

(3) If, moreover, the assumptions of Theorem 8.4 are satisfied, then the second commutation formula holds,

$$(\hat{L}K_{\Lambda^n}(h)\varphi)(x) = (L(x, p; 0) - ihR_1 \varphi)(x) + h^2\chi(x, h), \tag{13.6}$$

where $\chi \in O_0(\mathbf{R}_x^n)$ (see §1); the explicit form of operator R_1 is given in Theorem 8.4.

Moreover, the canonical operator, with accuracy to $O(h)$, is independent of the choice of a canonical atlas, of focal coordinates in the charts, and of a partition of unity (Theorem 8.1). Since we are going to construct a concrete approximate eigenfunction, the atlas, the partition of unity as well as the focal coordinates will be fixed.

In § 8 we constructed the canonical operator under the assumptions:
(1) $\oint_\gamma \langle p, dx \rangle = 0$ for any 1-cycle on Λ^n;
(2) ind $\gamma = 0$ for any 1-cycle on Λ^n.
What are the consequences when these assumptions are violated? First we consider an example and then formulate the final result.

Let Λ^n admit a diffeomorphic projection on $\mathbf{R}_x^n$ and let the fundamental group $\pi_1(\Lambda^n) \approx \mathbf{Z}$ ($\mathbf{Z}$ is the group of integers). It is sufficient to consider the case $n = 2$ with the projection of Λ^2 on $\mathbf{R}_x^2$ being the ring $C: r < |x| < R$. Then Λ^2 is given by the equation

$$p = p(x), \qquad x \in C,$$

where $p(x)$ is a smooth single-valued vector function. Since Λ^2 is a Lagrangian manifold, we have locally, i.e. in a small simply connected neighbourhood $U(x^0)$ of any point $x^0 \in C$,

$$p(x) = \frac{\partial S(x)}{\partial x}, \quad x \in U(x^0),$$

where S is a smooth function, so that $\langle p(x), dx \rangle = dS(x)$. However, the 'function'

$$S(x) = \int_{x^0}^{x} \langle p(x), dx \rangle$$

need not be single-valued, if the form $\omega^1 = \langle p, dx \rangle$ has a non-zero period

$$J = \int_{\gamma^0} \langle p, dx \rangle \neq 0.$$

Here γ_0 is the circle $|x| = \rho$, $r < \rho < R$; it is evident that J does not depend on ρ. Curve γ_0 is a generator of the homology group $H_1(\Lambda^2, \mathbf{Z})$. Moreover let us remark that ind $\gamma_0 = 0$. We fix the point $r^0 = (x^0, p^0) \in \Lambda^2$ as well as the volume element $d\sigma(x)$ on Λ^2, and consider the expression

$$(K_{\Lambda^2}(h)\varphi)(x) = \sqrt{\left|\frac{d\sigma(x)}{dx}\right|} \exp\left(\frac{i}{h} \int_{r^0}^{r(x)} \langle p, dx \rangle\right) \qquad (13.7)$$

(where $\varphi \in C_0^\infty(\Lambda^2)$) which is constructed à la precanonical operator in a non-singular chart (see (6.2)). This expression is, in general, a multi-valued function of x (with parameter h being fixed). In fact, all values of the integral in the exponent, considered at one point $r(x) = (x, p(x))$, are given by

$$\int_{r^0}^{r(x)} \langle p, dx' \rangle = \int_{\tilde{\gamma}_0} \langle p, dx' \rangle + kJ, \quad k = 0, \pm 1, \pm 2, \ldots,$$

where $\tilde{\gamma}_0$ is a fixed path connecting the points x^0, x and lying in C. One may assume that $J > 0$. Formula (13.7) determines a single-valued function, if and only if

$$J = 2\pi m h,$$

where $m > 0$ is an integer, i.e. for discrete values of parameter h,

$$h_m = \frac{J}{2\pi m}, \quad m = 1, 2, \ldots. \tag{13.8}$$

Thus, for discrete values $h \in \{h_m\}$ there is an operator family $\{K_{\Lambda^2}(h_m)\}$. The commutation formulae obtained in §8 remain true with the only difference that $h \to 0$ via the sequence $\{h_m\}$.

However, take up again our example. Let there be a family of Lagrangian manifolds $\{\Lambda^2(E)\}$, $E \in \Delta$ (where Δ is an interval), smoothly depending on parameter E, and let $\pi_x \Lambda^2(E)$ coincide with the ring C as before. Then the period $J = J(E)$, and the condition of single-valuedness of the integral $\int \langle p, dx \rangle$ takes the form

$$J(E) = 2\pi m h. \tag{13.9}$$

Let $J(E) > 0$, $dJ(E)/dE > 0$ for $E \in I$. Then for each fixed $h \in (0, h_0)$, Equation (13.9) has a family of solutions $\{E_m(h)\}$, where m runs through the set of integers $M(h) : m_0(h) < m < m_1(h)$, depending on h. The number of elements of the set $M(h)$ linearly increases (as h^{-1}), if h decreases. If h is fixed, and $E = E_m(h) m \in M(h)$, then formula (13.7) determines the operator $K_{\Lambda^2 E_m(h)}(h)$; we denote it shortly by $K(E_m(h))$. Thus we obtain a discrete set of the operators

$$K(E_m(h)) : C_0^\infty(\Lambda^2_{E_m(h)}) \to C_0^\infty(C), \quad m \in M(h).$$

for each fixed h.

Now we shall treat the general case. Suppose Λ^n is a Lagrangian C^∞-manifold in the phase space, compact and without boundary. We are interested in the case when Λ^n is not simply connected. Let $\dim H_1(\Lambda^n, \mathbf{Z}) = p_1$ and $\{\gamma_j\}$, $1 \leq j \leq p_1$, be the basis of the first homology group $H_1(\Lambda^n, \mathbf{Z})$. We set

$$J_j = \int_{\gamma_j} \langle p, dx \rangle, \quad l_j = \text{ind } \gamma_j$$

(here J_j are the periods of the 1-form $\omega^1 = \langle p, dx \rangle$). Let $\tilde{r} = (\tilde{x}, \tilde{p}) \in \Lambda^n$ be a non-singular point of Ω_j. We shall construct the operator $K_{\Lambda^n}(h)$ formally according to formula (13.3); then $(K_{\Lambda^n}(h)\varphi)(x)$ will be a multi-valued function of x. If supp φ is concentrated in a small neighbourhood of point

$\tilde{r}$, then all values of the function $(K_{\Lambda^n}(h)\varphi)(\tilde{x})$ are given by the formula

$$(K_{\Lambda^n}(h)\varphi)(\tilde{x}) = \sqrt{\left|\frac{d\sigma}{dx}\right|} \exp\left[\frac{i}{h}\int_\gamma \langle p, dx \rangle - \frac{i\pi}{2} l\right] \varphi(\tilde{x}, p(\tilde{x})),$$

$$l = \text{ind } \gamma,$$

where γ is an arbitrary path connecting the points $r^0, \tilde{r}$ and lying on Λ^n. If path γ_0 if fixed, then the path γ is homologous to the linear combination

$$\gamma \sim \gamma_0 + \sum_{j=1}^{p_1} m_j \gamma_j,$$

where m_j are integers. Consequently, the expression $(K_{\Lambda^n}(h)\varphi)(\tilde{x})$ is defined up to a factor of the form

$$\exp\left[i \sum_{j=1}^{p_1} m_j \left(\frac{J_j}{h} - \frac{\pi}{2} l_j\right)\right]; \tag{13.10}$$

The same is true for any non-singular point of manifold Λ^n. Thus the expression $(K_{\Lambda^n}(h)\varphi)(x)$ is a single-valued function (for the time being for $x \notin \pi_x \Sigma(\Lambda^n)$), if and only if the relations

$$\frac{J_j}{h} - \frac{\pi}{2} l_j = 2\pi k_j, \quad 1 \leq j \leq p_1, \tag{13.11}$$

hold, where k_j are integers. It is easily seen that relations (13.11) do not depend on the choice of a basis $\{\gamma_j\}$ of the homology group $H_1(\Lambda^n, \mathbf{Z})$. In fact, if $\{\tilde{\gamma}_j\}$ is another such basis, then

$$\tilde{\gamma}_j = \sum_{l=1}^{p_1} C_{jl} \gamma_j, \quad 1 \leq j \leq p_1,$$

where C_{jl} are integers. Since the matrix $C = (C_{jl})$ is invertible and its inverse C^{-1} is integral, too, and since we denote

$$\tilde{\mathbf{J}}_j = \sum_{l=1}^{p_1} C_{jl} \mathbf{J}_j, \quad \tilde{l}_j = \sum_{l=1}^{p_1} C_{jl} l_j,$$

relations (13.11) are equivalent to the relations

$$\frac{\tilde{\mathbf{J}}_j}{h} - \frac{\pi}{2} \tilde{l}_j = 2\pi \tilde{k}_j, \quad 1 \leq j \leq p_1.$$

It is not hard to verify that if $p_1 \geq 2$, relations (13.11) can be unsatisfied for any value of h. For instance, if $l_1 = l_2 = 0$, the relations

$$\mathbf{J}_1 = 2\pi h k_1, \quad \mathbf{J}_2 = 2\pi h k_2$$

imply that the periods $\mathbf{J}_1, \mathbf{J}_2$ are commensurable.

Thus the canonical operator $K_{\Lambda^n}(h)$ on a given Lagrangian manifold Λ^n can be undefined for all values of h.

Relations (13.11) can obviously be fulfilled, if there exists a p_1-parameter family of Lagrangian manifolds (in the above example we have studied the case $p_1 = 1$). Namely, let there be a family $\{\Lambda^n(\alpha)\}$ of Lagrangian C^∞-manifolds, $\alpha = (\alpha_1, \ldots, \alpha_{p_1}) \in U$ being a domain in $\mathbf{R}^{p_1}$ such that the following conditions are true:

$\Lambda 13.1$. For each fixed $\alpha \in U$, $\Lambda^n(\alpha)$ is a compact C^∞-manifold without boundary, and the family $\{\Lambda^n(\alpha)\}$ depends smoothly on parameter $\alpha \in U$.

To distinct values of $\alpha \in U$ there correspond non-intersecting manifolds $\Lambda^n(\alpha)$; the union $\bigcup_{\alpha \in U} \Lambda^n(\alpha)$ is a C^∞-manifold Λ^{n+p_1} of the dimension $n + p_1$.

$\Lambda 13.2$. The dimension of the cycle of singularities $\Sigma(\Lambda^n(\alpha))$ is not larger than $n - 1$ for $\alpha \in U$, and the cycles of singularities depend smoothly on α.

$\Lambda 13.3$. On each manifold $\Lambda^n(\alpha)$, $\alpha \in U$, there exists an n-dimensional volume element $d\sigma^n(\alpha)$ of class C^∞ which is normalized by the condition

$$\int\limits_{\Lambda^n(\alpha)} d\sigma^n(\alpha) = 1.$$

Notice that the *number of parameters* $(\alpha_1, \ldots, \alpha_{p_1})$ *is equal to the dimension of the first homology group of the manifolds* $\Lambda^n(\alpha)$.

Under these assumptions, if the domain U is sufficiently small, we can choose canonical atlases $\{\Omega_j(\alpha)\}$, partitions of unity $\{e_j(\alpha)\}$ of class $C^\infty(\Lambda^n(\alpha) \times U)$, where the number of charts is independent of α, as well as the bases $\{\gamma_j(\alpha)\}$ depending smoothly on α. Then the periods

$$\mathbf{J}_j(\alpha) = \int\limits_{\gamma_j(\alpha)} \langle p, dx \rangle$$

are functions of class $C^\infty(U)$, and the values of ind $\gamma_j(\alpha) = l_j$ do not depend on α. Relation (13.11) now takes the form

$$h^{-1}\mathbf{J}_j(\alpha) = \frac{\pi}{2}l_j + 2\pi m_j, \quad 1 \leq j \leq p_1. \tag{13.12}$$

There is an obvious sufficient condition of compatibility of this system: the independence of the periods $J_j(\alpha)$ as functions of α. We introduce the notation:

$$J(\alpha) = (J_1(\alpha), \ldots, J_{p_1}(\alpha)),$$

$$l = (l_1, \ldots, l_{p_1}), \quad M = (m_1, \ldots, m_{p_1}).$$

LEMMA 13.2.   *Let* $\alpha^0 \in U$. $\det(\partial J(\alpha^0)/\partial \alpha) \neq 0$ *and* $0 < h < h_0$, $h_0 > 0$ *being sufficiently small. Then the system* (13.12) *has solutions* $\{\alpha_M(h)\}$ *for each* $h \in (0, h_0)$, *where the integer-component vectors M form a cubic lattice with edge length of the order* const. h^{-1} *for small h.*

Proof. If U_0 is a sufficiently small neighbourhood of point α^0, then its image under the map $J = (1/2\pi) J(\alpha) - \frac{1}{2} hl$ contains a neighbourhood V_0 of the point $J(\alpha^0)$; one may assume that V_0 is a cube. The desired set of solutions consists of all points $\tilde{J} \in V_0$ such that $h^{-1}\tilde{J}$ is an integer-component vector; for sufficiently small h, this set is non-void.

Moreover, $\alpha_M(h)$ depends smoothly on h for fixed M, provided point $\alpha_M(h)$ lies in domain U_0, q.e.d.

Suppose the conditions $\Lambda 13.1$–$\Lambda 13.3$ and the assumptions of Lemma 13.2 are satisfied, $U_0 \subset U$ is a sufficiently small domain specified in the proof of Lemma 13.2, and function φ is defined on the family $\{\Lambda^n(\alpha)\}$, $\alpha \in U_0$, and belongs to class C^∞. The families of canonical atlases $\{\Omega_j(\alpha)\}$ as well as of the focal coordinates in these charts, of partitions of unity $\{e_j(\alpha)\}$, and of bases $\{\gamma_j(\alpha)\}$ of the first homology groups $H_1(\Lambda^n(\alpha), \mathbf{Z})$ are chosen as before. We fix $h \in (0, h_0)$, an integer-component vector M and a solution $\alpha_M(h)$ of system (13.12), $\alpha_M(h) \in U_0$.

Consider the integral

$$\tilde{S}(r, \alpha, h) = \frac{1}{h} \int\limits_{r^0(\alpha)}^{r} \langle p, dx \rangle - \frac{\pi}{2} l[\gamma(r^0(\alpha), r)],$$

where the integration path γ lies on manifold $\Lambda^n(\alpha)$. If $\alpha = \alpha_M(h)$, the periods of integral $\tilde{S}$ are multiples of 2π, so that the expression $\exp[i\tilde{S}(r, \alpha_M(h), h)]$ is a single-valued function on the manifold $\Lambda^n(\alpha_M(h))$. Consequently, we may construct the family of canonical operators $\{K_{\Lambda^n(\alpha)}, \alpha = \alpha_M(h)\}$ as in § 8.

If the vector M is fixed, then this family is well defined for those h for which $\alpha_M(h) \in U_0$. Evidently, $\alpha_M(h) \to 0$ for $h \to 0$, so that the point $\alpha_M(h)$ leaves the domain U_0. On the other hand, for each fixed (and sufficiently small) h there is a finite set $\mathfrak{M}(h)$ of integer-component vectors M such that the point $\alpha_M(h) \in U_0$ for $M \in \mathfrak{M}(h)$.

Thus, for each fixed h, the *family of Lagrangian manifolds* $\{\Lambda^n(\alpha)\}$ is *quantized* – there is a *discrete family* of manifolds ($\alpha = \alpha_M(h)$) on which the canonical operator is defined.

Let us remark that commutation formulae (13.5) and (13.6) remain valid for operators $K_{\Lambda^n(\alpha)}$, $\alpha = \alpha_M(h)$, and that the estimates of the remainders in formulae (13.4)–(13.6) are uniform with respect to $\alpha \in U_0$. This follows from the stationary phase method for parameter-dependent integrals (Theorem 1.4) and from the assumptions $\Lambda 13.1$–$\Lambda 13.3$.

3. The Asymptotic Series of Eigenvalues

Consider a h^{-1}-p.d. operator $A(h)$ which is real and formally self-adjoint. This operator can be represented in the form

$$A(h) = \tfrac{1}{2}[L(\overset{2}{x}, h\overset{1}{D}_x; ih) + L(\overset{1}{x}, h\overset{2}{D}_x; ih)] \tag{13.13}$$

with a real-valued symbol $L(x, p; ih)$. We introduce the assumptions:

L13.1. Symbol $L(x, p; ih) \in T_+^m$ for some m.

L13.2. The minimal operator $A^0(h)$ corresponding to the expression $A(h)$ can be extended to a self-adjoint operator $A(h): L_2(\mathbf{R}^n) \to L_2(\mathbf{R}^n)$ with purely discrete spectrum for each fixed $h \in (0, h_0)$.

Let g^t be the displacement along trajectories of the Hamilton system associated with symbol $L(x, p; 0)$. The asymptotic series of the eigenvalues which will be constructed below, is generated by a family of Lagrangian manifolds invariant under displacements g^t. We suppose:

$\Lambda 13.4$. There exists a family of Lagrangian manifolds $\{\Lambda^n(\alpha)\}$, $\alpha \in U$, satisfying the conditions $\Lambda 13.1$–$\Lambda 13.3$ and such that

(1) $g^t \Lambda^n(\alpha) = \Lambda^n(\alpha)$ for all $t \in \mathbf{R}$;

(2) $L(x, p; 0) \equiv E(\alpha)$, $(x, p) \in \Lambda^n(\alpha)$.

Thus each manifold $\Lambda^n(\alpha)$ is invariant with respect to displacements along trajectories of the Hamilton system and the symbol $L|_{h=0} \equiv \text{const.}$ on each of the manifolds of the family (the constant can depend on α).

There exists still one classical object related to operator $A(h)$ – the differential operator

$$R = \frac{1}{i}\left(\frac{d}{dt} + \frac{\partial L}{\partial(ih)}\bigg|_{h=0}\right), \tag{13.14}$$

where d/dt is the derivative with respect to the Hamilton system. Since operator $A(h)$ is self-adjoint and real, the function $(1/i)(\partial L/\partial(ih))\big|_{h=0}$ is real. We fix $\alpha \in U_0$; then, due to the assumptions stated above, R is a self-adjoint operator in the space $L_{2,\sigma^n(\alpha)}(\Lambda^n(\alpha))$. Consequently, its spectrum is real.

We assume that operator R has a family of eigenfunctions φ_α, smooth with respect to $\alpha \in U$, and corresponding to a smooth family of eigenvalues $\mu(\alpha)$:

$$R\varphi_\alpha = \mu(\alpha)\varphi_\alpha, \quad \alpha \in U. \tag{13.15}$$

In particular, if symbol L does not depend on h, we may put

$$\mu(\alpha) \equiv 0, \quad \varphi_\alpha \equiv 1. \tag{13.16}$$

Now we construct approximate eigenfunctions of the family $\{A(h)\}$. We fix $h \in (0, h_0)$ and find the collection $\{\alpha_M(h)\}$ of the solutions of system (13.12) (Lemma 13.2). We look for an approximate eigenfunction of the form

$$u_M(x, h) = K_{\Lambda^n(\alpha)}\varphi_\alpha, \quad (\alpha = \alpha_M(h)). \tag{13.17}$$

By applying commutation formula (13.6) and taking into account that $L(x, p; 0) \equiv E_0(\alpha)$ on $\Lambda^n(\alpha)$, we get $(\alpha = \alpha_M(h))$

$$\begin{aligned}A(h)u_M &= K_{\Lambda^n(\alpha)}(L(x, p; 0) + R)\varphi_\alpha + O(h^2) = \\ &= (E(\alpha) + h\mu(\alpha))K_{\Lambda^n(\alpha)}\varphi_\alpha + O(h^2) = \\ &= (E(\alpha) + h\mu(\alpha))u_M + O(h^2).\end{aligned} \tag{13.18}$$

For the remainder we have the estimate

$$|O(h^2)| \leq Ch^2, \tag{13.19}$$

where $x \in \mathbf{R}^n, \alpha \in U_0, h \in (0, h_0)$, and a constant C is independent of x, α, h. Furthermore, there exists a constant $C_1 > 0$ such that

$$\| K_{\Lambda^n(\alpha)}\varphi_\alpha \|_{L_2(\mathbf{R}^n)} \geq C_1, \tag{13.20}$$

$\alpha = \alpha_M(h)$. This follows from the fact that, in a non-singular chart, the canonical operator is reduced to the multiplication by a non-vanishing function, the modulus of which is independent of h (see (13.41)).

By virtue of Lemma 13.1, we have (for fixed h, M)

$$d(E(\alpha) + h\mu(\alpha)) \leq \frac{C}{C_1}h^2, \quad (\alpha = \alpha_M(h)), \tag{13.21}$$

where $d(\rho)$ is the distance of point ρ from the spectrum $\sigma(A(h))$. This proves the main result of this paragraph.

THEOREM 13.3. *Let $\{A(h)\}, h \in (0, h_0)$, be a family of h^{-1}-p.d. operators and let there exist a family of Lagrangian manifolds $\{\Lambda^n(\alpha)\}$ fulfilling the assumptions $\Lambda 13.1 - \Lambda 13.4$.*

Then the operator $A(h)$ has a series of eigenvalues of the form

$$E_M(h) = E(\alpha) + h\mu(\alpha) + O(h^2) \tag{13.22}$$

for each fixed $h \in (0, h_1)$ and for $h_1 > 0$ sufficiently small. Here $\alpha = \alpha_M(h)$, $|O(h^2)| \leqq C_0 h^2$, constant C_0 does not depend on h, M.

The approximate eigenfunction $u_M(x, h)$ corresponding to the eigenvalue (13.22) is given by formula (13.17).

Now we shall discuss Theorem 13.3. Let $p_1 = 2$ for simplicity, and $E_0 < E(\alpha) < E_1$ for $\alpha \in U$. We fix $h > 0$ and then we find a discrete collection of solutions $\alpha(h, m_1, m_2)$ of the system (13.12)

$$J_1(\alpha_1, \alpha_2) = \frac{\pi h}{2} l_1 + 2\pi m_1 h,$$

$$J_2(\alpha_1, \alpha_2) = \frac{\pi h}{2} l_2 + 2\pi m_2 h,$$

for which $E_0 < E(\alpha(h, m_1, m_2)) < E_1$.

Formula (13.22) determines the series of eigenvalues depending on two integral parameters m_1, m_2:

$$E_{m_1, m_2}(h) = E(\alpha(h, m_1, m_2)) + h \, \mu(\alpha(h, m_1, m_2)) + O(h^2).$$

Here the set of admissible pairs (m_1, m_2) depends on parameter h.

Let us consider a classical example – the eigenvalue problem for the one-dimensional Schrödinger operator on the whole real axis

$$- h^2 \psi'' + V(x)\psi = E\psi.$$

Potential $V(x) \in C^\infty(\mathbf{R})$ is real, $V(\pm \infty) = \pm \infty$, and for simplicity $V(x)$ is assumed to behave at infinity as a polynomial. Then, among other things, function $V(x)$ decreases (increases) strictly monotonically for $x < -R$ $(x > R)$, provided $R > 0$ is sufficiently large.

The equation

$$p^2 + V(x) = \alpha$$

for $\alpha \geqq \alpha_0 \gg 1$ determines a one-dimensional Lagrangian manifold $\Lambda^1(\alpha)$, diffeomorphic to the circle S^1, so that $p_1 = 1$. The period $J(\alpha)$ is equal to

$$J(\alpha) = \int\limits_{\Lambda^1(\alpha)} p\,dx = 2 \int\limits_{x_-(\alpha)}^{x_+(\alpha)} \sqrt{\alpha - V(x)}\,dx,$$

where $x_+(\alpha)$ are the roots of the equation $V(x) = \alpha, x_-(\alpha) < x_+(\alpha)$. The curve $\Lambda^1(\alpha)$ is oriented clockwise. It is not difficult to see that $dJ(\alpha)/d\alpha > 0$. Further the function $H = p^2 + V(x)$ turns out to be a first integral of the corresponding Hamilton system so that $E(\alpha) \equiv \alpha$. Consequently, all assumptions $\Lambda 13.1 - \Lambda 13.4$, $L13.1 - L13.2$ are fulfilled. Finally, ind $\Lambda^1(\alpha) = +2$. We take $\mu(\alpha) = 0$, $\varphi(\alpha) \equiv 1$ on $\Lambda^1(\alpha)$ as in (13.16). Then, as follows from Theorem 13.3, there exists the series of eigenvalues

$$E_m(h) = \alpha_m(h) + O(h^2),$$

where $\alpha_m(h)$ is determined by the equation

$$J(\alpha_m) = 2\pi m h + \pi h,$$

i.e. $\alpha_m(h)$ satisfies the equation

$$\int\limits_{x_-(\alpha)}^{x_+(\alpha)} \sqrt{\alpha - V(x)}\,dx = \pi h(m + \tfrac{1}{2}).$$

This formula coincides with the *Bohr quantization formula*, well known from the physical literature.

The corresponding approximate eigenfunctions ψ_m have the form

$$\psi_m(x) = (K_{\Lambda^1(\alpha_m)} 1)(x), \qquad \alpha_m = \alpha_m(h)$$

(for details see [59] and others), where 1 is the function identically equal to unity on curve Λ^1.

REMARK 13.4. Let the family $\{\Lambda(E)\}$ of compact Lagrangian manifolds without boundary smoothly depended on E for $E \in \varepsilon = [E^0 - \varepsilon, E^0 + \varepsilon]$ and the conditions (1), (2) from Λ 13.4 are fulfilled. Let the quantization conditions (13.11) holds for some set $\{E^i\} \subset \varepsilon$ depending on h. Then operator A has the set of eigenvalues

$$\lambda^i = E^i + h\mu(E^i) + O(h^2)$$

(see [59], [60]).

REMARK 13.5. Let $u_M(x, h)$ be an approximate eigenfunction (see (13.17)) and $h > 0$ small enough. Then this function is concentrated in a neighbourhood of the projection $V = \pi_x \operatorname{supp} K_{\Lambda^n(\alpha)}$, $\alpha = \alpha_M(h)$, since outside this domain $K_{\Lambda^n(\alpha)}(h)\varphi_\alpha = O(h^\infty)$. More precisely, if x lies outside some neighbourhood of the set U, then

$$|u_M(x, h)| \leq C_N h(1 + |x|)^{-N}$$

for any N (see (13.4)).

4. Asymptotic Series of Eigenvalues of the Laplace–Beltrami Operator on the Sphere

In this section we review the results obtained in [39]. Let S^{n-1} be the unit sphere $|x| = 1$ in the Euclidean space $\mathbf{R}^n$. Consider the eigenvalue problem

$$(\Delta_\theta + k^2)u(x) = 0, \qquad x \in S^{n-1}, \tag{13.23}$$

where Δ_θ is the angular part of the Laplace operator $\Delta = \sum_{j=1}^n (\partial^2 / \partial x_j^2)$. We introduce the spherical coordinates on the sphere S^{n-1},

$$x_1 = \sin \theta_{n-1} \ldots \sin \theta_2 \sin \theta_1,$$
$$x_2 = \sin \theta_{n-1} \ldots \sin \theta_2 \cos \theta_1,$$
$$\ldots\ldots\ldots\ldots\ldots\ldots\ldots\ldots\ldots$$
$$x_{n-1} = \sin \theta_{n-1} \cos \theta_{n-2},$$
$$x_n = \cos \theta_{n-1},$$

where $0 \leqq \theta_1 < 2\pi$, $0 \leqq \theta_j < \pi$ $(j \neq 1)$. Operator Δ_θ has the form

$$\Delta_\theta = \frac{1}{\sin^{n-2}\theta_{n-1}} \frac{\partial}{\partial\theta_{n-1}} \sin^{n-2}\theta_{n-1} \frac{\partial}{\partial\theta_{n-1}} +$$

$$+ \frac{1}{\sin^2\theta_{n-1}} \frac{1}{\sin^{n-3}\theta_{n-2}} \frac{\partial}{\partial\theta_{n-2}} \sin^{n-3}\theta_{n-2} \frac{\partial}{\partial\theta_{n-2}} +$$

$$+ \ldots + \frac{1}{\sin^2\theta_{n-1} \ldots \sin^2\theta_2} \frac{\partial^2}{\partial\theta_1^2}. \tag{13.24}$$

The eigenvalue problem (13.23) is exactly solvable. The eigenvalues of the operator Δ_θ have the form $k_l^2 = l(l + n - 2)$, $l = 0, 1, 2, \ldots$, and the eigenfunctions corresponding to the eigenvalue k_l^2 are restrictions of homogeneous harmonic polynomials of degree l on the sphere S^{n-1}. With the help of Theorem 13.3 we shall construct the asymptotics of some series of eigenvalues and eigenfunctions of operator Δ_θ.

The metric in $\mathbf{R}^n$ induces the metric

$$ds^2 = d\theta_{n-1}^2 + \sin^2 \theta_{n-1}\, d\theta_{n-2}^2 +$$
$$+ \ldots + \sin^2 \theta_{n-1} \ldots \sin^2 \theta_{n-2}\, d\theta_1^2, \qquad (13.25)$$

on sphere S^{n-1} so that the sphere S^{n-1} equipped with metric (13.25) becomes an $(n-1)$-dimensional compact Riemannian manifold. We recall the definition of the Laplace–Beltrami operator on a Riemannian manifold M^s [32]. Let Ω be a chart of manifold M^s, $x = (x_1, \ldots, x_s)$ be the local coordinates in Ω and $g_{ij}(x)$ be the metric tensor. We introduce, as usual, the notation $g = |\det(g_{ij})|$, and (g^{ij}) the matrix inverse to the matrix (g_{ij}). Any scalar function $f \in C^\infty(M^s)$ generates the vector field grad f (the gradient of function f) on M^s, the restriction of which on Ω is of the form

$$\operatorname{grad} f = \sum_{i,j=1}^{s} g^{ij} \frac{\partial f}{\partial x_i} \frac{\partial}{\partial x_j}.$$

It is not hard to verify that the expression on the right-hand side of the last equality does not depend on the choice of coordinates in Ω. Further, a vector field X on M^s generates the scalar function div X (the divergence of field X), the restriction of which on Ω is given by

$$\operatorname{div} X = \frac{1}{\sqrt{g}} \sum_{i=1}^{s} \frac{\partial}{\partial x_i} (\sqrt{g}\, X_i).$$

The right-hand side of this equality does not depend on the choice of coordinates in Ω. The Laplace–Beltrami operator is defined by

$$\Delta f = \operatorname{div} \operatorname{grad} f, \quad (f \in C^\infty(M^s))$$

or, in the local coordinates,

$$\Delta f = \frac{1}{\sqrt{g}} \sum_{i=1}^{s} \frac{\partial}{\partial x_i} \left(\sum_{j=1}^{s} g^{ij} \sqrt{g} \frac{\partial}{\partial x_j} \right) f. \qquad (13.26)$$

It follows from (13.25) that, for the sphere S^{n-1}, we have in the coordinates $(\theta_1, \ldots, \theta_{n-1})$)

$$g_{n-1,n-1} = 1, \quad g_{n-2,n-2} = \sin^2 \theta_{n-1}, \ldots,$$
$$g_{11} = \sin^2 \theta_{n-1} \ldots \sin^2 \theta_{n-2},$$
$$g = \sin^{2(n-2)} \theta_{n-1} \ldots \sin^2 \theta_{n-2}, \qquad (13.27)$$

and $g_{ij} = 0, i \neq j$. Inserting (13.27) into (13.26) we obtain (13.23) so that Δ_θ is the Laplace–Beltrami operator on the unit sphere S^{n-1}.

The Hamilton–Jacobi equation associated with equation (13.23) has the form

$$H(\theta, p) = 1. \tag{13.28}$$

Here

$$p = \frac{\partial S(\theta)}{\partial \theta},$$

$$H(\theta, p) = \left(p_{n-1}^2 + \frac{1}{\sin^2 \theta_{n-1}} \left(p_{n-2}^2 + \frac{1}{\sin^2 \theta_{n-2}} \left(p_{n-3}^2 + \cdots \right. \right. \right.$$

$$\left. \left. \left. \cdots + \frac{1}{\sin^2 \theta_3} \left(p_2^2 + \frac{1}{\sin^2 \theta_2} p_1^2 \right) \cdots \right) \right) \right) =$$

$$= \sum_{i,j=1}^{n-1} g^{ij} p_i p_j. \tag{13.29}$$

The Hamilton system has the form

$$\frac{d\theta_i}{dt} = 2 \sum_{j=1}^{n-1} g^{ij} \frac{\partial H}{\partial p_i}, \qquad \frac{dp_i}{dt} = -\frac{\partial H}{\partial \theta_i}. \tag{13.30}$$

We shall find the first integrals of this system. For this purpose we use the formula

$$\dot{h}(\theta, p) = \{h, H\},$$

where $\dot{h}$ is the derivative with respect to the Hamilton system with the Hamiltonian $H(\theta, p)$, and $\{h, H\}$ is the Poisson bracket of functions h, H (§4). Consequently, the necessary and sufficient condition for function h to be the first integral of the Hamilton system is $\{h, H\} = 0$. In this manner we find $n - 1$ first integrals of the system (13.30):

$$h_1 = p_1, \quad h_2 = p_2^2 + \frac{c_1^2}{\sin^2 \theta_2}, \dots, \quad h_{n-1} = p_{n-1}^2 + \frac{c_{n-2}^2}{\sin^2 \theta_{n-1}}, \tag{13.31}$$

where $c_1, \dots, c_{n-2}$ are arbitrary constants. These first integrals are in involution, i.e. $\{h_i, h_j\} = 0$ for all i, j; it is easy to see that the integrals are independent. As a consequence, the system (13.30) is completely integrable.

Consider the set $\Lambda^{n-1} = \Lambda^{n-1}(c_1, \dots, c_{n-2})$ in the phase space $\mathbf{R}_{\theta,p}^{2(n-1)}$

which is given by the equations

$$h_1 = c_1, \quad h_2 = c_2^2, \dots, h_{n-2} = c_{n-2}^2, \quad h_{n-1} = 1, \tag{13.32}$$

where $0 < c_1 < c_2 < \dots < c_{n-2} < 1$. The variables θ are defined within the intervals $0 \le \theta_1 < 2\pi, 0 \le \theta_j < \pi \, (j \ge 2)$, but, for shortness, the phase space will be denoted by $\mathbf{R}_{\theta,p}^{2(n-1)}$. Since the set Λ^{n-1} is compact and connected, and the system (13.30) is completely integrable, then, according to the Liouville theorem [4], the set $\Lambda^{n-1}(c_1, \dots, c_{n-2})$ is diffeomorphic to the $(n-1)$-dimensional torus T^{n-1}. It is not difficult to establish this fact directly. Indeed, the equation

$$p^2 + \frac{a^2}{\sin^2 \theta} = b^2, \quad 0 < a < b,$$

defines a smooth curve γ in the domain $0 < \theta < \pi, p \in \mathbf{R}$, which is diffeomorphic to the circle. The points in which the tangent to curve γ is vertical, have the coordinates $(\theta_0, 0), (\pi - \theta_0, 0)$, where $\theta_0 = \arcsin(a/b)$. Further, the projection of Λ^{n-1} on the plane (p_1, θ_1) has the form $p_1 = c_1$, $0 \le \theta_1 \le 2\pi$, i.e. of the circle. Since $h_j = h_j(\theta_j, p_j)$, our considerations imply that manifold Λ^{n-1} is diffeomorphic to the torus T^{n-1}. The manifold Λ^2 is drawn in Figure 16.

The cycle of singularities $\Sigma(\Lambda^{n-1})$ turns out to be the set of points on Λ^{n-1} for which $p_2 = \dots = p_{n-2} = 0$. In particular for $n = 3$, the cycle of singularities consists of two circles (Figure 16) given by the equations

$$\{p_1 = c_1, p_2 = 0, \theta_2 = \theta_2^0, 0 \le \theta_1 \le 2\pi\},$$
$$\{p_1 = c_1, p_2 = 0, \theta_2 = \pi - \theta_2^0, 0 \le \theta_1 \le 2\pi\},$$

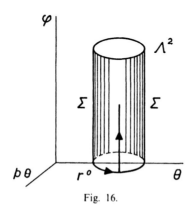

Fig. 16.

where $\theta_2^0 = \arcsin(c_1/c_2)$. For $n \geqq 4$ the cycle of singularities consists of 2^{n-3} tori T^{n-2}.

We choose a basis $\gamma_1, \ldots, \gamma_{n-1}$ of the homology group $H_1(\Lambda^{n-1}, \mathbf{Z})$ in the following way: as $\gamma_j, j \geqq 2$, we take the intersection of Λ^{n-1} with the plane (θ_j, p_j). As γ_1 we take the circle $0 \leqq \theta_1 \leqq 2\pi$, with the remaining variables θ_j, p_j fixed; moreover we suppose that γ_1 does not lie on the cycle of singularities (Figure 16). Now we compute the periods $\mathbf{J}_s$ of the 1-form $\omega^1 = \langle p, d\theta \rangle$ and the indexes $l_j = \text{ind } \gamma_j$.

We have $l_1 = 0$ by construction (the curve γ_1 does not intersect the cycle of singularities). Further, the curve $\gamma_s, s \geqq 2$, intersects the cycle of singularities exactly in two points in which $p_s = 0$; it is not hard to see that $l_s = 2$ for $s \geqq 2$. Furthermore,

$$\mathbf{J}_1 = \int_{\gamma_1} p_1 \, d\theta_1 = \int_0^{2\pi} c_1 \, d\theta_1 = 2\pi c_1,$$

$$\mathbf{J}_2 = \int_{\gamma_2} p_2 \, d\theta_2 = 2 \int_{\alpha_2}^{\pi - \alpha_2} \sqrt{c_2^2 - \frac{c_1^2}{\sin^2 \theta_2}} \, d\theta_2,$$

where $\alpha_2 = \arcsin(c_1/c_2)$. Consequently, $\mathbf{J}_2 = 2\pi(c_2 - c_1)$. The other periods are computed analogously, so that

$$\mathbf{J}_1 = 2\pi c_1, \mathbf{J}_s = 2\pi(c_s - c_{s-1}), \qquad (s \geqq 2), \tag{13.33}$$

where $c_{n-1} = 1$. Equations (13.12) in our case have the form

$$kc_1 = m_1,$$
$$k(c_2 - c_1) = m_2 + \tfrac{1}{2},$$
$$\ldots\ldots\ldots\ldots\ldots\ldots\ldots \tag{13.34}$$
$$k(c_{n-2} - c_{n-3}) = m_{n-2} + \tfrac{1}{2},$$
$$k(1 - c_{n-2}) = m_{n-1} + \tfrac{1}{2},$$

where m_j are integers. Summing up these equations and setting $l = m_1 + m_2 + \ldots + m_{n-1}$, we obtain

$$k = l + \frac{n-2}{2}. \tag{13.35}$$

We select the numbers $c_1, \ldots, c_{n-2}, k$ in such a way that relations (13.34) hold. Then the canonical operator $K_{\Lambda^{n-1}}(h), h = k^{-1}$, is defined on the Lagrangian manifold $\Lambda^{n-1}(c_1, \ldots, c_{n-2})$. As in (3.16) we set $\mu \equiv 0$, $\varphi = 1$, i.e. we take the function $\psi = K_{\Lambda^{n-1}}(h)1$ as an approximate eigenfunction. Then

$$\left(\frac{1}{k^2}\Delta\right)\psi = \left(-1 + O\left(\frac{1}{k^2}\right)\right)\psi =$$

$$= \left(-1 + O\left(\frac{1}{l^2}\right)\right)\psi, \qquad (l \to +\infty),$$

since $k \sim l$. The last relation and Lemma 13.1 yield the existence of the asymptotic series of eigenvalues of the Laplace–Beltrami operator on the sphere:

$$k_l = l + \frac{n-2}{2} + O(l^{-1}), \qquad (l \to +\infty). \tag{13.36}$$

Since the eigenvalue problem for the Laplace–Beltrami operator on the sphere is exactly solvable, and even by simpler means, the obtained formula (13.36) for the series of eigenvalues is not very interesting by itself. However, more important are the asymptotic formulae for the corresponding eigenfunctions, though their rigorous derivation is missing at present.

We shall consider here the case $n = 3$, i.e. the sphere S^2 in the three-dimensional space; for the general case see [39]. The first integrals and the quantization conditions have the form

$$p_\varphi^2 = c_1^2, \qquad p_\theta^2 + \frac{c_1^2}{\sin^2\theta} = 1, \tag{13.37}$$

$$kc_1 = m_1, \qquad k(1-c_1) = m_2 + \tfrac{1}{2}, \tag{13.38}$$

where $\theta_1 = \varphi, \theta_2 = 0$. We find from here

$$l_1 = l + \tfrac{1}{2} + O(l^{-1}), \qquad c_1 = 2m_1/(2l+1),$$

where $l = m_1 + m_2$. To the eigenvalue k_l, there corresponds the set of eigenfunctions $u_{m_1}(\theta, \varphi)$, where $0 < \delta \leq (|m_1|/l) \leq 1 - \delta, \delta > 0$ being a fixed number. This restriction arises because the numbers $c_1, 1 - c_2$ must be separated from zero. It is clear that $N(l) \approx 2l$, where $N(l)$ is the dimensionality of the invariant subspace of operator Δ corresponding to the eigenvalue k_l.

The Equations (13.37) determine a family of invariant Lagrangian

tori T^2, smoothly depending on the parameter c_1. We choose the point
$\{\varphi = 0, \theta = \arcsin c_1, p_\theta = 0\}$ as the initial point r^0, and the large circle
$\theta = \pi/2$ as the initial manifold T_0^1. The measure on T_0^1 induces the invariant measure $d\sigma$ on the torus T^2.

We calculate the action

$$S(\varphi, \theta) = \int_{r^0}^{r} (p_\varphi \, d\varphi + p_\theta \, d\theta)$$

in non-singular points. This integral is path-independent. Therefore,
as curve on T^2 connecting points r^0 and $r = (\varphi, \theta, p_\theta)$, we can take the
curve consisting of the arc of the curve $p_\theta^2 + (c_1^2/\sin^2 \theta) = 1$ along which
$\varphi = 0$ and which connects the projections of points r^0, r, and of the
segment on which θ, p_θ are constant (Figure 16). Then

$$S(\varphi, \theta) = \int_0^\varphi p_\varphi \, d\varphi + \int_{\theta_0}^\theta p_\theta \, d\theta = \int_0^\varphi c_1 \, d\varphi + \int_{\theta_0}^\theta \sqrt{1 - \frac{c_1^2}{\sin^2 \theta}} \, d\theta =$$

$$= c_1 \varphi + \tan^{-1} \frac{\sqrt{\sin^2 \theta - c_1^2}}{\cos \theta} - c_1 \tan^{-1} \frac{\sqrt{\sin^2 \theta - c_1^2}}{c_1 \cos \theta}$$

(all square roots are arithmetic). However, there are two points on the
torus T^2 projected into the point (φ, θ): if we denote the second of them by
$r^2(\varphi, \theta)$, then the action will be evaluated according to the same formula
with the only difference that $p_\theta = -\sqrt{1 - (c_1^2/\sin^2 \theta)}$ along the integration
path. Finally, we obtain two values $S_\pm$ for S:

$$S_\pm(\varphi, \theta) = c_1 \varphi \pm S_1(\theta),$$

$$S_1(\theta) = \tan^{-1} \frac{\sqrt{\sin^2 \theta - c_1^2}}{\cos \theta} - c_1 \tan^{-1} \frac{\sqrt{\sin^2 \theta - c_1^2}}{c_1 \cos \theta}. \qquad (13.39)$$

Furthermore

$$\left| \frac{D\theta}{D\sigma} \right| = \sqrt{\frac{\sin^2 \theta - c_1^2}{1 - c_1^2}}. \qquad (13.40)$$

The canonical atlas on T^2 can be composed of four charts. The cycle
of singularities consists of two circles on which $p_\theta = 0$, and its projection
on the plane (θ, p_θ) consists of two points $(\theta_0, 0), (\pi - \theta_0, 0)$. We cover the

curve $p_\theta^2 + (c_1^2/\sin^2 \theta) = 1$ by four charts U_j, two of which are the neighbourhoods of the points $(\theta_0, 0)$, $(\pi - \theta_0, 0)$ (we denote them U_1, U_2), U_3 lies in the half-plane $p_\theta > 0$, and U_4 in the half-plane $p_\theta < 0$. We choose a point $r^1 \neq r^0$ in which $p_\theta > 0$ as the distinguished point in the chart U_1. Then the curve l_1 connecting the points r^1 and $r = (\varphi, \theta, p_\theta)$, where $p_\theta > 0$, does not intersect the cycle of singularities, and $\mathrm{ind}\, l_1 = 0$; if $p_\theta < 0$, then the curve l_1 intersects the cycle of singularities and $\mathrm{ind}\, l_1 = 1$. Consequently, in the non-singular point (θ, φ) we have

$$K_{T^2}(1)(\theta, \varphi) = \sqrt{\left|\frac{D\sigma}{D\theta}\right|} (e^{ikS_+ + (i\pi/2)\alpha_+} + e^{ikS_- + (i\pi/2)\alpha_-}),$$

where $\alpha_+ = 0, \alpha_- = \pi/2$. By multiplying this expression by $e^{-i\pi/4}$ and by taking relations (13.37)–(13.40) into account, we obtain the following asymptotic formula for the approximate eigenfunction (in the non-focal points):

$$u_m(\theta, \varphi) = e^{im\varphi} \left(\frac{1 - \dfrac{m^2}{k^2}}{\sin^2 \theta - \dfrac{m^2}{k^2}}\right)^{1/4} \cos\left(kS_1 - \frac{\pi}{4}\right), \qquad (13.41)$$

$$S_1(\theta) = \tan^{-1} \frac{\sqrt{\sin^2 \theta - \dfrac{m^2}{k^2}}}{\cos \theta} - \frac{m}{k} \tan^{-1} \frac{k\sqrt{\sin^2 \theta - \dfrac{m^2}{k^2}}}{m \cos \theta}.$$

An analogous formula was derived for spheres of arbitrary dimensions in [39].

Formula (13.41) was obtained under the assumption that the ratio $|m/k|$ is separated from zero, and, in particular, that $m \neq 0$. If we put $m = 0, k = l + \frac{1}{2}$ in the formula and multiply its both sides by the corresponding normalization factor, then we obtain the known asymptotic formula for the Legendre polynomials:

$$u_l(\varphi, \theta) = \sqrt{\frac{2}{\pi l \sin \theta}} \sin\left[(l + \tfrac{1}{2})\theta + \frac{\pi}{4}\right].$$

The approximate eigenfunctions constructed above (see (13.41)) are localized in a neighbourhood of the projection of torus T^2 on the sphere S^2, i.e. in a neighbourhood of the strip $\theta_0 \leq \theta \leq \pi - \theta_0, \theta_0 = \arcsin(m/k)$. This strip contracts to the equator if $m \to 0$.

For the Laplace–Beltrami operator it is possible to prove that the approximate eigenfunctions constructed above are close to the exact eigenfunctions. This is connected with the fact that the spectrum is exactly known in the considered case: the eigenvalues k_l^2 have the form

$$k_l^2 = l(l + n - 2), \quad l = 1, 2, \ldots. \tag{13.42}$$

The existence of asymptotic series of eigenvalues of the form (13.36) was proved above. For $l \gg 1$ the consecutive eigenvalues k_l^2 are distributed with the mutual distances of order nl, so that formula (13.36) contains all eigenvalues of the Laplace–Beltrami operator (with those values of l for which it was derived). Further, the approximate eigenfunctions u_l were constructed (for simplicity, we label them by one index) for which

$$\|u_l\| \gg C_1, \qquad \|(\Delta + k_l^2)u_l\| \leq C_2,$$

for $l \gg 1$, where the constants C_j do not depend on l.

Finally, we shall use the following theorem from the spectral theory. Let A be a self-adjoint operator in a Hilbert space $\mathscr{H}$, for which the point $\lambda_0 = 0$ is an isolated eigenvalue with a finite multiplicity. Let an element $f \in D(A)$,

$$\|f\| \geq C_3, \qquad \|Af\| \leq C_4 \tag{13.43}$$

and let the estimate

$$C_4/C_3 < d_0,$$

hold, where d_0 is the distance of the point $\lambda_0 = 0$ from the rest of the spectrum $\sigma(A) \setminus \{0\}$.

Then

$$\|f - P_0 f\| \leq C_4 (d_0 C_3)^{-1}, \tag{13.44}$$

where P_0 is the orthogonal projector on the invariant subspace $\mathscr{H}_0$ of operator A which corresponds to the eigenvalue $\lambda_0 = 0$.

Application of this theorem to the Laplace–Beltrami operator yields the estimate

$$\|u_l - P_l u_l\| \leq C l^{-1} \tag{13.45}$$

for $l \gg 1$. Here constant C does not depend on l and P_l is the orthogonal projector on the invariant subspace of the Laplace–Beltrami operator corresponding to the eigenvalue k_l^2.

§ 14. Semi-Classical Approximations for the Relativistic Dirac Equation

We shall construct the semi-classical asymptotic solution of the Cauchy problem for the relativistic Dirac equation. It will be shown that the spin polarization has a classical limit for $h \to +0$.

1. The Dirac Equation

The motion of a relativistic charged particle of spin $\frac{1}{2}$ in an external electromagnetic field is described by the Dirac equation [7], [12]

$$- ih \frac{\partial \psi}{\partial t} = \hat{H}\psi. \tag{14.1}$$

The wave function $\psi = \psi(t, x_1, x_2, x_3)$ is a 4-vector (a column $\psi = (\psi_1, \psi_2, \psi_3, \psi_4)$, and the Hamiltonian has the form [76]

$$H = \alpha(c\hat{p} - eA) - e\varphi + \beta mc^2. \tag{14.2}$$

Here e is the charge, m the mass, c the light velocity in the vacuum, h the Planck constant, $A = (A_1, A_2, A_3)$ and φ – the vector and the scalar potentials of the electromagnetic field. Further,

$$\alpha\hat{p} = \alpha_1 \hat{p}_1 + \alpha_2 \hat{p}_2 + \alpha_3 \hat{p}_3,$$
$$\alpha A = \alpha_1 A_1 + \alpha_2 A_2 + \alpha_3 A_3,$$

where $\hat{p}_j = - ih(\partial/\partial x_j)$. The matrices α_j, β have the form

$$\alpha_j = \begin{pmatrix} 0 & \sigma_j \\ \sigma_j & 0 \end{pmatrix}, \quad \beta = \begin{pmatrix} 1 & 0 \\ 0 & -1 \end{pmatrix},$$

where 0 and 1 are the zero and the unit (2×2)-matrices, respectively, and σ_j are the Pauli spin matrices:

$$\sigma_1 = \begin{pmatrix} 0 & 1 \\ 1 & 0 \end{pmatrix}, \quad \sigma_2 = \begin{pmatrix} 0 & -i \\ i & 0 \end{pmatrix}, \quad \sigma_3 = \begin{pmatrix} 1 & 0 \\ 0 & -1 \end{pmatrix}.$$

As usual, we shall assume that all functions $A_j(t, x)$, $\varphi(t, x)$ are real-valued and infinitely differentiable.

Notice that all matrices $\alpha_j, \sigma_j, \beta$ are Hermitian. The symbol of operator

$\hat{H}$ is given by the Hermitian matrix

$$H(t, x, p) = \begin{pmatrix} A & B \\ B & C \end{pmatrix},$$

$$A = (-e\varphi + mc^2)\begin{pmatrix} 1 & 0 \\ 0 & 1 \end{pmatrix}, B = \begin{pmatrix} B_3 & \bar{B}_2 \\ B_2 & -B_3 \end{pmatrix},$$ (14.3)

$$C = (-e\varphi - mc^2)\begin{pmatrix} 1 & 0 \\ 0 & 1 \end{pmatrix},$$

with the notation

$$B_3 = -eA_3 + cp_3, \quad B_2 = -e(A_1 + iA_2) + c(p_1 + ip_2).$$ (14.4)

We take the Cauchy problem

$$\psi|_{t=0} = \psi^0(x) \exp\left[\frac{i}{h} S_0(x)\right],$$ (14.1')

where $\psi^0 \in C_0^\infty(\mathbf{R}^3)$, $S_0 \in C^\infty(\mathbf{R}^3)$, function S_0 is real-valued, and then we investigate the asymptotic solution of the problem (14.1), (14.1') for $h \to 0$.

2. The Hamilton–Jacobi and the Transport Equations

We first calculate the eigenvalues of matrix H. Owing to the block structure of this matrix we get

$$\det(H - hI_4) = \det((A - hI_2)(C - hI_2) - $$
$$- (A - hI_2)B(A - hI_2)^{-1} B) = $$
$$= \det([(e\varphi + h)^2 - m^2c^4]I_2 - B^2),$$

where I_k is the unit $(k \times k)$-matrix. Since $B^2 = (|B_2|^2 + B_3^2)I_2$, the eigenvalues of matrix H have the form

$$h_\pm = -e\varphi \mp \sqrt{D}.$$ (14.5)

Here

$$D = (cp - eA)^2 + m^2c^4 = |B_2|^2 + B_3^2 + m^2c^4.$$ (14.6)

Consequently, the matrix H has two eigenvalues $h_\pm$ of multiplicity 2 and of class C^∞ for any real t, x, p. The corresponding two Hamilton–Jacobi equations are

$$\frac{\partial S_\pm}{\partial t} = \pm c \sqrt{\left(\nabla S - \frac{e}{c}A\right)^2 + m^2c^2} + e\varphi.$$ (14.7)

The eigenvectors $f_j^\pm, j = 1, 2$, associated with the eigenvalues $h_\pm$ can be chosen as follows:

$$f_1^+ = \begin{pmatrix} -B_3 \\ -B_2 \\ mc^2 + \sqrt{D} \\ 0 \end{pmatrix}, \quad f_2^+ = \begin{pmatrix} -\bar{B}_2 \\ B_3 \\ 0 \\ mc^2 + \sqrt{D} \end{pmatrix},$$

$$f_1^- = \begin{pmatrix} mc^2 + \sqrt{D} \\ 0 \\ B_3 \\ B_2 \end{pmatrix}, \quad f_2^- = \begin{pmatrix} 0 \\ mc^2 + \sqrt{D} \\ \bar{B}_2 \\ -B_3 \end{pmatrix}. \tag{14.8}$$

Vectors $f_j^\pm$ form an orthogonal basis in $\mathbf{R}^4$ (the scalar product is Hermitian) and have the same length:

$$|f_j^\pm|^2 = 2\sqrt{D}\,(mc^2 + \sqrt{D}).$$

Moreover, the following identities are true:

$$\langle f_1^+, \alpha_j f_2^+ \rangle \equiv 0, \qquad j = 1, 2, 3, \tag{14.9}$$

and analogous identities hold for vectors f^-.

The space $\mathbf{R}^4$ decomposes (for any fixed t, x, p) into an orthogonal direct sum of two invariant subspaces $\mathbf{R}^4 = \mathbf{R}_+^2 \oplus \mathbf{R}_-^2$ of matrix H. The bases of the subspaces $\mathbf{R}_+^2$ and $\mathbf{R}_-^2$ are formed by the vectors $f_{1,2}^+$ and $f_{1,2}^-$, respectively. Correspondingly, the vector function entering the Cauchy data (14.1') can be represented in the form

$$\psi^0 = \psi_+^0 + \psi_-^0, \quad \text{where} \quad \psi_+^0 \in \mathbf{R}_+^2, \quad \psi_-^0 \in \mathbf{R}_-^2.$$

We shall assume that $\psi^0 \in \mathbf{R}_+^2$.

The formal asymptotic solution of the Cauchy problem (14.1), (14.1') will be sought in the usual form

$$\psi(t, x) = \exp\left[\frac{i}{h} S(t, x)\right] \sum_{k=0}^{N} \left(\frac{h}{i}\right)^k \psi^k(t, x). \tag{14.10}$$

We shall evaluate the leading term of the asymptotics. For the phase S, we obtain the Hamilton–Jacobi equation

$$\frac{\partial S}{\partial t} + h_+ = 0, \qquad S|_{t=0} = S_0(x).$$

The vector function $\psi^0(t, x)$ can be written in the form

$$\psi^0(t, x) = \omega_1(t, x) f_1^+ + \omega_2(t, x) f_2^+. \tag{14.11}$$

The unknown functions $\omega_{1,2}$ are determined from the system of transport equations

$$\frac{d\omega_1}{at} = a_{11}\omega_1 + a_{12}\omega_2, \tag{14.12}$$

$$\frac{d\omega_2}{dt} = a_{21}\omega_1 + a_{22}\omega_2.$$

The Cauchy data are given by

$$\omega_j\big|_{t=0} = \omega_j^0(x)$$

where ω_j^0 are found from the relation

$$\psi^0(x) = \omega_1^0 f_1^+\big|_{t=0} + \omega_2^0 f_2^+\big|_{t=0}.$$

By virtue of Lemma 11.5 we have

$$a_{jk} = \langle f_j^+, f_j^+ \rangle^{-1} \left[\left(\left\langle f_j^+, \frac{\partial f_k^+}{\partial t} \right\rangle + \frac{1}{2}\left(\frac{\partial^2 h_+}{\partial x \, \partial p} - \frac{d}{dt} \ln J_+ \right) \delta_{jk} + \right.\right.$$

$$\left.\left. + \left\langle f_j^+, \left(\frac{\partial H}{\partial p} - \frac{\partial h_+}{\partial p} \right) \frac{\partial f_k^+}{\partial x} \right\rangle \right]. \tag{14.13}$$

The factor $\langle f_j^+, f_j^+ \rangle^{-1}$ appears since f_j^+ are not unit vectors. The summation over conjugate variables is understood, i.e.

$$\frac{\partial h_+}{\partial p} \frac{\partial f}{\partial x} = \sum_{j=1}^{3} \frac{\partial h_+}{\partial p_j} \frac{\partial f}{\partial x_j},$$

etc. Here J_+ is the Jacobian corresponding to the Hamilton system

$$\frac{dx}{dt} = \frac{\partial h_+}{\partial p}, \qquad \frac{dp}{dt} = -\frac{\partial h_+}{\partial x}, \tag{14.14}$$

i.e.

$$J_+ = \det \frac{\partial x(t, y)}{\partial y},$$

where $(x(t, y), p(t, y))$ is the solution of system (14.14) with the Cauchy data

$$x\big|_{t=0} = y, \qquad p\big|_{t=0} = \frac{\partial S_0(y)}{\partial y}.$$

Due to orthogonality of the vectors f_1^+, f_2^+, the identities (14.9) and reality of h_+, H, we have the identity

$$a_{11} \equiv -\bar{a}_{12} \tag{14.15}$$

However, the direct calculation of the coefficients a_{jk} according to formula (14.13) leads to cumbersome expressions. This is connected, roughly speaking, with the fact that we separate the variable t. Another derivation of the transport equation is given in the next section.

3. The Squared Dirac Equation and the Transport Equations

The symbol of the Dirac operator is equal to $-p_0 I_4 + H(x_0, x, p)$, where $\hat{p}_0 = (h/i)(\partial/\partial t), x_0 = t$. We shall find the matrix $H_1(x_0, x, p)$ such that

$$(-p_0 I_4 + H_1)(-p_0 I_4 + H) = (p_0 - h_+)(p_0 - h_-)I_4, \tag{14.16}$$

where $h_\pm$ are the eigenvalues of matrix H. It is easy to check that

$$H_1 = -e\varphi I_4 - \alpha(cp - eA) - \beta mc^2. \tag{14.17}$$

The matrices $-p_0 I_4 + H_1$, $-p_0 I_4 + H$ are associated by virtue of (14.16), up to a scalar factor.

Let $\hat{L}_1$ be the Dirac operator, $\hat{L}_2$ the operator with the symbol $-p_0 I_4 + H_1(x_0, x, p)$. Then [76]

$$\hat{L}_2 \hat{L}_1 = [(\hat{p}_0 + e\varphi)^2 - (c\hat{p} - eA)^2 - m^2 c^4] I_4 + \frac{h}{i}(iec\sigma'H + c\alpha E). \tag{14.18}$$

Here

$$E = -\frac{\partial A}{\partial t} - \nabla\varphi, \qquad H = \operatorname{rot} A, \tag{14.19}$$

$$\sigma' = (\sigma'_1, \sigma'_2, \sigma'_3), \qquad \sigma'H = \sigma'_1 H_1 + \sigma'_2 H_2 + \sigma'_3 H_3,$$

and σ'_j are (4×4)-matrices

$$\sigma'_j = \begin{pmatrix} \sigma_j & 0 \\ 0 & \sigma_j \end{pmatrix}, \tag{14.20}$$

σ_j the Pauli spin matrices. The operator

$$S = \frac{h}{2}\sigma' \tag{14.21}$$

is said to be the *spin angular momentum of the electron.*

The operator $\hat{Q} = \hat{L}_2 \hat{L}_1$ is called the *squared Dirac operator*, and the equation

$$\hat{Q}\psi = 0 \tag{14.22}$$

is called the *squared Dirac equation*.

We set $x = (x_0, x_1, x_2, x_3), p = (p_0, p_1, p_2, p_3)$. The symbol Q of the operator $\hat{Q}$ is equal to (see (14.18))

$$Q = Q_0 + \frac{h}{i} Q_1, \tag{14.23}$$

$$Q_0 = [(p_0 + e\varphi)^2 - (cp - eA)^2 - m^2 c^4] I_4 = \\ = (p_0 - h_+)(p_0 - h_-) I_4, \tag{14.24}$$

$$Q_1 = B - \frac{1}{2} \frac{\partial^2 Q_0}{\partial x \, \partial p}. \tag{14.25}$$

Here B is the matrix

$$B = iec\sigma' H - c\alpha E. \tag{14.26}$$

Conjugate indices are summed over, i.e.

$$\frac{\partial^2 Q_0}{\partial x \, \partial p} = \sum_{j=0}^{3} \frac{\partial^2 Q_0}{\partial x_j \, \partial p_j}$$

etc.

Now we shall consider not the squared Dirac equation, but an arbitrary h^{-1}-p.d. operator with the symbol of the form (14.23), where Q_0, Q_1 are $(N \times N)$-matrices independent of h, matrix Q_0 being scalar.

We are going to derive the f.a. solution of Equation (14.22) with accuracy to $O(h)$. We give only formal arguments and do not mention the standard assumption on the symbol.

We seek the f.a. solution in the form

$$\psi(x) = \sum_{j=0}^{N} \left(\frac{h}{i} \right)^j \psi^j(x) \exp \left[\frac{i}{h} S(x) \right].$$

Inserting ψ into the equation and setting the coefficients at powers of h equal to zero, we obtain the equations

$$R_0 \psi^0 = 0, \qquad R_0 \psi^1 + R_1 \psi^0 = 0,$$

etc. Here

$$R_0 \psi = Q_0 \psi,$$

$$R_1 \psi = \frac{\partial Q_0}{\partial p} D_x \psi + \frac{1}{2} \frac{\partial^2 Q_0}{\partial p^2} \frac{\partial^2 S}{\partial x^2} \psi + Q_1 \psi, \qquad (14.27)$$

and summation over the conjugate variables is understood.

According to the assumption, the matrix Q_0 is scalar; we shall write it in the form $Q_0 I_N$, where Q_0 is a scalar function. Then the equations for ψ^0, ψ^1 take the form

$$Q_0(x, p) = 0, \qquad (p = \nabla S(x)), \qquad (14.28)$$

$$R_1 \psi^0 = 0. \qquad (14.29)$$

The Hamilton system

$$\frac{dx}{d\tau} = \frac{\partial Q_0}{\partial p}, \qquad \frac{dp}{d\tau} = -\frac{\partial Q_0}{\partial x} \qquad (14.30)$$

is associated with the Hamilton–Jacobi equation (14.28). By using the Liouville formula, the operator R_1 is transformed into the form

$$R_1 = \left[\frac{d}{d\tau} + \frac{1}{2} \left(\frac{d}{d\tau} \ln J - \frac{\partial^2 Q_0}{\partial x \, \partial p} \right) \right] I_N + Q_1 \qquad (14.31)$$

(see § 3). As a consequence, the following proposition holds:

PROPOSITION 14.1.   *The solution of the transport equation* (14.29) *has the form*

$$\psi^0(\tau) = \sqrt{\frac{J(0)}{J(\tau)}} \exp\left(\frac{1}{2} \int_0^\tau \frac{\partial^2 Q_0}{\partial x \, \partial p} \, d\tau \right) \chi^0(\tau), \qquad (14.32)$$

where $\chi^0(\tau)$ is the solution of the Cauchy problem

$$\frac{d\chi^0}{d\tau} + Q_1 \chi^0 = 0, \qquad \chi^0|_{\tau=0} = \psi^0|_{\tau=0}. \qquad (14.33)$$

In formula (14.32), as usual, $x = x(\tau), p = p(\tau)$ is the solution of the system (14.29), and $J(\tau) = \det(\partial x(\tau)/\partial x(0))$.

We shall consider the evolution equation (14.22), i.e.

$$Q_j = Q_j(t, x, p_0, p),$$

where $t \in \mathbf{R}, x \in \mathbf{R}^n$. Let the equation

$$Q_0(t, x, p_0, p) = 0 \tag{14.34}$$

with respect to the variable p_0 have a solution of class C^∞

$$p_0 = h(t, x, p) \tag{14.35}$$

(for simplicity) for all real t, x, p, and let this root be simple. Construction of the f.a. solution of the equation $\hat{Q}\psi = 0$ leads to the Hamilton–Jacobi equation (14.34) (here $p_0 = \partial S/\partial t, p = \partial S/\partial x$); we select the solution which satisfies Equation (14.35).

The Hamilton system

$$\frac{dt}{d\tau} = \frac{\partial Q_0}{\partial p_0}, \qquad \frac{dp_0}{d\tau} = -\frac{\partial Q_0}{\partial t},$$

$$\frac{dx}{d\tau} = \frac{\partial Q_0}{\partial p}, \qquad \frac{dp}{d\tau} = -\frac{\partial Q_0}{\partial x}, \tag{14.36}$$

is associated with Equation (14.34), and the truncated Hamilton system

$$\frac{dx}{d\tau} = -\frac{\partial h}{\partial p}, \qquad \frac{dp}{d\tau} = \frac{\partial h}{\partial x} \tag{14.37}$$

is associated with Equation (14.35). We shall establish the connections between the solutions of these systems and between the Jacobians J.

We take the Cauchy problems in the form

$$t|_{\tau=0} = 0, \qquad p_0|_{\tau=0} = E^0,$$

$$x|_{\tau=0} = x^0, \qquad p|_{\tau=0} = p^0 \tag{14.38}$$

and

$$x|_{\tau=0} = x^0, \qquad p|_{\tau=0} = p^0 \tag{14.39}$$

for the systems (14.36) and (14.37), respectively, where the data x^0, p^0 are the same. We assume that both problems are Lagrangian, i.e. $p^0 = = \partial S_0(x^0)/\partial x^0$. Our discussion will be local.

The solution of the problem (14.37), (14.39) will be denoted by x^+, p^+, to distinguish it from the solution of the problem (14.36), (14.38). We introduce the Jacobians

$$J = \frac{D(t, x)}{D(\tau, x^0)}, \qquad J^+ = \frac{Dx^+}{Dx^0}$$

(recall that $t = t(\tau, x^0), x = x(\tau, x^0)$ in the case of the problem (14.36), (14.38)). We assume that $J \neq 0, J^+ \neq 0$. Then (at least for small t) one may consider the solution $x(\tau, x^0)$ as a function of t, x^0.

LEMMA 14.2.   Let the Cauchy data (14.3) be such that $E^0 = h(0, x^0, p^0)$. Then for small t

$$x(t, x^0) = x^+(t, x^0), \qquad p(t, x^0) = p^+(t, x^0), \qquad (14.40)$$

$$J = \left. \frac{\partial Q_0}{\partial p_0} \right|_{p_0 = h} J^+.$$

Proof. The function Q_0 is the first integral of system (14.36); since it vanishes for $\tau = 0$, we have $Q_0 \equiv 0$ on the trajectories of system (14.36). Since the root $p_0 = h$ of Equation (14.34) is isolated, and since we have $p_0 = h$ for $\tau = 0$, equation (14.35) is satisfied identically along trajectories of the system (14.36).

By differentiating identity (14.34) with respect to p we get

$$\frac{\partial h}{\partial p} = -\frac{\partial Q_0}{\partial p} \left(\frac{\partial Q_0}{\partial p_0} \right)^{-1},$$

where $p_0 = h$. From the system (14.36) we find

$$\frac{dx}{dt} = \frac{\partial Q_0}{\partial p} \left(\frac{\partial Q_0}{\partial p_0} \right)^{-1} = -\frac{\partial h}{\partial p}.$$

One proves similarly that $dp/dt = \partial h/\partial x$, so the solution $x(t, x^0), p(t, x^0)$ of the problem (14.36), (14.38) satisfies the system (14.37). Since the Cauchy data coincide, we have the identities $x = x^+, p = p^+$. Furthermore,

$$dt \wedge dx_1 \wedge \ldots \wedge dx_n =$$

$$= dt \wedge \left(\frac{\partial x_1}{\partial t} dt + \sum_{k=1}^{n} \frac{\partial x_1}{\partial x_k^0} dx_k^0 \right) = J^+ dt \wedge dx_1^0 \wedge \ldots \wedge dx_n^0,$$

from which the last of the identities in (14.40) follows.

Consider the Cauchy problem (14.1') for the evolution equation $\hat{Q}\psi = 0$. Suppose symbol Q has the form (14.23) (recall that matrix Q_0 is scalar), and the assumptions of Lemma 14.2 hold. Then we have

PROPOSITION 14.3.   The Cauchy problem (14.22), (14.1') has the *formal asymptotic solution of the form*

$$\psi(t, x) = \sqrt{\frac{J^+(0)}{J^+(t)} \frac{\partial Q_0(0)}{\partial p_0} \left(\frac{\partial Q_0(\tau)}{\partial p_0}\right)^{-1}} \exp\left\{\frac{i}{h} S(t, x) + \right.$$

$$\left. + \frac{1}{2} \int_0^t \left[\sum_{j=1}^n \frac{\partial^2 Q_0}{\partial x_j \partial p_j} + \frac{\partial^2 Q_0}{\partial t \partial p}\right] \left(\frac{\partial Q_0}{\partial p_0}\right)^{-1} d\tilde{\tau}\right\} \chi^0(t, x).$$

$$(14.41)$$

Here S is the solution of the Hamilton–Jacobi Equation (14.35) *with the Cauchy data* $S|_{t=0} = S_0(x)$ *and* χ^0 *is the solution of the Cauchy problem*

$$\frac{d\chi^0}{dt} + \left(\frac{\partial Q_0}{\partial p_0}\right)^{-1} Q_1 \chi^0 = 0, \qquad \chi^0|_{t=0} = \psi^0. \qquad (14.42)$$

The proof is a consequence of Proposition 14.1, Lemma 14.2 and the relation $dt/d\tau = \partial Q_0/\partial p_0$. We set $p_0 = h$ in the formulae (14.41), (14.42).

Now we shall apply this method to the Cauchy problem (14.1′) for the Dirac equation. This problem is equivalent to the Cauchy problem

$$\psi|_{t=0} = \psi^0(x) \exp\left(\frac{i}{h} S(x)\right),$$

$$\left.\frac{\partial \psi}{\partial t}\right|_{t=0} = \frac{i}{h} \tilde{H} \psi^0(x) \exp\left(\frac{i}{h} S(x)\right) \qquad (14.43)$$

for the squared Dirac equation. Actually, $\hat{Q} = \hat{L}_2 \hat{L}_1$, and since, by virtue of (14.33), $\hat{L}_1 \psi|_{t=0} = 0$, then $\hat{L}_1 \psi \equiv 0$ for all t. The second of the conditions (14.43) leads to the relation

$$\left.\left(I_4 \frac{\partial S}{\partial t} - H\right)\right|_{t=0} \psi^0(x) = 0,$$

i.e. $(\partial S/\partial t)|_{t=0}$ is the eigenvalue, $\psi^0(x)$ the eigenvector of the matrix $H|_{t=0}$. According to the previous convention, $\psi^0 \in \mathbf{R}_2^+$, i.e. $H|_{t=0} \psi^0 = h_+|_{t=0} \psi^0$.

THEOREM 14.4. *The Cauchy problem* (14.1′) *for the Dirac equation*

(14.1), *where $\psi^0(x) \in \mathbf{R}_2^+$, has the formal asymptotic solution*

$$\psi(t, x) = \sqrt{\frac{J^+(0, y)}{J^+(t, y)}} \sqrt{\frac{[(cp - eA)^2 + m^2c^4]|_{t=0}}{[(cp - eA)^2 + m^2c^4]|_t}} \times$$

$$\times \exp\left(\frac{i}{h} S_+(t, x)\right) \chi^0(t, x). \tag{14.44}$$

Here χ^0 is the solution of the Cauchy problem

$$\frac{d\chi^0}{dt} = -\frac{1}{2}(ie c\sigma' H - c\alpha E)((cp - eA)^2 + m^2c^4)^{-1/2},$$

$$\chi^0|_{t=0} = \psi^0(x). \tag{14.45}$$

Function S_+ is the solution of Equation (14.7) with the Cauchy data $S|_{t=0} = S_0(x)$.

Proof follows from Proposition 14.3 and the formulae (14.25), (14.26). In formula (14.44) $x = x(t, y)$, $p = p(t, y)$ as usual (the solution of the Cauchy problem (14.37), (14.39) where $x|_{t=0} = y$).

Similarly, if $\psi|_{t=0} \in \mathbf{R}_-^2$, then there exists the f.a. solution of the form (14.44), where $J^+ \to J^-, S_+ \to S_-$. Thus the asymptotic solution of the Cauchy problem can be evaluated for arbitrary Cauchy data (14.1'), provided they satisfy the standard conditions of smoothness and of compact support.

Since the spin operator $\frac{1}{2} ie c\sigma' H$ enters the transport equation (14.45), the semi-classical approximation for the Dirac equation contains the spin angular momentum.

Formulae (14.44) and (14.45) give the f.a. solution of the Cauchy problem in the small. With the help of the canonical operator (see Theorem 11.10) one can construct the f.a. solution for any finite time interval.

The asymptotic formula (14.44) can be derived as follows. Let the scalar and vector electromagnetic potentials be real-valued functions of class $C^\infty(\mathbf{R}^3)$ not depending on x and bounded together with their derivatives up to the second order. Then the minimal symmetric operator $\hat{H}_0$ in $\hat{L}_2(\mathbf{R}_x^3)$ corresponding to the Dirac system 14.1 admits an extension to a self-adjoint operator $\hat{H}$.

Theorem 11.10 and Proposition 10.3 yield the estimate

$$\| \psi(t, x) - \psi^0(t, x) \| \leq Ch,$$

where ψ^0 is the right-hand side of formula (14.44).

REFERENCES

1. *Analytic Methods in the Theory of Wave Diffraction and Propagation*, Scientific Conference on Acoustics, (Academy of Sciences of USSR. Ministry of Radiotechnical Industry of USSR). Moscow 1970 (Russian).
2. Arnold, V. I.: 'A Characteristic Class Entering in Quantization Conditions', *Funct. Analys. and its Appl.* **1** (1967), 1–14 (Russian).
3. Arnold, V. I.: 'Normal Forms of Functions in Neighbourhoods of Degenerate Critical Points, the Weyl Groups of A_k, D_k, E_k, and Lagrangian Singularities', *Funct. Analys. and its Appl.* **6** (1972), 3–26 (Russian).
4. Arnold, V. I.: *Mathematical Methods of Classical Mechanics*, Springer-Verlag, Berlin 1978.
5. Babich, V. M. and Buldyrev, V. S.: *Asymptotic Methods in the Problems of Short-Wave diffraction; the Method of Reference Problems*, Nauka, Moscow 1972 (Russian).
6. Birkhoff, G. D.: 'Quantum Mechanics and Asymptotic Series', *Bull. Am. Math. Soc.* **39** (1933), 681–700.
7. Bogoliubov, N. N. and Shirkov, D. V.: *Introduction to the Theory of Quantized Fields*, Interscience, New York 1959.
8. Brekhovskikh, L. M.: *Waves in Layered Media*, Nauka, Moscow 1973 (Russian).
9. Buslayev, V. S.: 'A Generating Integral and Maslov's Canonical Operator in the WKB Method', *Funct. Analys. and its Appl.* **3** (1969), 17–31 (Russian).
10. Cartan, E.: *Leçons sur les invariants intégraux*, Herman, Paris 1921.
11. Courant, R.: *Partial Differential Equations*, New York–London 1962.
12. Dirac, P. A. M.: *The Principles of Quantum Mechanics*, 4th ed., Clarendon Press, Oxford 1958.
13. Duistermaat, J. J.: 'Oscillatory Integrals, Lagrange Immersions and Unfolding of Singularities', *Commun. Pure Appl. Math.* **27** (1974), 207–281.
14. Dunford, N. and Schwartz, J. T.: *Linear Operators*, Part I, Interscience, New York 1958.
15. Fedoriuk, M. V.: 'The Stationary Phase Method for Multi-Dimensional Integrals', *J. Comp. Math.* **2** (1962), 145–150 (Russian).
16. Fedoriuk, M. V.: 'Asymptotics of Green's Function for $t \to +0, |x| \to \infty$, for Petrovsky-Correct Equations with Constant Coefficients, and Correctness Classes of Solutions of the Cauchy Problem', *Mat. Sbornik* **62** (1963), No. 4, 397–468 (Russian).
17. Fedoriuk, M. V.: 'The Stationary Phase Method. Near Saddle Points in Multi-Dimensional Case', *J. Comp. Math.* **4** (1964), 671–682 (Russian).
18. Fedoriuk, M. V.: 'One-Dimensional Scattering Problem in the Semi-Classical Approximation, I', *Differential Equations* **1** (1965), 631–646 (Russian).
19. Fedoriuk, M. V.: 'One-Dimensional Scattering Problem in the Semi-Classical Approximation, II', *Differential Equations* **1** (1965), 1525–1536 (Russian).
20. Fedoriuk, M. V.: 'Stationary Phase Method and Pseudo-Differential Operators',

Uspekhi Mat. Nauk **26** (1971), No. 1, 67–112 (Russian).

21. Feynman, R. P. and Hibbs, A. R.: *Quantum Mechanics and Path Integrals*, McGraw-Hill, New York 1965.

22. Fock, V. A.: 'On the Canonical Transformation in Classical and Quantum Mechanics', *Vestnik LGU* **16** (1959), 67–71 (Russian).

23. Fock, V. A.: *Problems of Electromagnetic Wave Diffraction and Propagation*, Soviet Radio, Moscow 1970 (Russian).

24. Frank, Ph. and Mises, R. von: *Die Differential- und Integralgleichungen der Mechanik und Physik*, I, II, Vieweg, Braunschweig 1935.

25. *Functional Analysis*, Mathematical Handbook Library, ed. by S. G. Krein, Nauka, Moscow 1964 (Russian).

26. Fuks, D. B.: 'On the Maslov–Arnold Characteristic Classes', *Dokl. Akad. Nauk SSSR*, *ser. mat.* **178** (1968), 303–306 (Russian).

27. Fuks, D. B., Fomenko, A. T., and Gutenmacher, V. L.: *Homotopic Topology*, Moscow State University, Moscow 1969 (Russian).

28. Gantmacher, F. R.: *Lectures on Analytical Mechanics*, Fizmatgiz, Moscow 1960 (Russian).

29. Gårding, L.: *Cauchy's Problem for Hyperbolic Equations*, Lecture Notes, University of Chicago, 1957.

30. Gelfand, I. M. and Fomin S. V.: *Calculus of Variations*, Prentice-Hall, Englewood Cliffs, N. J. 1963.

31. Goldstein, H.: *Classical Mechanics*, Addison–Wesley, Reading, Mass., 1959.

32. Helgason, S.: *Differential Geometry and Symmetric Spaces*, Academic Press, New York 1962.

33. Hörmander, L.: 'Pseudo-Differential Operators and Hypoelliptic Equations', *Am. Math. Soc. Symp. Pure Math.* **10** (1966), 138–183. (Symp. on Singular Integral Operators).

34. Hörmander, L.: 'Fourier Integral Operators, I', *Acta Math.* **127** (1971), 79–183.

35. Hörmander, L. and Duistermaat, J. J.: 'Fourier Integral Operators, II', *Acta Math.* **128** (1972), 183–269.

36. Hörmander, L.: *Linear Partial Differential Operators*, Springer-Verlag, Berlin 1963.

37. Keller, J. B.: 'Corrected Bohr–Sommerfeld Quantum Conditions for Nonseparable Systems', *Ann. Phys.* **4** (1958), 180–188.

38. Keller, J. B. and Rubinow, S.: 'Asymptotic Solution of Eigenvalue Problems', *Ann. Phys.* **9** (1960), 24–75.

39. Kogan, V. R.: 'Asymptotics of the Laplace–Beltrami Operator on the Unit Sphere S^m', *Izv. Vyssh. Ucheb. Zaved., Radiophysics* **12** (1969), 1675–1680 (Russian).

40. Kogan, V. R.: 'Asymptotics of the Laplace–Beltrami Operator in the Unit Ball E^m', *Izv. Vyssh. Uche. Zaved., Radiophysics* **12** (1969), 1681–1689 (Russian).

41. Kohn, J. J. and Nirenberg, L.: 'On the Algebra of Pseudo-Differential Operators', *Comm. Pure Appl. Math.* **18** (1965), 269–305.

42. Kucherenko, V. V.: 'Semi-Classical Asymptotics of the Point Source Function for the Stationary Schrödinger Equation', *Theor. and Math. Phys.* **1** (1969), 384–406 (Russian).

43. Kucherenko, V. V.: 'Asymptotic Solution of the System $A(x, -ih(\partial/\partial x))u = 0$ for $h \to 0$ in the Case of Characteristics with Variable Multiplicity', *Izv. Akad. Nauk SSR*, *ser. mat.* **38** (1974), 625–662 (Russian).

44. Kucherenko, V. V.: 'Commutation Formula of a h^{-1}-Pseudo-Differential Operator with a Rapidly Oscillating Exponential', *Mat. Sbornik* **94** (1974) No. 1, 89–113 (Russian).
45. Kucherenko, V. V.: 'Asymptotic Solutions of an Equation with Complex Characteristics', *Mat Sbornik* **95** (1974), No. 2, 163–213 (Russian).
46. Landau, L. D. and Lifshitz, E. M. : *Mechanics*, Addison–Wesley. Reading, Mass, 1969.
47. Landau. L. D. and Lifshitz, E. M. : *The Classical Theory of fields*, Addison–Wesley. Reading. Mass., 1951.
48. Landau, L. D. and Lifshitz, E. M. : *Quantum Mechanics (The Non-Relativistic Theory)*, Addison–Wesley, Reading, Mass. 1958.
49. Lax, P.D. : 'Asymptotic Solutions of Oscillatory Initial Value Problems', *Duke Math. J.* **24**, (1957), 627–646.
50. Leray, J. : *Lectures on Hyperbolic Equations with Variable Coefficients*, Mimeographed Notes, Institute for Advanced Study, Princeton 1952.
51. Leray, J.: *Solutions asymptotiques des équations aux dérivées partielles*, Conv. Intern. Phys. Math., Roma 1972.
52. Leray, J.: 'Solutions asymptotiques et groupe symplectique', in *Fourier Integral Operators and Partial Differential Equations*, Ed. by J. Chazarain. Lecture Notes in Mathematics, Vol. 459, Springer-Verlag, Berlin 1975, pp. 73–97.
53. Leray, J., Gårding, L., and Kotake, T. : 'Problème de Cauchy', *Bull. Soc. Math. France* **92** (1964), 263–361.
54. Ludwig. D.: 'Exact and Asymptotic Solutions of the Cauchy Problem', *Commun. Pure Appl. Math.* **13** (1960), 473–508.
55. Malgrange, B. : *Opérateurs de Fourier d'après Hörmander et Maslov*, Séminaire Bourbaki, No. 411, 24-e année, 1971/72.
56. Maslov, V.P.: 'Semi-Classical Asymptotics of Solutions of Some Problems of Mathematical Physics, I', *J. Comp. Math.* **1** (1961), 113–128 (Russian).
57. Maslov, V.P.: 'Semi-Classical Asymptotics of Solutions of Some Problems of Mathematical Physics, II', *J. Comp. Math.* **1** (1961), 638–663 (Russian).
58. Maslov, V. P.: 'The Scattering Problem in Semi-Classical Approximation', *Dokl. Akad. Nauk SSSR, ser. mat.* **151** (1963), 306–309 (Russian).
59. Maslov, V. P.: *Théorie des perturbations et méthodes asymptotiques*, Dunod, Paris 1972 (contains transl. of [2], [9]).
60. Maslov, V. P.: 'The WKB Method in Multi-Dimensional Case', Appendix in the Russian edition of: Heading, J.: *An Introduction to Phase-Integral Methods* (J. Wiley, New York 1962) Mir, Moscow 1965 (Russian).
61. Maslov, V.P.: 'On the Regularization of the Cauchy Problem for Pseudodifferential equations', *Dokl. Akad. Nauk SSSR, ser. mat.* **177** (1967), 1277–1280 (Russian).
62. Maslov, V. P. : 'On the Stationary Phase Method for Feynman's Continual Integrals', *Theor. Math. Phys.* **2** (1970), 30–40 (Russian).
63. Maslov, V. P. :'The Characteristics of Pseudo-Differential Operators and Difference Schemes,' Actes Congrès Intern. Math., Vol. 2, 1970, pp. 755–769.
64. Maslov, V. P.: *Operational Methods*, Mir, Moscow 1973.
65. Maslov, V. P. and Fedoriuk, M. V. : *The Canonical Operator (Real Case)*. Contemporary Problems of Mathematics, Vol. I, VINITI Publ., Moscow 1973 (Russian).
66. Milnor, J. W. : *Morse Theory*, Annals of Mathematics Studies, Vol. 51, Princeton University Press, Princeton, N. J., 1963.

293

67. Mishchenko, A. S. and Sternin, B. Yu.: *The Method of Canonical Operator in Applied Mathematics*, Pt. 1, MIEM, Moscow 1974 (Russian).
68. Mishchenko, A. S., Sternin, B. Yu., and Shatalov, V. E.: *The Method of Maslov's Canonical Operator, Complex Theory*, MIEM, Moscow 1974 (Russian).
69. Morse, M.: *The Calculus of Variations in the Large*, Amer. Math. Soc. Coll. Publ. 18, New York. 1934.
70. Novikov, S. P.: 'Algebraic Construction and Properties of Hermitian Analogues of the K-Theory over the Fields with Involution, from the Viewpoint of the Hamilton Formalism. Some Applications to Differential Topology and the Theory of Characteristic Classes, I', *Izv. Akad. Nauk SSSR, ser. mat.* **34** (1970), 253–288 (Russian).
71. Novikov, S. P.: 'Algebraic Construction and Properties of Hermitian Analogues of the K-Theory over the Fields with Involution, from the Viewpoint of the Hamilton Formalism. Some Applications to Differential Topology and the Theory of Characteristic Classes, II', *Izv. Akad. Nauk SSSR, ser. mat.* **34** (1970), 475–500 (Russian).
72. Petrovsky, I. G.: *Über das Cauchysche Problem für System von partiellen Differentialgleichungen*, Mat. Sbornik (1973), No. 2, pp. 815–870.
73. Pokrovskij, V. L. and Khalatnikov, I. M.: 'On the Above-Barrier Reflection of High Energy Particles', *JETP* **40** (1961), 1713–1719 (Russian).
74. Rashevskij, P. K.: *Geometrical Theory of Partial Differential Equations*, Gostekhizdat, Moscow 1947 (Russian).
75. Schaeffer, D. and Guillemin, V.: *Maslov Theory and Singularities*, Mimeographed Notes, M.I.T., Cambridge, Mass., 1972.
76. Schiff, L. I.: *Quantum Mechanics*, 2nd ed., McGraw-Hill, New York 1955.
77. Smirnov, V. I.: *A Course of Higher Mathematics*, Vol. 4, Fizmatgiz, Moscow 1958 (Russian).
78. Sobolev, S. L.: *Some Applications of Functional Analysis in Mathematical Physics*, Leningrad State University, Leningrad 1950 (Russian).
79. Souriau, J.-M.: 'Indice de Maslov des variétés lagrangiennes orientables', *Compt. Rend. Acad. Sci. Paris, sér. A.* **276** (1973), 1025–1026.
80. Sternberg, S.: *Lectures on Differential Geometry*, Prentice-Hall, Englewood Cliffs, N. J., 1964.
81. Vishik, M. I. and Liusternik, L. A.: 'Regular Degeneracy and Boundary Layer for Linear Differential Equations with a Small Parameter', *Usp. Mat. Nauk* **12** (1957) No. 5, 3–122 (Russian).
82. Vladimirov, V. S.: *Equations of Mathematical Physics*, Nauka, Moscow 1975 (Russian).
83. Weinberg, B. R.: 'On Analytic Properties of the Resolvent for One Class of Operator Fields, *Mat. Sbornik* **77** (1968), No. 2, 259–296 (Russian).
84. Weinberg, B. R.: 'On Short-Wave Asymptotics of Solutions of Stationary Problems and the Asymptotics of Non-Stationary Problems for $t \to \infty$', *Usp. Mat. Nauk* **30** (1975), No. 2, 3–55 (Russian).
85. Weinstein, L. A.: *Open Resonators and Open Wave-Guides*, Soviet Radio, Moscow 1966 (Russian).
86. Weinstein, A.: 'On Maslov's Quantization Condition', in *Fourier Integral Operators and Partial Differential Equations*, Ed. by J. Chazarain, Lecture Notes in Mathematics, Vol. 459, Springer-Verlag, Berlin 1975, pp. 341–372.
87. Yosida, K.: *Functional Analysis*, Springer-Verlag, Berlin 1968.

88. Zhdanova, G. V. and Fedoriuk, M. V.: 'Asymptotic Theory of the Systems of Second Order Ordinary Differential Equations and the Scattering Problem', *Trudy Mosk. Mat. Obshch.* 1976 (Russian).
89. Le Vu An: 'Classical Asymptotics of the Free Schrödinger Equation for Calculating the Corrections in the Stationary Phase Method', *Theor. and Math. Phys.* **25** (1975), 270–274. (Russian).
90. Belov, V. V. and Vorobjev, E. M.: *Collection of Problems on Mathematical Physics*, additional chapters, Vysshaja Shkola, Moscow, 1978 (Russian).

SUBJECT INDEX

INDEX OF ASSUMPTIONS, THEOREMS ETC.

300

MATHEMATICAL PHYSICS AND
APPLIED MATHEMATICS

Editors:

M. FLATO (*(Université de Dijon, Dijon, France*)
R. RĄCZKA (*Institute of Nuclear Research, Warsaw, Poland*)